T0350119

Nonconventional Limit Theorems and Random Dynamics

Nonconventional Limit Theorems and Random Dynamics

Yeor Hafouta
Hebrew University of Jerusalem, Israel

Yuri Kifer
Hebrew University of Jerusalem, Israel

World Scientific

NEW JERSEY · LONDON · SINGAPORE · BEIJING · SHANGHAI · HONG KONG · TAIPEI · CHENNAI · TOKYO

Published by

World Scientific Publishing Co. Pte. Ltd.
5 Toh Tuck Link, Singapore 596224
USA office: 27 Warren Street, Suite 401-402, Hackensack, NJ 07601
UK office: 57 Shelton Street, Covent Garden, London WC2H 9HE

Library of Congress Cataloging-in-Publication Data
Names: Hafouta, Yeor, author. | Kifer, Yuri, 1948–   author.
Title: Nonconventional limit theorems and random dynamics / by Yeor Hafouta
    (Hebrew University of Jerusalem, Israel), Yuri Kifer (Hebrew University of Jerusalem, Israel).
Description: New Jersey : World Scientific, 2018. | Includes bibliographical references and index.
Identifiers: LCCN 2018000570 | ISBN 9789813235007 (hardcover : alk. paper)
Subjects: LCSH: Limit theorems (Probability theory) | Random dynamical systems.
Classification: LCC QA273.67 .H335 2018 | DDC 519.2--dc23
LC record available at https://lccn.loc.gov/2018000570

British Library Cataloguing-in-Publication Data
A catalogue record for this book is available from the British Library.

For any available supplementary material, please visit
http://www.worldscientific.com/worldscibooks/10.1142/10849#t=suppl

Desk Editor: Benny Lim

Printed in Singapore

Preface

Nonconventional ergodic theorems dealt with the limits of expressions having the form $1/N \sum_{n=1}^{N} T^{q_1(n)} f_1 \cdots T^{q_\ell(n)} f_\ell$ where T is a measure preserving transformation, f_i's are bounded measurable functions and q_i's are linear or polynomial functions taking on integer values on integers. These results were used in the ergodic theory proof of Szemerédi's theorem on arithmetic progressions (see [21] and [22]) but since then their various extensions became a topic of its own interest.

From the probabilistic point of view ergodic theorems can be regarded as laws of large numbers and once they are established it is natural to inquire about other limit theorems of probability such as the central limit theorem, Poisson limit theorem, large deviations etc. During the last decade these questions were studied in a number of papers (see, for instance, [29], [30], [41]-[43] and [45]-[47]). These limit theorems were studied for nonconventional sums of the form

$$S_N = \sum_{n=1}^{N} F(\xi_{q_1(n)}, ..., \xi_{q_\ell(n)})$$

where $\{\xi_n, n \geq 0\}$ is a sufficiently fast mixing process with some stationarity properties, F is a Borel function with some regularity properties and q_j's are functions taking on integer values on integers and satisfying certain conditions (for instance, linear or polynomial ones). These results hold true, in particular, when $\xi_1, \xi_2, \xi_3, ...$ form a Markov chain satisfying some form of the Doeblin condition while in the dynamical systems setup these results are applicable to topologically mixing subshifts of finite type considered with appropriate Gibbs measures with implications to systems having corresponding symbolic representations such as C^2 Axiom A diffeomorphism (in particular, Anosov) in a neighborhood of an attractor or an expanding C^2 endomorphism of a Riemannian manifold.

Some of number theoretic (combinatorial) applications of such limit theorems can be described in the following way. For each point $\omega \in [0,1)$ consider its base m or continued fraction expansions with digits $\xi_k(\omega)$, $k = 1, 2, \ldots$. Next, count the number $S_N(\omega)$ of those ℓ-tuples $q_1(n), \ldots, q_\ell(n)$, $n \leq N$ for which, say, $\xi_{q_j(n)}(\omega) = a_j$, $j = 1, \ldots, \ell$ for some fixed integers $a_1, \ldots, a_\ell$ with the most notable q_j's forming an arithmetic progression $q_j(n) = jn$, $j = 1, \ldots, \ell$. Now we can write

$$S_N(\omega) = \sum_{n=1}^{N} \prod_{j=1}^{\ell} \delta_{a_j \xi_{q_j(n)}(\omega)}$$

where $\delta_{ks} = 1$ if $k = s$ and $= 0$ otherwise, arriving at the above setup. To make ξ_k's random variable, we supply the unit interval with an appropriate probability measure such as the Lebesgue measure for base m expansions and the Gauss measure for continued fraction expansions. We mention also another application to limit theorems for numbers of certain patterns in random sets. Namely, define a random set Γ in positive integers via a sequence of random variables $\xi_1, \xi_2, \ldots$ taking on values 0 or 1 by saying that $n \in \Gamma$ if and only if $\xi_n = 1$. Then, for instance,

$$S_N = \sum_{n=1}^{N} \prod_{j=1}^{\ell} \xi_{jn}$$

counts the number of arithmetic progressions of length ℓ in Γ starting at n and having the step n where n is between 1 and N.

Most of previous papers dealing with the central limit theorem for the above sums relied on the well known martingale approximation method and in Chapter 1 we return to this question showing how Stein's method can be successfully applied in the nonconventional situation. The advantage of this method is that we obtain simultaneously the speed of convergence in the corresponding central limit theorem improving estimates from [30], so that the central limit theorem and a version of the Berry-Esseen theorem come together.

In Chapter 2 we return to the study of the nonconventional local (central) limit theorem (LLT) which was considered first in [29]. The method of [29] was restricted to Markov chains with certain regularity of their transition probability (a version of the Doeblin condition) which excluded applications to important classes of dynamical systems such as subshifts of finite type. In order to overcome this restriction we had to extend the Ruelle-Perron-Frobenius (RPF) type theorems to random complex operators which is developed in Part 2 of the manuscript. After establishing this LLT, we present in Section 2.10 what seems to be a new approach to certain type of random dynamical systems which stems from the proof of the nonconventional LLT and is based on certain regularity conditions of the system around periodic points.

In Chapter 3 we consider nonconventional arrays of the form

$$S_N = \sum_{n=1}^{N} F(\xi_{q_1(n,N)}, ..., \xi_{q_\ell(n,N)})$$

where now the summands themselves depend on the number N of summands. Ergodic theorems for such arrays were recently studied in [48]. Here we restrict ourselves to the linear case $q_j(n,N) = p_j n + q_j N$ and obtain under certain conditions the strong law of large number, the central limit theorem and the Poisson limit theorem for such expressions.

Though our motivation for the results in Part 2 came from the need to extend the nonconventional local limit theorem to additional important classes of processes, the RPF type theory for random complex operators developed there is certainly interesting by its own. For instance, in Chapter 7 we apply it in order to obtain, for the first time, a version of the Berry-Esseen and the local limit theorems for processes in random dynamical environment, where the proof of the latter involves ideas from the proof of the nonconventional LLT, as well. Extensions of the nonconventional LLT (and CLT) for such processes will also be discussed, and this theory should find additional applications. The RPF theorem for products of real (random and deterministic) operators is a well-studied topic being the main step in the thermodynamic formalism constructions for many families of operators (see [10], [37], [40] and [53]), but this is the first exposition of the RPF theorem for random complex operators which emerge naturally in the study of the nonconventional LLT. The RPF theorem for a single deterministic complex operator was established, for instance, in [28] and [54], where both works rely on quasi compactness of a fixed real operator and the perturbation theory (see [36]). Recently, a complex version of the Hilbert metric was introduced and the corresponding theory of cones contractions was established yielding the complex deterministic RPF theorem, as well (see [18], [19] and [58]). In Part 2 we adapt this theory to the situation of complex random operators and as an application we produce the random complex thermodynamic formalism type constructions.

Contents

Thermodynamic Formalism for Random Complex Operators and applications 171

PART 1
Nonconventional Limit Theorems

Chapter 1

Stein's method in the nonconventional setup

1.1 Introduction: local (strong) dependence structure

Let $\xi = \{\xi_n : n \geq 0\}$ be a sequence of random variables and let $\ell \in \mathbb{N}$. Consider the sums $S_N = \sum_{n=1}^{N} F(\Xi_n)$, $N \in \mathbb{N}$, where $\Xi_n = (\xi_n, \xi_{2n}, ..., \xi_{\ell n})$ and F is a function which satisfies some regularity conditions. The summands in S_N are long range dependent even when ξ is a sequence of independent random variables. For instance, Ξ_n and Ξ_{kn} are strongly dependent for any n and $1 \leq k \leq \ell$. Still, each Ξ_n can depend only on the random vectors Ξ_m for m's which are members of $\{\frac{in}{j} : 1 \leq i, j \leq \ell\}$. When this type of dependence structure occurs, it is convenient to think about the indexes $1, 2, ..., N$ as vertices of a graph, and from this point of view the random variables $\{F(\Xi_n) : 1 \leq n \leq N\}$ are "locally dependent". More precisely, we will say that n and m are "connected" if $in = jm$ for some $1 \leq i, j \leq \ell$. Then each n can be connected to at most ℓ^2 integers m, and the random vectors Ξ_n and $\tilde{\Xi}_n = \{\Xi_m : m$ and n are not connected$\}$ are independent. When ξ_n's are not independent but instead satisfy some mixing (weak dependence) conditions then we fix some $l \geq 1$, which is small relatively to N, and say that n and m are "connected" if $|in - jm| < l$ for some $1 \leq i, j \leq \ell$. Then each n is connected to at most $\ell^2(2l+1)$ integers m, and the random vectors Ξ_n and $\tilde{\Xi}_n$ defined in a similar to the above way are "weakly dependent".

We see then that the summands of S_N are "locally (strongly) dependent" in the sense that each summand $F(\Xi_n)$ can strongly depend only on the summands indexed by members from the neighborhood of n which is the set of all m's connected with n. This local dependence structure is one of the main situations in which Stein's method is effective, and, in fact, is not restricted to the above specific arithmetic progression structure of the indexes $n, 2n, ..., \ell n$. For instance, such local dependence structure occurs also for sums of the form $\sum_{n=1}^{N} F(\xi_{q_1(n)}, \xi_{q_2(n)}, ..., \xi_{q_\ell(n)})$ where q_i's are polynomials with integer coefficients whose leading coefficients are positive, and, more generally, when the q_i's satisfy some growth conditions. In this situation we will say that n and m are connected if $|q_i(n) - q_j(m)| < l$ for some $1 \leq i, j \leq \ell$.

In this chapter we will use the local dependence structure described above in order to obtain a central limit theorem (CLT) for the normalized sums $Z_N = N^{-\frac{1}{2}} S_N$ when the sequence $\xi_1, \xi_2, \xi_3, ...$ satisfies some mixing and moment conditions and when F satisfies some regularity properties. Under rather general assumptions we will obtain almost optimal convergence rate of order $N^{-\frac{1}{2}} \ln^2(N+1)$. These results improve the rates obtained in [26] and [30]. Our proofs require adaptation of the arguments of [15] to the situation when Ξ_n and $\tilde{\Xi}_n$ defined above are not independent but only weakly dependent and the following section is devoted to formulation of such results. We will also obtain a functional CLT for the random functions $\mathcal{Z}_N(t) = N^{-\frac{1}{2}} S_{[Nt]}$, and the necessary background for Stein's method in the situation of random functions is presented in Section 1.5.

We will use two main probability metrics. Let X and Y be two random variables. The Wasserstein distance between their laws $\mathcal{L}(X)$ and $\mathcal{L}(Y)$ is defined by
$$d_W(\mathcal{L}(X), \mathcal{L}(Y)) = \sup\{|\mathbb{E}h(X) - \mathbb{E}h(Y)| : h \in Lip_1\} \qquad (1.1.1)$$
where Lip_1 is the class of Lipschitz functions with constant 1. The Kolmogorov (uniform) distance between $\mathcal{L}(X)$ and $\mathcal{L}(Y)$ is defined by
$$d_K(\mathcal{L}(X), \mathcal{L}(Y)) = \sup_{x \in \mathbb{R}} |P(X \leq x) - P(Y \leq x)|. \qquad (1.1.2)$$
Then by Theorem 3.1 in [3] (see also [24]) we have
$$d_K(\mathcal{L}(X), \mathcal{L}(Y)) \leq (1 + \sup |G'|)\sqrt{d_W(\mathcal{L}(X), \mathcal{L}(Y))} \qquad (1.1.3)$$
if Y has a distribution function G with density G'.

We will also use the following.

Lemma 1.1.1. *Let X and Y be two random variables defined on the same probability space. Let Z be a random variable with density ρ bounded from above by some constant $c > 0$. Then,*
$d_K(\mathcal{L}(Y), \mathcal{L}(Z)) \leq 3d_K(\mathcal{L}(X), \mathcal{L}(Z)) + 4c\|X - Y\|_{L^\infty}$ *and for any $b \geq 1$,*
$$d_K(\mathcal{L}(Y), \mathcal{L}(Z)) \leq 3d_K(\mathcal{L}(X), \mathcal{L}(Z)) + (1 + 4c)\|X - Y\|_{L^b}^{1-\frac{1}{b+1}}.$$

The second inequality is proved in Lemma 3.3 in [30], while the proof of the first inequality goes in the same way as the proof of that Lemma 3.3, taking in (3.2) from there $\delta = \|X - Y\|_{L^\infty}$.

1.2   Stein's method for normal aproximation

In the past 50 years Stein's method has become one of the main methods for approximating a distribution on the real line by the standard normal distribution and in particular for proving central limit theorems with some convergence rate. For readers' convenience, we begin this section with a short introduction to the method.

1.2.1   *A short introduction*

Consider the operator $\mathcal{A}$ which acts on differentiable functions $f : \mathbb{R} \to \mathbb{R}$ by the formula $\mathcal{A}f(w) = f'(w) - wf(w)$. Then, (see [3] and [62]) a random variable W has the standard normal distribution if and only if

$$\mathbb{E}\mathcal{A}f(W) = 0$$

for any bounded function f with a bounded derivative. The operator $\mathcal{A}$ is often referred to as the Stein operator associated with the standard normal distribution. The idea behind Stein's method is to quantify this characterization of the normal distribution. Let h be a Liphshitz function with constant 1, and consider the ordinary differential equation

$$\mathcal{A}f(w) = h(w) - \mathbb{E}h(Z) \tag{1.2.1}$$

where Z is a standard normal random variable. Then (see [3] and [62]) there exists a twice differentiable function f_h which solves (1.2.1) and (in particular) satisfies

$$\max\left(\sup|f_h|, \sup|f_h'|, \sup|f_h''|\right) \leq 4$$

where as usual $\sup|g|$ stands for the supremum of the absolute value of a real valued function g. Let W be a random variable whose first absolute moment $\mathbb{E}|W|$ is finite. Plugging in $w = W$ in (1.2.1) we arrive at

$$\mathbb{E}h(W) - \mathbb{E}h(Z) = \mathbb{E}\mathcal{A}f_h(W)$$

and therefore the following inequality holds true:

$$d_W(\mathcal{L}(W), \mathcal{N}) \tag{1.2.2}$$
$$\leq 4\sup\left\{|\mathbb{E}[f'(W) - Wf(W)]| : \max(\sup|f|, \sup|f'|, \sup|f''|) \leq 1\right\}$$

where $\mathcal{N}$ stands for the standard normal law. We see then that estimating the difference of expectations of Lipschitz functions requires estimation of expectations of the form $|\mathbb{E}\mathcal{A}f|$ for some class of differentiable functions.

Many situations require estimation of the Kolmogorov distance between $\mathcal{L}(W)$ and $\mathcal{N}$. Such estimates, of course, follow from (1.1.3) and (1.2.2), but this may cause a loss of accuracy. For instance (see [3]) when applied to sums of n i.i.d. random variable with finite third absolute moments, Stein's method yields optimal upper bound of order $n^{-\frac{1}{2}}$ in the Wasserstein metric which does not lead to an optimal upper bound in the Kolmogorov metric.

The idea behind obtaining more accurate estimates of the Kolmogorov distance is to approximate the indicator function $\mathbb{I}(w \leq x)$ of the ray $(\infty, x]$ by a piecewise linear function h and then approximating $|\mathbb{E}\mathcal{A}f_h|$ (see (1.4.17) and (1.4.19)). Such approximations, often referred to as "smoothing inequalities", are used in the proof of (1.1.3), but the proof in [3] does not take into account the specific form of the function f_h. This function satisfies some fine regularity properties (see (1.4.20) and (1.4.21)), and exploiting these properties requires some concentration inequality (see Proposition 1.4.2) which means controlling probabilities of the form $P(a \leq W \leq b)$. The proof of Theorem 1.2.2 begins with a general description of this concentration inequality approach, so the readers who are interested to see some of the technical details are referred to the beginning of this proof. We also refer the readers to a more detailed discussion about this concentration inequality approach in [3].

1.2.2 *Normal approximation for graphical indexation*

This section is devoted to formulation of abstract normal-approximation theorems in the situation of graphical indexation, which will be effective in the situation of local (strong) dependence. There are many ways to quantify the amount of dependence, and in Section 1.2.4 we will see that our results can be reformulated in terms of familiar mixing (weak dependence) coefficients. For the sake of readability, the results are not formulated under the most general (moment) assumptions, and the readers who are interested to see more general versions of these results are referred to (1.4.19) and (1.4.38) in Section 1.4.

Let $G = (E, V)$ be a finite graph (where E is the set of edges and V is the set of vertices) and let $d_G(\cdot, \cdot)$ be its associated shortest path distance (when v and u are not connected by a path we set $d_G(v, u) = \infty$). Denote by $B(v, r)$ a ball of radius $r \geq 0$ around $v \in V$. For any $v \in V$ set

$$N_v = B(1, v) = \{v\} \cup \{u \in V : (u, v) \in E\}. \qquad (1.2.3)$$

Let $\{X_v, v \in V\}$ be a collection of centered random variables defined on some probability space $(\Omega, \mathcal{F}, P)$ and set $W = \sum_{v \in V} X_v$. For any $A \subset V$ set $X_A = \{X_i : i \in A\}$ and $A^c = V \setminus A$. We denote by $\mathcal{N}$ the standard normal law on $\mathbb{R}$ and by $|\Gamma|$ the cardinality of a set Γ.

Theorem 1.2.1. *Let* $\gamma_4 > 0$, $c_0, D \geq 1$ *and* $1 \leq \rho \leq 3$ *be such that* $\mathbb{E}|X_v|^4 \leq \gamma_4^4$, $|N_v| \leq D$ *and*

$$|B(v,3)| \leq c_0 D^\rho$$

for any $v \in V$. *Then*

$$d_W(\mathcal{L}(W), \mathcal{N}) \leq 12c_0^{\frac{1}{2}} D^{\max(2, \frac{\rho+1}{2})}(\gamma_4^3|V| + \gamma_4^2|V|^{\frac{1}{2}}) + 4\sum_{i=1}^{3} \delta_i + 4|\mathbb{E}W^2 - 1|$$

where

$$\delta_1 = \sup\Big\{\big|\sum_{v \in V} \mathbb{E}X_v g\big(\sum_{u \in N_v^c} X_u\big)\big| : \sup|g| \leq 1\Big\}, \quad \delta_2 = \big|\sum_{v \in V}\sum_{u \in N_v^c} \mathbb{E}X_v X_u\big|,$$

$$(\delta_3)^2 = \big|\sum_{(v_1, u_1, v_2, u_2) \in \Gamma} Cov(X_{v_1} X_{u_1}, X_{v_2} X_{u_2})\big|$$

and $\Gamma = \{(v_1, u_1, v_2, u_2) : d(v_1, v_2) > 3 \text{ and } u_i \in N_{v_i}, i = 1, 2\} \subset V^4 = V \times V \times V \times V$ *which satisfies* $|\Gamma| \leq |V|^2 D^2$.

Note that we can always take $c_0 = 1$ and $\rho = 3$, but for certain graphs the cardinality of balls $B(v, 3)$ of radius 3 is of order D, and in this case we can choose $\rho = 1$.

Relying on (1.1.3), Theorem 1.2.1 yields estimates of the Kolmogorov distance between $\mathcal{L}(W)$ and $\mathcal{N}$, as well, but (as mentioned earlier) this cannot yield close to optimal upper bounds. For instance, when W is a sum of i.i.d. random variables and $\mathbb{E}W^2 = 1$ then considering a graph G with no edges we have $N_v = \{v\}$, and in this situation all δ_i's vanish and we get an upper bound of order $|V|^{-\frac{1}{2}}$, which only implies that $d_K(\mathcal{L}(W), \mathcal{N}) \leq C|V|^{-\frac{1}{4}}$ for an appropriate constant C.

Theorem 1.2.2. *Let* $\gamma_4 > 0$, $c_0, D \geq 1$ *and* $1 \leq \rho \leq 3$ *be as in Theorem 1.2.1. Then*

$$d_K(\mathcal{L}(W), \mathcal{N}) \leq 16c_0 D^{1+\rho}(|V|\gamma_4^3 + |V|^{\frac{1}{2}}\gamma_4^2) + 4|\mathbb{E}W^2 - 1| + 8\sum_{i=1}^{5} \delta_i + 3\delta_6^{\frac{1}{2}}$$

where δ_1, δ_2 *and* δ_3 *are defined in Theorem 1.2.1,*

$$\delta_4 = \sum_{v \in V} Cov(|W - Z_v|, |X_v \min(Y_v^2, 1)|) \quad \text{and for } i = 5, 6, \quad (1.2.4)$$

$$\delta_i = \big|\sum_{v \in V}\sum_{u: d(v,u)>3} \big(\mathbb{E}X_v X_u g_i(Y_v, Y_u) - \mathbb{E}X_v X_u^* g_i(Y_v, Y_u^*)\big)\big| \quad (1.2.5)$$

where $g_5(a, b) = \mathbb{I}_{\{(a,b):ab \geq 0\}} \min(|a|, |b|, 1)$, $g_6(a, b) = \big(g_5(a, b)\big)^2$, $\mathbb{I}_A$ *stands for the indicator function of a set* A, $Y_v = \sum_{u \in N_v} X_u$, $Z_v = \sum_{k: d(v,k) \leq 2} X_k$ *and* (X_u^*, Y_u^*) *is a copy of* (X_u, Y_u) *which is independent of* (X_v, Y_v).

Both theorems follow from the arguments in the proof of Theorem 2.1 in [15] and their proofs are postponed until Section 1.4.

1.2.3 *Weak local dependence coefficients*

The terms δ_i, $i = 1, ..., 6$ in Theorem 1.2.2 vanish when X_A and X_B are independent for some specific pairs of sets A and B which are not connected by a single edge. In general, these terms are small when X_A and X_B are "weakly dependent" for the appropriate A's and B's. In the following we will estimate δ_i's in terms of various dependence coefficients. First, for any $A, B \subset V$ and $1 \leq p, q \leq \infty$ we will measure the dependence between X_A and X_B via the quantities

$$\varepsilon_{p,q}(A, B)$$
$$= \sup\big\{|\mathrm{Cov}\big(g(X_A), h(X_B)\big)| : \max(\|g(X_A)\|_{L^p}, \|h(X_B)\|_{L^q}) \leq 1\big\},$$
$$\mathcal{E}_1(A, B) = \sup\big\{|\mathbb{E}G(X_A, X_B) - \mathbb{E}G(X_A, X_B^*)| : \mathbb{E}|G(X_A, X_B^*)| \leq 1\big\}$$
$$\text{and } \mathcal{E}_\infty(A, B) = \sup\big\{|\mathbb{E}G(X_A, X_B) - \mathbb{E}G(X_A, X_B^*)| : \|G\|_\infty \leq 1\big\}$$
$$(1.2.6)$$

where X_B^* is a copy of X_B which is independent of X_A, and $\|G\|_\infty = \|G\|_{\infty, A, B}$ stands here for essential supremum of the absolute value of a real valued function $G(\cdot, \cdot)$ with respect to the sum of the laws of (X_A, X_B) and (X_A, X_B^*). We remark that $\varepsilon_{p,q}$ is nonincreasing in both p and q and that $\mathcal{E}_\infty \leq \mathcal{E}_1$.

The first approximation of the δ_i's is given in the following simple lemma.

Lemma 1.2.3. *Suppose that* $\max_{v \in V} \|X_v\|_{L^r} \leq \gamma_r$ *for any* $1 \leq r \leq \infty$ *with some* $0 < \gamma_r \leq \infty$. *Then*

$$\delta_1 \leq \sum_{v \in V} \gamma_p \varepsilon_{p,\infty}(\{v\}, N_v^c), \quad \delta_2 \leq \sum_{v \in V} \sum_{u \in N_v^c} \gamma_p \gamma_q \varepsilon_{p,q}(\{u\}, \{v\}), \quad (1.2.7)$$

$$(\delta_3)^2 \leq \sum_{(v_1, u_1, v_2, u_2) \in \Gamma} \gamma_p^2 \gamma_q^2 \varepsilon_{\frac{p}{2}, \frac{q}{2}}(\{v_1, u_1\}, \{v_2, u_2\}), \quad (1.2.8)$$

$$\text{and } \delta_4 \leq \sum_{v \in V} (\|W\|_{L^p} + D^2 \gamma_p) \gamma_q \varepsilon_{p,q}(V \setminus B(v, 2), N_v) \quad (1.2.9)$$

for any $p, q \geq 2$, *and*

$$\delta_5 \leq \sum_{v \in V} \sum_{u:d(v,u)>3} D\gamma_3^3 \mathcal{E}_1(N_v, N_u) \text{ and} \quad (1.2.10)$$

$$\delta_6 \leq \sum_{v \in V} \sum_{u:d(v,u)>3} D^2 \gamma_4^4 \mathcal{E}_1(N_v, N_u). \quad (1.2.11)$$

Proof. Let $u, v \in V$. The inequalities in (1.2.7) follow directly from our assumption about the γ_r's. By the Hölder inequality for any $r \geq 2$,

$$\|X_v X_u\|_{L^{\frac{r}{2}}} \leq \|X_v\|_{L^r} \|X_u\|_{L^r} \leq \gamma_r^2.$$

Using this estimate with $r = p$ and $r = q$ and the definition of $\varepsilon_{\frac{p}{2}, \frac{q}{2}}$ we obtain (1.2.8). Next, (1.2.9) follows from the estimates

$$\|X_v \min(Y_v^2, 1)\|_{L^q} \leq \|X_v\|_{L^q} \quad \text{and} \quad \|W - Z_v\|_{L^p} \leq \|W\|_{L^p} + D^2 \gamma_p,$$

our assumption about the γ_r's and the assumption that $|N_v| \leq D$.

Finally, by the definitions of g_5 and g_6, the Hölder inequality and the inequality $|N_v| \leq D$,

$$\|X_v X_u^* g_5(Y_v, Y_u^*)\|_{L^1} \leq \|X_v X_u^* Y_v\|_{L^1} \leq \|X_v\|_{L^3} \|X_u\|_{L^3} \|Y_v\|_{L^3} \leq D\gamma_3^3$$

and

$$\|X_v X_u^* g_6(Y_v, Y_u^*)\|_{L^1} \leq \|X_v X_u^* Y_v^2\|_{L^1} \leq \|X_v X_u\|_{L^2} \|Y_v^2\|_{L^2}$$
$$\leq \|X_v\|_{L^4} \|X_u\|_{L^4} \|Y_v\|_{L^4}^2 \leq \gamma_4^4 D^2$$

and inequalities (1.2.10) and (1.2.11) follow. $\qquad\square$

Observe that $\mathcal{E}_\infty(A, B) \leq \mathcal{E}_1(A, B)$. In certain situations it will be easier to estimate $\mathcal{E}_\infty(N_v, N_v)$ when $d(u, v) > 3$, and in this case the following lemma will be useful.

Lemma 1.2.4. *Let $p > 2$ and $\gamma_p > 0$ be such that $\|X_v\|_{L^p} \leq \gamma_p$ for any $v \in V$. Then,*

$$\max(\delta_5, \delta_6) \leq 5\gamma_p^2 \sum_{v, u \in V : d(v, u) > 3} \left(\mathcal{E}_\infty(N_v, N_u)\right)^{1 - \frac{2}{p}}. \qquad (1.2.12)$$

Proof. For any random variable X and $R > 0$ set $X^{(R)} = X \mathbb{I}_{\{|X| \leq R\}}$, where $\mathbb{I}_{\{|X| \leq R\}}$ is the random variable which equals 1 when $|X| \leq R$ and 0 otherwise. Then by the Hölder and Markov inequalities, for any two random variables X and X_0,

$$\mathbb{E}|XX_0 - X^{(R)} X_0^{(R)}| \leq \mathbb{E}|X(X_0 - X_0^{(R)})| + \mathbb{E}|X_0^{(R)}(X - X^{(R)})|$$
$$= \mathbb{E}|XX_0 \mathbb{I}_{\{|X_0| \leq R\}}| + \mathbb{E}|X_0^{(R)} X \mathbb{I}_{\{|X| \leq R\}}|$$
$$\leq \|XX_0\|_{L^{\frac{p}{2}}} (P(|X_0| > R))^{1 - \frac{2}{p}} + \|XX_0\|_{L^{\frac{p}{2}}} (P(|X| > R))^{1 - \frac{2}{p}}$$
$$\leq \left(\|X\|_{L^p} \|X_0\|_{L^p}^{p-1} + \|X\|_{L^p}^{p-1} \|X_0\|_{L^p}\right) R^{-(p-2)}.$$

Therefore, taking into account that $\sup |g_5| \leq 1$, for any $v, u \in V$ and

$R > 0$ we have

$$\mathbb{E}|X_v X_u g_5(Y_v, Y_u) - X_v^{(R)} X_u^{(R)} g_5(Y_v, Y_u)|$$
$$\leq \mathbb{E}|X_v X_u - X_v^{(R)} X_u^{(R)}| \leq 2\gamma_p^p R^{-(p-2)} \quad \text{and}$$
$$\mathbb{E}|X_v X_u^* g_5(Y_v, Y_u^*) - X_v^{(R)} (X_u^*)^{(R)} g_5(Y_v, Y_u^*)| \leq 2\gamma_p^p R^{-(p-2)}.$$

Next, by the definition of $\mathcal{E}_\infty(N_v, N_u)$,

$$\left|\mathbb{E}X_v^{(R)} X_u^{(R)} g_5(Y_v, Y_u) - \mathbb{E}X_v^{(R)} (X_u^*)^{(R)} g_5(Y_v, Y_u^*))\right| \leq R^2 \mathcal{E}_\infty(N_v, N_u).$$

Taking $R = R_{v,u} = \gamma_p (\mathcal{E}_\infty(N_v, N_u))^{-\frac{1}{p}}$, we derive from the above estimates that

$$\left|\mathbb{E}X_v X_u g_5(Y_v, Y_u) - \mathbb{E}X_v X_u^* g_5(Y_v, Y_u^*)\right| \leq R^2 \mathcal{E}_\infty(N_v, N_u) \quad (1.2.13)$$
$$+ 4\gamma_p^p R^{-(p-2)} = 5\gamma_p^2 (\mathcal{E}_\infty(N_v, N_u))^{1-\frac{2}{p}}.$$

Note that such choice of R is possible only when $\mathcal{E}_\infty(N_v, N_u) > 0$, but when $\mathcal{E}_\infty(N_v, N_u) = 0$ then X_{N_v} and X_{N_u} are independent which implies that the left-hand side of (1.2.13) vanishes, and so (1.2.13) trivially holds true. We conclude from the above estimates and the definition of δ_5 that

$$\delta_5 \leq 5\gamma_p^2 \sum_{v,u \in V : d(v,u) > 3} (\mathcal{E}_\infty(N_v, N_u))^{1-\frac{2}{p}}.$$

Repeating the above arguments with the function g_6 in place of g_5 we obtain the same upper bound for δ_6, and the lemma follows. □

1.2.4   *Relations with more familiar mixing coefficients*

For any $A \subset V$ let $\mathcal{G}_A = \sigma\{X_A\} = \sigma\{X_v : v \in A\}$ be the σ-algebra generated by X_A. We first recall that the α, ϕ and ψ mixing (dependence) coefficients associated with two sub-σ-algebras $\mathcal{G}$ and $\mathcal{H}$ of $\mathcal{F}$ are defined by the formulas

$$\alpha(\mathcal{G}, \mathcal{H}) = \sup\left\{|P(\Gamma \cap \Delta) - P(\Gamma)P(\Delta)| : \Gamma \in \mathcal{G}, \Delta \in \mathcal{H}\right\}, \quad (1.2.14)$$

$$\phi(\mathcal{G}, \mathcal{H}) = \sup\left\{\left|\frac{P(\Gamma \cap \Delta)}{P(\Gamma)} - P(\Delta)\right| : \Gamma \in \mathcal{G}, \Delta \in \mathcal{H}, P(\Gamma) > 0\right\} \quad (1.2.15)$$

and

$$\psi(\mathcal{G}, \mathcal{H}) = \sup\left\{\left|\frac{P(\Gamma \cap \Delta)}{P(\Gamma)P(\Delta)} - 1\right| : \Gamma \in \mathcal{G}, \Delta \in \mathcal{H}, P(\Gamma)P(\Delta) > 0\right\}.$$
$$(1.2.16)$$

By Theorem A.5, Corollary A.1 and Corollary A.2 in [27] for any $A, B \subset V$ and $p > 1$,

$$\varepsilon_{\infty,\infty}(A,B) \leq 4\alpha(\mathcal{G}_A,\mathcal{G}_B), \quad \varepsilon_{p,\infty}(A,B) \leq 6\big(\alpha(\mathcal{G}_A,\mathcal{G}_B)\big)^{1-\frac{1}{p}} \quad (1.2.17)$$

$$\text{and} \quad \varepsilon_{p,q}(A,B) \leq 8\big(\alpha(\mathcal{G}_A,\mathcal{G}_B)\big)^{1-\frac{1}{p}-\frac{1}{q}}$$

for any $q > 1$ such that $\frac{1}{p} + \frac{1}{q} < 1$. In fact, it is clear that $\varepsilon_{\infty,\infty}(A,B) \geq \alpha(\mathcal{G}_A,\mathcal{G}_B)$ which makes these coefficients equivalent. Moreover, by Theorem A.6 in [27],

$$\varepsilon_{p,q}(A,B) \leq 2\big(\phi(\mathcal{G}_A,\mathcal{G}_B)\big)^{\frac{1}{p}}$$

for any $1 < q, p \leq \infty$ such that $\frac{1}{p} + \frac{1}{q} = 1$.

In our application to nonconventional sums we will bound directly $\mathcal{E}_\infty(A,B)$ for appropriate A's and B's, but still it is important to estimate these weak dependence coefficients in terms of the more familiar ones. Indeed, taking expectations in Lemma 3.2 in [26] we obtain that

$$\mathcal{E}_\infty(A,B) \leq 2\phi(\mathcal{G}_A,\mathcal{G}_B).$$

Similarly, taking expectations in Lemma 3.1 in [39] we deduce that

$$\mathcal{E}_1(A,B) \leq 2\psi(\mathcal{G}_A,\mathcal{G}_B).$$

Remark 1.2.5. In [56] a general upper bound on $d_K(\mathcal{L}(W),\mathcal{N})$ was obtained when the summands X_v's are bounded while the arguments in [61] lead to such estimates assuming existence of finite eighth moments where in both situations the X_v's were normalized in a special way. In the resulting upper bounds obtained in either [56] or [61] the terms δ_5 and δ_6 are replaced with expressions which can be written as sums of covariances of the form $\text{Cov}(g(X_A), h(X_B))$ for appropriate A's and B's which are not connected by a single edge and functions g and h with some polynomial growth. Thus, in these circumstances it is possible to obtain similar estimates to the ones in Theorem 1.2.2 which only involve the coefficients $\varepsilon_{p,q}(A,B)$. Using (1.2.17), the resulting upper bounds will involve only mixing coefficients of the form $\alpha(\mathcal{G}_A,\mathcal{G}_B)$.

1.3 Nonconventional CLT with convergence rates

1.3.1 *Assumptions and main results*

Our setup consists of a $\wp$-dimensional stochastic process $\{\xi_n, n \geq 0\}$ on a probability space $(\Omega, \mathcal{F}, P)$ and a family of sub-σ-algebras $\mathcal{F}_{k,l}$, $-\infty \leq k \leq$

$l \leq \infty$ such that $\mathcal{F}_{k,l} \subset \mathcal{F}_{k',l'} \subset \mathcal{F}$ if $k' \leq k$ and $l' \geq l$. We will impose restrictions on the mixing coefficients

$$\phi_n = \sup\{\phi(\mathcal{F}_{-\infty,k}, \mathcal{F}_{k+n,\infty}) : k \in \mathbb{Z}\} \qquad (1.3.1)$$

where $\phi(\cdot, \cdot)$ was defined in (1.2.15).

In order to ensure some applications, in particular, to dynamical systems we will not assume that ξ_n is measurable with respect to $\mathcal{F}_{n,n}$ but instead impose restrictions on the approximation rate

$$\beta_{q,r} = \sup_{k \geq 0} \|\xi_k - \mathbb{E}[\xi_k | \mathcal{F}_{k-r,k+r}]\|_{L^q}. \qquad (1.3.2)$$

We do not require stationarity of the process $\{\xi_n, n \geq 0\}$, assuming only that the distribution of ξ_n does not depend on n and that the joint distribution of (ξ_n, ξ_m) depends only on $n - m$ which we write for further reference by

$$\xi_n \overset{d}{\sim} \mu \quad \text{and} \quad (\xi_n, \xi_m) \overset{d}{\sim} \mu_{m-n} \qquad (1.3.3)$$

where $Y \overset{d}{\sim} \mu$ means that Y has μ for its distribution.

For each $\theta > 0$, set

$$\gamma_\theta^\theta = \|\xi_n\|_{L^\theta}^\theta = \int |x|^\theta d\mu. \qquad (1.3.4)$$

Let $F = F(x_1, ..., x_\ell)$, $x_j \in \mathbb{R}^\wp$ be a function on $(\mathbb{R}^\wp)^\ell$ such that for some $K, \iota > 0$, $\kappa \in (0,1]$ and all $x_i, z_i \in \mathbb{R}^\wp$, $i = 1, ..., \ell$, we have

$$|F(x) - F(z)| \leq K[1 + \sum_{i=1}^{\ell}(|x_i|^\iota + |z_i|^\iota)] \sum_{i=1}^{\ell} |x_j - z_j|^\kappa \qquad (1.3.5)$$

and

$$|F(x)| \leq K[1 + \sum_{i=1}^{\ell} |x_i|^\iota] \qquad (1.3.6)$$

where $x = (x_1, ..., x_\ell)$ and $z = (z_1, ..., z_\ell)$. In fact, if ξ_n is measurable with respect to $\mathcal{F}_{n,n}$ then our results will follow with any Borel function F satisfying (1.3.6) without imposing (1.3.5), since the latter is needed only for approximation of ξ_n by conditional expectations $\mathbb{E}[\xi_n | \mathcal{F}_{n-r,n+r}]$ using (1.3.2). To simplify formulas we assume the centering condition

$$\bar{F} := \int F(x_1, ..., x_\ell) d\mu(x_1) \ldots d\mu(x_\ell) = 0 \qquad (1.3.7)$$

which is not really a restriction since we can always replace F by $F - \bar{F}$.

For each $N \in \mathbb{N}$ set

$$S_N = \sum_{n=1}^{N} F(\xi_n, \xi_{2n}, ..., \xi_{\ell n}) \quad \text{and} \quad Z_N = N^{-\frac{1}{2}} S_N.$$

The main goal in this section is to prove a central limit theorem (CLT) with close to optimal convergence rates for the normalized sums Z_N via Stein's method. Stein's method yields a functional CLT for the random function $\mathcal{Z}_N(\cdot)$ defined by

$$\mathcal{Z}_N(t) = N^{-\frac{1}{2}} \sum_{n=1}^{[Nt]} F(\xi_n, \xi_{2n}, ..., \xi_{\ell n}),$$

as well. These type of results require introduction of appropriate notations, and are delayed until Section 1.6.

Our results will rely on the following assumption.

Assumption 1.3.1. There exist $b \geq 2$, $q \geq 1$ and $m > 0$ such that

$$\frac{1}{b} > \frac{\iota}{m} + \frac{\kappa}{q}, \quad \gamma_m < \infty \text{ and } \gamma_{\iota b} < \infty. \tag{1.3.8}$$

We will also need either

Assumption 1.3.2. There exist $d \geq 1$ and $c \in (0,1)$ such that for any $n \geq 0$,

$$\phi_n + \beta_{q,n}^{\kappa} \leq dc^n \tag{1.3.9}$$

where $\beta_{q,n}^{\kappa} = (\beta_{q,n})^{\kappa}$,

or the following weaker

Assumption 1.3.3. There exist $d \geq 1$ and $\theta > 2$ such that for any $n \in \mathbb{N}$,

$$\phi_n + \beta_{q,n}^{\kappa} \leq dn^{-\theta} \tag{1.3.10}$$

where $\beta_{q,n}^{\kappa} = (\beta_{q,n})^{\kappa}$.

1.3.2 *Asymptotic variance*

The following theorem follows from arguments in [30].

Theorem 1.3.4. *Suppose that Assumption 1.3.1 holds true and that*

$$\Theta(b, q, \kappa) := \sum_{n=0}^{\infty} (n+1)\phi_n^{1-\frac{1}{b}} + \sum_{n=0}^{\infty} (n+1)\beta_{q,n}^{\kappa} < \infty. \tag{1.3.11}$$

Then the limit $\sigma^2 = \lim_{N \to \infty} \mathbb{E} Z_N^2$ exists and for some $c_\ell > 0$ which depends only on ℓ,

$$|\mathbb{E} Z_N^2 - \sigma^2| \le c_\ell C_0 N^{-\frac{1}{2}} \tag{1.3.12}$$

for any $N \in \mathbb{N}$, where $C_0 = K^2(1 + \gamma_m^\iota)^2 \Theta(b, q, \kappa)$. Moreover, $\sigma^2 > 0$ if and only if there exists no stationary in the wide sense process $\{V_n : n \ge 0\}$ such that

$$F(\xi_n^{(1)}, \xi_{2n}^{(n)}, ..., \xi_{\ell n}^{(\ell)}) = V_{n+1} - V_n, \quad P\text{-}a.s.$$

for any $n \in \mathbb{N}$, where $\xi^{(i)}$, $i = 1, ..., \ell$ are independent copies of the process $\xi = \{\xi_n : n \ge 0\}$ and a.s. stands for almost surely.

Note that (1.3.11) holds true under Assumption 1.3.2. In fact, it holds true when Assumption 1.3.3 is satisfied with some $\theta > \frac{2b}{b-1}$, for instance, when $\theta > \frac{5}{2}$.

Remark 1.3.5. Set $U_n = F(\xi_n^{(1)}, \xi_{2n}^{(n)}, ..., \xi_{\ell n}^{(\ell)})$, $n \in \mathbb{N}$. Then the process $U = \{U_n : n \ge 1\}$ is stationary in the wide sense and under Assumptions 1.3.1 and 1.3.3 the limit

$$s^2 = \lim_{N \to \infty} N^{-1} \mathrm{Var}\Big(\sum_{n=1}^N U_n \Big)$$

exists and the above characterization of positivity of σ^2 is equivalent to the statement that $s^2 > 0$ (see [11]). We refer the readers to [30] for the exact details. When, in addition, ξ is stationary then U is stationary and so by Theorem 18.2.1 in [33] (applied with the process U) this characterization can be replaced with the condition that there exists no square integrable function g such that

$$F = g \circ T_\ell - g, \quad \mu^\ell - \text{ a.s.}$$

where T is the measure preserving map generating ξ, $T_\ell = T \times T^2 \times \cdots \times T^\ell$ and $\mu^\ell = \mu \times \mu \times \cdots \times \mu$. This condition generalizes the usual coboundary condition in the "conventional" case ($\ell = 1$) to nonconventional sums above.

Remark 1.3.6. When $\xi_1, \xi_2, \xi_3, ...$ forms a stationary and sufficiently fast mixing Markov chain then under additional general conditions the asymptotic variance σ^2 is positive, unless F vanishes μ^ℓ-almost surely. In particular, this characterization of positivity of σ^2 is valid when ξ_n's are independent. See Theorem 2.4 in [29].

1.3.3 CLT with convergence rate

We denote by $\mathcal{N}$ the standard normal law on $\mathbb{R}$. Our main result is the following Berry-Esseen type theorem for nonconventional sums.

Theorem 1.3.7.

(i) Suppose that Assumptions 1.3.1 and 1.3.2 hold true with $b = 5$ and that $\sigma^2 > 0$. Then there exists $z_1 > 0$ which depends only on ℓ such that for any $N \in \mathbb{N}$,

$$\max\big(d_W(\mathcal{L}(\sigma^{-1}Z_N), \mathcal{N}),\, d_K(\mathcal{L}(\sigma^{-1}Z_N), \mathcal{N})\big)$$
$$\leq z_1 C_1^{\frac{3}{2}}(1 + R)\max(R^{\frac{5}{6}}, R^2)N^{-\frac{1}{2}}\ln^2(N+1)$$

where $R = \sigma^{-1}K(1 + \gamma_m^{\ell})$ and

$$C_1 = 1 + d^2 + (-\ln c)^{-2} + d(1 - c^{\frac{4}{5}})^{-2}.$$

(ii) Suppose that Assumptions 1.3.1 and 1.3.3 hold true with some θ and $b \geq 5$ such that $\theta \geq \frac{2b}{b-4}$. Further assume that $\sigma^2 > 0$. Then there exists $z_2 > 0$ which depends only on ℓ such that for any $N \in \mathbb{N}$,

$$\max\big(d_W(\mathcal{L}(\sigma^{-1}Z_N), \mathcal{N}),\, d_K(\mathcal{L}(\sigma^{-1}Z_N), \mathcal{N})\big)$$
$$\leq z_2 C_2^{\frac{3}{2}}(1 + R)\max(R^{1 - \frac{1}{b+1}}, R^2)N^{-(\frac{1}{2} - 2\zeta)}$$

where R is defined as above, $\zeta = \frac{3b}{\theta(b-4)+4b}$ and with $\theta_b = \frac{\theta(b-1)}{b}$,

$$C_2 = 1 + d^2 + \beta_{q,0}^{\kappa} + d(\theta_b - 2)^{-1}.$$

Remark that under the conditions of Theorem 1.3.7 (ii) we have $\theta > \frac{2b}{b-1}$ and so the conditions of Theorem 1.3.4 hold true. We also note that by the contraction of conditional expectations, $\beta_{q,0} \leq 2\gamma_q$. Moreover, when $\beta_{r_0} = 0$ for some r_0, then Theorem 1.3.7 holds true when F is only a Borel function satisfying (1.3.6) and (1.3.7), namely, it is unnecessary to require any type of continuity of F. In particular, this takes place when our mixing assumptions hold true with the σ-algebras $\mathcal{F}_{m,n} = \sigma\{\xi_m, ..., \xi_n\}$ and $r_0 = 0$ which yields the same convergence rates assuming only (1.3.6) and (1.3.7).

Remark 1.3.8. The purpose of Theorem 1.3.7 is to obtain close to optimal convergence rate in the CLT for the normalized sums Z_N, under sufficient moment and mixing conditions. The assumption that $b \geq 5$ in this theorem serves this cause, and the readers who are only interested to see under which moment conditions the CLT holds true are referred to Theorem 1.6 in which a functional CLT is established, in particular, when $b = 2 + \delta$ and $\theta > 4 + \frac{8}{\delta}$ for some $\delta > 0$.

1.3.4   The associated strong dependency graphs

Before proving Theorem 1.3.7 we will describe the graph associated with nonconventional sums discussed in Section 1.1. For any $x, y \in \mathbb{R}$ set

$$d_\ell(x,y) = \min_{1 \le i,j \le \ell} |ix - jy| = \min_{1 \le i,j \le \ell} |q_i(x) - q_j(y)| \qquad (1.3.13)$$

where $q_i(x) = ix, i = 1, ..., \ell$, and for any $\Gamma, \Lambda \subset \mathbb{R}$ set

$$d_\ell(\Gamma, \Lambda) = \inf\{d_\ell(x,y) : x \in \Gamma, y \in \Lambda\}. \qquad (1.3.14)$$

Let $A \subset \mathbb{R}$ and set

$$\Gamma_A = \{ja : a \in A, 1 \le j \le \ell\} = \bigcup_{i=1}^{\ell} q_i(A). \qquad (1.3.15)$$

Then by the definition of d_ℓ, for any $A, B \subset \mathbb{R}$ we have

$$dist(\Gamma_A, \Gamma_B) = d_\ell(A, B) \qquad (1.3.16)$$

where

$$dist(E_1, E_2) = \inf\{|e_1 - e_2| : e_1 \in E_1, e_2 \in E_2\}$$

for any two sets $E_1, E_2 \subset \mathbb{R}$.

Let $N \in \mathbb{N}$, $l \ge 1$ and set $V = V_N = \{1, ..., N\}$. Consider the graph $G = G_\ell(N, l)$ given by

$$G = (V_N, E_{N,l}) \text{ where } E_{N,l} = \{(n,m) \in V_N \times V_N : d_\ell(n,m) < l\}. \quad (1.3.17)$$

Namely, we will say that n and m are connected if $d_\ell(n,m) < l$. Denote by d_G the shortest path distance on the graph and by $B(v,r)$ a ball of radius r around $v \in V$ with respect to d_G. Then the sets $N_v = B(v,1), v \in V$ satisfy

$$N_v = \{u \in V : d_\ell(v,u) < l\} = \bigcup_{1 \le i,j \le \ell} V \cap I_{i,j}(v) \qquad (1.3.18)$$

where $I_{i,j}(v) = (\frac{i}{j}v - \frac{l}{j}, \frac{i}{j}v + \frac{l}{j})$, and therefore,

$$|N_v| \le 3\ell^2 l := D \qquad (1.3.19)$$

since the length of each $I_{i,j}(v)$ does not exceed $2l$ and $l \ge 1$. The following lemma shows that the cardinality of balls of radius 3 is of order D.

Lemma 1.3.9. *There exists $c_0 > 1$ which depends only on ℓ such that for any $v \in V$,*

$$|B(v,3)| = |\{u \in V : d_G(v,u) \le 3\}| \le c_0 D.$$

Proof. Let $v = v_0 \in V$. Then

$$B(v_0, 3) = \bigcup_{v_2 \in B(v_0, 2)} N_{v_2} = \bigcup_{v_1 \in N_{v_0}} \bigcup_{v_2 \in N_{v_1}} N_{v_2} \qquad (1.3.20)$$

since $N_u = B(u, 1)$ for any $u \in V$. Consider the linear maps $L_{i,j}^{(1)}$ and $L_{i,j}^{(2)}$, $1 \le i, j \le \ell$, given by

$$L_{i,j}^{(1)}(x) = \frac{ix - l}{j} \quad \text{and} \quad L_{i,j}^{(2)}(x) = \frac{ix + l}{j}.$$

For each $1 \le i_0, j_0 \le \ell$, the interval $I_{i_0,j_0}(v_0)$ in (1.3.18) can be written in the form

$$I_{i_0,j_0}(v_0) = (L_{i_0,j_0}^{(1)}(v_0), L_{i_0,j_0}^{(2)}(v_0)) := (v_{i_0,j_0,1}, v_{i_0,j_0,2})$$

and hence the set of all neighbors of members of $I_{i_0,j_0}(v_0)$ is contained in the union

$$\bigcup_{1 \le i_1, j_1 \le \ell} (L_{i_1,j_1}^{(1)}(v_{i_0,j_0,1}), L_{i_1,j_1}^{(2)}(v_{i_0,j_0,2})).$$

Applying the same reasoning to the latter intervals, we deduce from (1.3.18) and (1.3.20) that the set $B(v_0, 3)$ is contained in the union

$$\bigcup_{1 \le i_0,j_0,i_1,j_1,i_2,j_2 \le \ell} I_{i_0,j_0,i_1,j_1,i_2,j_2}(v_0)$$

of the intervals given by

$$I_{i_0,j_0,i_1,j_1,i_2,j_2}(v_0) = (L_{i_2,j_2}^{(1)} \circ L_{i_1,j_1}^{(1)} \circ L_{i_0,j_0}^{(1)}(v_0), L_{i_2,j_2}^{(2)} \circ L_{i_1,j_1}^{(2)} \circ L_{i_0,j_0}^{(2)}(v_0))$$
$$(1.3.21)$$

where $\circ$ stands for composition of maps. The length of each of these ℓ^6 intervals does not exceed $6\ell^2 l$, since each of them has the form $(av_0 - b, av_0 + b)$ for some $0 \le b \le l(1 + \ell + \ell^2)$. Since $l \ge 1$, we conclude that

$$|B(v_0, 3)| \le \ell^6 (6\ell^2 l + 1) \le 7\ell^8 l$$

and the lemma follows. $\qquad \qquad \square$

We conclude that the conditions in Theorems 1.2.1 and 1.2.2 about the cardinality of balls of radius 1 and 3 are satisfied with $D = 2\ell^2 l$, $\rho = 1$ and the above c_0.

1.3.5 *Expectation estimates*

We first recall that (see [11], Ch. 4) for any two sub-σ-algebras $\mathcal{G}, \mathcal{H} \subset \mathcal{F}$,

$$2\phi(\mathcal{G}, \mathcal{H}) = \sup\{\|\mathbb{E}[g|\mathcal{G}] - \mathbb{E}g\|_{L^\infty} : g \in L^\infty(\Omega, \mathcal{H}, P), \|g\|_{L^\infty} \leq 1\} \tag{1.3.22}$$

where $\phi(\mathcal{G}, \mathcal{H})$ is defined by (1.2.15). The following result does not seem to be new but for readers' convenience and completeness we will sketch its proof here.

Lemma 1.3.10. *Let $\mathcal{G}_1, \mathcal{G}_2 \subset \mathcal{F}$ be two sub-σ-algebras of $\mathcal{F}$ and for $i = 1, 2$ let V_i be a $\mathbb{R}^{d_i}$-valued random $\mathcal{G}_i$-measurable vector with distribution μ_i. Set $d = d_1 + d_2$, $\mu = \mu_1 \times \mu_2$, denote by κ the distribution of the random vector (V_1, V_2) and consider the measure $\nu = \frac{1}{2}(\kappa + \mu)$. Let $\mathcal{B}$ be the Borel σ-algebra on $\mathbb{R}^d$ and $H \in L^\infty(\mathbb{R}^d, \mathcal{B}, \nu)$. Then $\mathbb{E}[H(V_1, V_2)|\mathcal{G}_1]$ and $\mathbb{E}H(v, V_2)$ exist for μ_1-almost any $v \in \mathbb{R}^{d_1}$ and*

$$|\mathbb{E}[H(V_1, V_2)|\mathcal{G}_1] - h(V_1)| \leq 2\|H\|_{L^\infty(\mathbb{R}^d, \mathcal{B}, \nu)}\phi(\mathcal{G}_1, \mathcal{G}_2), \; P - a.s. \tag{1.3.23}$$

where $h(v) = \mathbb{E}H(v, V_2)$ and a.s. stands for almost surely.

Proof. Clearly H is bounded μ and κ almost surely. Thus, $\mathbb{E}[H(V_1, V_2)|\mathcal{G}_1]$ exists and existence of $\mathbb{E}H(v, V_2)$ (μ_1-a.s.) follows from the Fubini theorem. Relying on (1.3.22), inequality (1.3.23) follows easily for functions of the form $G(v_1, v_2) = \sum_i \mathbb{I}_{\{v_1 \in A_i\}} g_i(v_2)$, where $\{A_i\}$ is a measurable partition of the support of μ_1 and $\mathbb{I}_{\{v_1 \in A_i\}} = 1$ when $v_1 \in A_i$ and equals 0 otherwise. Any uniformly continuous function H is a uniform limit of functions of the above form, which implies that (1.3.23) holds true for uniformly continuous functions. Finally, by Lusin's theorem (see [57]), any function $H \in L^\infty(\mathbb{R}^d, \mathcal{B}, \nu)$ is an L^1 (and a.s.) limit of a sequence $\{H_n\}$ of continuous functions with compact support satisfying $\|H_n\|_{L^\infty(\mathbb{R}^d, \mathcal{B}, \nu)} \leq \|H\|_{L^\infty(\mathbb{R}^d, \mathcal{B}, \nu)}$ and (1.3.23) follows for any $H \in L^\infty(\mathbb{R}^d, \mathcal{B}, \nu)$. $\qquad\square$

Next, let U_i, $i = 1, 2, ..., k$ be d_i-dimensional random vectors defined on the probability space $(\Omega, \mathcal{F}, P)$ from Section 1.3.1, and $\{\mathcal{C}_j : 1 \leq j \leq s\}$ be a partition of $\{1, 2, ..., k\}$. Consider the random vectors $U(\mathcal{C}_j) = \{U_i : i \in \mathcal{C}_j\}$, $j = 1, ..., s$, and let

$$U^{(j)}(\mathcal{C}_i) = \{U_i^{(j)} : i \in \mathcal{C}_j\}, \; j = 1, ..., s$$

be independent copies of the $U(\mathcal{C}_j)$'s. For each $1 \leq i \leq k$ let $a_i \in \{1, ..., s\}$ be the unique index such that $i \in \mathcal{C}_{a_i}$, and for any bounded Borel function $H : \mathbb{R}^{d_1 + d_2 + ... + d_k} \to \mathbb{R}$ set

$$\mathcal{D}(H) = \left|\mathbb{E}H(U_1, U_2, ..., U_k) - \mathbb{E}H(U_1^{(a_1)}, U_2^{(a_2)}, ..., U_k^{(a_k)})\right|. \tag{1.3.24}$$

The following result is a consequence of Lemma 1.3.10.

Corollary 1.3.11. *Suppose that each* U_i *is* $\mathcal{F}_{m_i,n_i}$-*measurable, where* $n_{i-1} < m_i \leq n_i < m_{i+1}$, $i = 1, ..., k$, $n_0 = -\infty$ *and* $m_{k+1} = \infty$. *Then, for any bounded Borel function* $H : \mathbb{R}^{d_1+d_2+...+d_k} \to \mathbb{R}$,

$$\mathcal{D}(H) \leq 4 \sup |H| \sum_{i=2}^{k} \phi_{m_i - n_{i-1}}$$

where $\sup |H|$ *is the supremum of* $|H|$. *In particular, when* $s = 2$ *then*

$$\alpha\big(\sigma\{U(\mathcal{C}_1)\}, \sigma\{U(\mathcal{C}_2)\}\big) \leq 4 \sum_{i=2}^{k} \phi_{m_i - n_{i-1}}$$

where $\sigma\{X\}$ *stands for the* σ-*algebra generated by a random variable* X *and* $\alpha(\cdot, \cdot)$ *is given by (1.2.14).*

Proof. Denote by μ_i the distribution of the random vector $U(\mathcal{C}_i)$, $i = 1, ..., s$. Then, we have to show that for any Borel function $H : \mathbb{R}^{d_1+d_2+...+d_k} \to \mathbb{R}$,

$$\left| \mathbb{E}H(U_1, U_2, ..., U_k) - \int H(u_1, u_2, ..., u_k) d\mu_1(u^{(\mathcal{C}_1)}) d\mu_2(u^{(\mathcal{C}_2)})...d\mu_s(u^{(\mathcal{C}_s)}) \right|$$

$$\leq 4 \sup |H| \sum_{i=2}^{k} \phi_{m_i - n_{i-1}} \qquad (1.3.25)$$

where $u^{(\mathcal{C}_i)} = \{u_j : j \in \mathcal{C}_i\}$.

In order to prove (1.3.25), denote by ν_i the distribution of U_i, $i = 1, 2, ..., k$. We first prove by induction on k that for any choice of H and U_i's with the required properties,

$$\left| \mathbb{E}H(U_1, U_2, ..., U_k) - \int H(u_1, u_2, ..., u_k) d\nu_1(u_1) d\nu_2(u_2)...d\nu_k(u_k) \right|$$

$$\leq 2 \sup |H| \sum_{i=2}^{v} \phi_{m_i - n_{i-1}}. \qquad (1.3.26)$$

Indeed, suppose that $k = 2$ and set $V_1 = U_1$, $V_2 = U_2$, $h(u_1) = \mathbb{E}H(u_1, U_2)$, $\mathcal{G}_1 = \mathcal{F}_{-\infty,n_1}$ and $\mathcal{G}_2 = \mathcal{F}_{m_2,\infty}$. Taking expectation in (1.3.23) yields

$$|\mathbb{E}H(U_1, U_2) - \mathbb{E}h(U_1)| \leq 2 \sup |H| \phi_{m_2 - n_1}$$

which means that (1.3.26) holds true when $k = 2$. Now, suppose that (1.3.26) holds true for any $k \leq j - 1$, $U_1, ..., U_k$ with the required properties and any bounded Borel function $H : \mathbb{R}^{e_1+...+e_{k-1}} \to \mathbb{R}$, where $e_1, ..., e_{k-1} \in$

N. In order to deduce (1.3.26) for $k = j$, set $V_1 = (U_1, ..., U_{j-1})$, $V_2 = U_j$, $h(v_1) = \mathbb{E}H(v_1, U_j)$, $v_1 = (u_1, ..., u_{j-1})$, $\mathcal{G}_1 = \mathcal{F}_{-\infty, n_{j-1}}$ and $\mathcal{G}_2 = \mathcal{F}_{m_j, \infty}$. Taking expectation in (1.3.23) yields

$$|\mathbb{E}H(U_1, U_2, ..., U_j) - \mathbb{E}h(U_1, U_2, ..., U_{j-1})| \leq 2 \sup |H| \phi_{m_j - n_{j-1}}.$$

Applying the induction hypothesis with the function h completes the proof of (1.3.26), since $\sup |h| \leq \sup |H|$. Next, we prove by induction on s that for any choice of k, H, U_i's with the required properties and $C_1, ..., C_s$,

$$\left| \int H(u_1, u_2, ..., u_k) d\mu_1(u^{(C_1)}) d\mu_2(u^{(C_2)}) ... d\mu_s(u^{(C_s)}) \right. \tag{1.3.27}$$
$$\left. - \int H(u_1, u_2, ..., u_k) d\nu_1(u_1) d\nu_2(u_2) ... d\nu_k(u_k) \right| \leq 2 \sup |H| \sum_{i=2}^{k} \phi_{m_i - n_{i-1}}.$$

For $s = 1$ this is just (1.3.26). Now suppose that (1.3.27) holds true for any $s \leq j - 1$, and any real valued bounded Borel function H defined on $\mathbb{R}^{d_1 + ... + d_k}$, where k and $d_1, ..., d_k$ are some natural numbers. In order to prove (1.3.27) for $s = j$, set $u^{(I)} = (u^{(C_1)}, u^{(C_2)}, ..., u^{(C_{s-1})})$ and let the function I be defined by

$$I(u^{(I)}) = \int H(u_1, u_2, ..., u_k) \prod_{j \in C_s} d\nu_j(u_j). \tag{1.3.28}$$

Then

$$\int H(u_1, u_2, ..., u_k) d\nu_1(u_1) d\nu_2(u_2) ... d\nu_k(u_k) = \int I(u^{(I)}) \prod_{j \notin C_s} d\nu_j(u_j). \tag{1.3.29}$$

Let the function J be defined by

$$J(u^{(I)}) = \int H(u_1, u_2, ..., u_k) d\mu_s(u^{(C_s)}). \tag{1.3.30}$$

Then by (1.3.26), for any $u^{(C_1)}, ..., u^{(C_{s-1})}$,

$$|I(u^{(I)}) - J(u^{(I)})| \leq 2 \sup |H| \sum_{i \in C_s} \phi_{m_i - n_{i-1}}. \tag{1.3.31}$$

It is clear that $\sup |J| \leq \sup |H|$. Applying the induction hypothesis with the function J (considered as a function of the variable u) and taking into account (1.3.29) and (1.3.31) we obtain (1.3.27) with $s = j$ and we complete the induction. Inequality (1.3.25) follows now by (1.3.26) and (1.3.27), and the proof of Corollary 1.3.11 is complete. $\qquad \square$

Remark 1.3.12. In the notations of Corollary 1.3.11, let $Z_i, i = 1, ..., s$ be a bounded $\sigma\{U(\mathcal{C}_i)\}$-measurable random variable. Then each Z_i has the form $Z_i = H_i(U(\mathcal{C}_i))$ for some function H_i which satisfies $\sup |H_i| \leq \|Z_i\|_{L^\infty}$. Considering the function $H(u) = \prod_{i=1}^{s} H_i(u^{(\mathcal{C}_i)})$, we obtain from (1.3.25) that,

$$|\mathbb{E} \prod_{i=1}^{s} Z_i - \prod_{i=1}^{s} \mathbb{E}Z_i| \leq 4\big(\prod_{i=1}^{s} \|Z_i\|_{L^\infty}\big) \sum_{j=2}^{k} \phi_{m_j - n_{j-1}}. \quad (1.3.32)$$

In general we can replace $\sup |H|$ in the right-hand side of (1.3.25) by some essential supremum of $|H|$ with respect to an appropriate measure which has a similar but more complicated form as κ in Lemma 1.3.10.

We also obtain the following result.

Corollary 1.3.13. *Suppose that each U_i is $\mathcal{F}_{m_i, n_i}$-measurable. Let $H :$ $\mathbb{R}^{d_1 + d_2 + ... + d_k} \to \mathbb{R}$ satisfies*

$$|H(u)| \leq K(1 + \sum_{i=1}^{k} |u_i|^\iota) \quad (1.3.33)$$

for some $\iota > 0$ and all $u = (u_1, u_2, ..., u_k) \in \mathbb{R}^{d_1 + ... + d_k}$. Suppose that $\|U_i^\iota\|_{L^b} \leq \Gamma_b$ for some $b \geq 1$, $\Gamma_b \in (0, \infty)$ and all $i = 1, 2, ..., k$. Then

$$\mathcal{D}(H) \leq 6\mathcal{T}_b \varphi_k^{1 - \frac{1}{b}} \quad (1.3.34)$$

where $\varphi_k = \sum_{i=2}^{k} \phi_{m_i - n_{i-1}}$, $\varphi_k^{1 - \frac{1}{b}} = (\varphi_k)^{1 - \frac{1}{b}}$ and $\mathcal{T}_b = K(1 + k\Gamma_b)$.

Note that even though H is not bounded, the growth condition (1.3.33) and the moment assumptions in Corollary 1.3.13 guarantee that $\mathcal{D}(H)$ given by (1.3.24) is well defined, namely both expectations exist.

The proof of Corollary 1.3.13 is a standard application of the Hölder and Markov inequalities and it goes as follows.

Proof. First, when $\varphi_k = 0$ then $U_1, ..., U_k$ are independent and the left-hand side of (1.3.34) vanishes, and so we assume without loss of generality that $\varphi_k > 0$. For any $R > 0$ let the function H_R be defined by $H_R(u) = H(u)I(|H(u)| \leq R)$. Then by (1.3.25),

$$\mathcal{D}(H_R) \leq 4R \sum_{i=2}^{k} \phi_{m_i - n_{i-1}} = 4R\varphi_k.$$

By the Hölder and Markov inequalities, for any random variable X we have

$$\mathbb{E}|X\mathbb{I}_{\{|X|>R\}}| \leq \|X\|_{L^b} \big(P(|X| > R)\big)^{1 - \frac{1}{b}} \leq \|X\|_{L^b}^b R^{-(b-1)}$$

where $\mathbb{I}_{\{|X|>R\}}$ is the random variable which equals 1 when $|X| > R$ and equals 0 otherwise. By (1.3.33) and since $\|U_i^\iota\|_{L_b} \leq \Gamma_b$, $i = 1, ..., k$, we have

$$\|H(U_1, U_2, ..., U_k)\|_{L^b} \leq K(1 + k\Gamma_b) = \mathcal{T}_b.$$

Applying the previous inequality with $X = H(U_1, U_2, ..., U_k)$ we deduce that

$$\|H(U_1, U_2, ..., U_k) - H_R(U_1, U_2, ..., U_k)\|_{L^1} \leq \mathcal{T}_b^b R^{-(b-1)}.$$

Using the same arguments with $H(U_1^{(a_1)}, U_2^{(a_2)}, ..., U_k^{(a_k)})$ in place of $H(U_1, U_2, ..., U_k)$ we obtain

$$\|H(U_1^{(a_1)}, U_2^{(a_2)}, ..., U_k^{(a_k)}) - H_R(U_1^{(a_1)}, U_2^{(a_2)}, ..., U_k^{(a_k)})\|_{L^1} \leq \mathcal{T}_b^b R^{-(b-1)}.$$

Taking $R = \varphi_k^{-\frac{1}{b}} \mathcal{T}_b$, we conclude that

$$\mathcal{D}(H) \leq \mathcal{D}(H_R) + 2\mathcal{T}_b^b R^{-(b-1)} \leq 4R\varphi_k + 2\mathcal{T}_b^b R^{-(b-1)} = 6\mathcal{T}_b \varphi_k^{1-\frac{1}{b}}$$

and the corollary follows.   □

In order to obtain an explicit constant C_0 which satisfies (1.3.12), we will need the following consequence of Corollary 1.3.13.

Corollary 1.3.14. *Let* $H : \mathbb{R}^{d_1+d_2+...+d_k} \to \mathbb{R}$ *satisfies (1.3.33) and (1.3.5) with H in place of F and with u_i's in place of x_i's. Let $q, b > 1$ and $m > 0$ be such that*

$$\frac{1}{b} > \frac{\iota}{m} + \frac{\kappa}{q}$$

and set $r_i = [\frac{1}{3}(m_i - n_{i-1})]$, $i = 2, ..., k$, $r_1 = r_2$ and $r_{k+1} = r_k$. Suppose that $\|U_i\|_{L^b} < \infty$ for any $1 \leq i \leq k$. Then

$$\mathcal{D}(H) \leq 6R_0 \Lambda_{b,\kappa} \tag{1.3.35}$$

where

$$\Lambda_{b,\kappa} = \left(\sum_{i=2}^{k} \phi_{r_i}\right)^{1-\frac{1}{b}} + \sum_{i=1}^{k} \left(\|U_i - \mathbb{E}[U_i | \mathcal{F}_{m_i-r_i, n_i+r_{i+1}}]\|_{L^q}\right)^{\kappa}$$

$R_0 = K(1 + k\gamma_m^\iota)$ *and* $\gamma_m = \max\{\|U_i\|_{L^m} : 1 \leq i \leq k\}$.

Proof. First, for any $i = 1, ..., k$ set $U_{i,r} = \mathbb{E}[U_i | \mathcal{F}_{m_i-r_i, n_i+r_{i+1}}]$ and for any $s > 0$ set

$$\Delta_{i,r,s} = \|U_i - U_{i,r}\|_{L^s} \text{ and } \gamma_s = \max\{\|U_i\|_{L^s} : 1 \leq i \leq k\}.$$

Then, by the Hölder inequality, for any $i = 1, ..., k$ we have

$$\left\||U_i - U_{i,r}|^{\kappa}\right\|_{L^b} \leq \Delta_{i,r,\kappa b}^{\kappa} \leq \Delta_{i,r,q}^{\kappa} \tag{1.3.36}$$

where in the second inequality we used that $q > \kappa b$. Let $1 \leq i, z \leq k$. Then by Lemma 3.1 *(i)* in [45] and our assumption that $\frac{1}{b} > \frac{\iota}{m} + \frac{\kappa}{q}$,

$$\left\| |U_z|^\iota |U_i - U_{i,r}|^\kappa \right\|_{L^b} \leq \gamma_m^\iota \Delta_{i,r,q}^\kappa \tag{1.3.37}$$

and observe that the above inequality holds true with $U_{z,r}$ in place of U_z, as well, since conditional expectation contracts L^p norms. We conclude from (1.3.5) and the above estimates that

$$\| H(U_1, U_2, ..., U_k) - H(U_{1,r}, U_{2,r}, ..., U_{k,r}) \|_{L^b} \tag{1.3.38}$$

$$\leq K(1 + 2k\gamma_m^\iota) \sum_{i=1}^{k} \Delta_{i,r,q}^\kappa.$$

Next, let $U_r^{(j)}(\mathcal{C}_j) = \{ U_{i,r}^{(j)} : i \in \mathcal{C}_j \}, j = 1, ..., s$ be independent copies of $U_r(\mathcal{C}_j) := \{ U_{i,r} : i \in \mathcal{C}_j \}, j = 1, ..., s$. Applying Corollary 1.3.13 we deduce that

$$|\mathbb{E}H(U_{1,r}, U_{2,r}, ..., U_{k,r}) - \mathbb{E}H(U_{1,r}^{(a_1)}, U_{2,r}^{(a_2)}, ..., U_{k,r}^{(a_k)})| \tag{1.3.39}$$

$$\leq 6 T_b \Big(\sum_{i=2}^{k} \phi_{r_i} \Big)^{1 - \frac{1}{b}}$$

where we recall that a_i is the unique index such that $i \in \mathcal{C}_{a_i}$, and we have used again the contraction of conditional expectations which imply that $\| U_{i,r}^\iota \|_{L^b} \leq \| U_i^\iota \|_{L^b} \leq \Gamma_b$ for each i. Considering the product of the laws of the random vectors

$$\mathcal{V}_j = \{ U(\mathcal{C}_j), U_r(\mathcal{C}_j) \} = \{ U_i, U_{i,r} : i \in \mathcal{C}_j \}, \ j = 1, ..., s$$

we can always assume that there exist (on a larger probability space) independent copies $\mathcal{V}_j^{(j)} = \{ U_i^{(j)}, U_{i,r}^{(j)} : i \in \mathcal{C}_j \}, j = 1, ..., s$ of the $\mathcal{V}_j$'s such that (1.3.36) holds true for any $1 \leq i, z \leq k$ with $U_i^{(a_i)}$, $U_z^{(a_z)}$, $U_{i,r}^{(a_i)}$ and $U_{z,r}^{(a_z)}$ in place of U_i, U_z, $U_{i,r}$ and $U_{z,r}$, respectively. The appropriate version of (1.3.37) holds true with $U_{z,r}^{(a_z)}$ in place of $U_z^{(a_z)}$, as well. Thus, similarly to (1.3.38) we obtain from (1.3.5) and (1.3.33) that

$$\left\| \mathbb{E}H(U_1^{(a_1)}, U_2^{(a_2)}, ..., U_k^{(a_k)}) - \mathbb{E}H(U_{1,r}^{(a_1)}, U_{2,r}^{(a_2)}, ..., U_{k,r}^{(a_k)}) \right\|_{L^b} \tag{1.3.40}$$

$$\leq K(1 + 2k\gamma_m^\iota) \sum_{i=1}^{k} \Delta_{i,r,q}^\kappa.$$

The corollary follows now from (1.3.38), (1.3.39) and (1.3.40). $\qquad\square$

Proof of Theorem 1.3.4. As claimed before its formulation, Theorem 1,3.4 follows from the arguments in [45], [44] and [30]. Indeed, the proof in [45] that σ^2 exists relies only on Lemma 4.3 from there which provides approximations of the left-hand side of (1.3.35), and so the proof that σ^2 exists in our circumstances proceeds in the same way as in [45], relying on Corollary 1.3.14 instead. Similarly, the proofs of the characterization of positivity of σ^2 and inequality (1.3.12) go exactly as in [30] and [44], respectively, relying on Corollary 1.3.14 instead of Lemma 3.4 in [30] and Proposition 3.2 (ii) in [44]. The appearance of an explicit constant C_0 satisfying (1.3.12) is guaranteed from (1.3.35), since, contrary to the above Proposition 3.2 (ii), the right-hand side of (1.3.35) includes only explicit constants. The exact form of C_0 given in Theorem 1.3.4 is easily recovered by keeping track of constants which come from the applications of Corollary 1.3.14. □

1.3.6   *Proof of Theorem 1.3.7*

Set

$$R = R_{\sigma,K,m,\iota} = \sigma^{-1}K(1 + \gamma_m^\iota).$$

Let $n, N \in \mathbb{N}$, $l \geq 1$ and set $r = [\frac{l}{3}]$,

$$\Xi_n = \left(\xi_n, \xi_{2n}, ..., \xi_{\ell n}\right) \text{ and } \Xi_{n,r} = \left(\xi_{n,r}, \xi_{2n,r}, ..., \xi_{\ell n,r}\right)$$

where for any $m \in \mathbb{N}$ and $r \geq 0$,

$$\xi_{n,r} = \mathbb{E}[\xi_m | \mathcal{F}_{m-r,m+r}].$$

Then with these notations we have $Z_N = N^{-\frac{1}{2}} \sum_{n=1}^{N} F(\Xi_n)$. Set

$$Z_{N,r} = N^{-\frac{1}{2}} \sum_{n=1}^{N} F(\Xi_{n,r}) \quad \text{and} \tag{1.3.41}$$

$$\bar{Z}_{N,r} = N^{-\frac{1}{2}} \sum_{n=1}^{N} \left(F(\Xi_{n,r}) - \mathbb{E}F(\Xi_{n,r})\right).$$

The first part of the proof is to approximate Z_N by these normalized sums. Let $1 \leq i, j \leq \ell$. Then by (1.3.38) applied with $k = \ell$, $H = F$ and the random vectors $U_i = \xi_{in}, i = 1, ..., \ell$,

$$\|F(\Xi_n) - F(\Xi_{n,r})\|_{L^b} \leq K\ell(1 + \ell\gamma_m^\iota)\beta_{q,r}^\kappa \leq \sigma\ell^2 R\beta_{q,r}^\kappa. \tag{1.3.42}$$

where $\beta_{q,r}^\kappa = (\beta_{q,r})^\kappa$, and therefore,

$$\|Z_N - Z_{N,r}\|_{L^b} \leq \sigma\ell^2 R N^{\frac{1}{2}}\beta_{q,r}^\kappa. \tag{1.3.43}$$

Next, we have

$$\|Z_{N,r} - \bar{Z}_{N,r}\|_{L^\infty} = N^{-\frac{1}{2}} |\sum_{n=1}^{N} \mathbb{E}F(\Xi_{n,r})|. \qquad (1.3.44)$$

In order to estimate the right-hand side in (1.3.44), let $n > l$, consider the random vectors $U_i = \xi_{in,r}$ and set $m_i = in - r$ and $n_i = in + r$, $i = 1, ..., \ell$. Then each U_i is $\mathcal{F}_{m_i, n_i}$ measurable and

$$m_i - n_{i-1} = n - 2r \geq l - 2r \geq \frac{l}{3} \geq r.$$

Using the triangle inequality and the estimate (1.3.42) and then applying Corollary 1.3.14 with $k = \ell$ and the sets $\mathcal{C}_i = \{in\}, i = 1, ..., \ell$ we deduce that

$$\left| \mathbb{E}F(\Xi_{n,r}) - \mathbb{E}F(\xi_n^{(1)}, \xi_{2n}^{(2)}, ..., \xi_{\ell n}^{(\ell)}) \right|$$
$$\leq \left| \mathbb{E}F(\Xi_{n,r}) - \mathbb{E}F(\Xi_n) \right| + \left| \mathbb{E}F(\Xi_n) - \mathbb{E}F(\xi_n^{(1)}, \xi_{2n}^{(2)}, ..., \xi_{\ell n}^{(\ell)}) \right|$$
$$\leq \sigma \ell^2 R \beta_{q,r}^\kappa + 6\sigma \ell^2 R(\phi_r^{1-\frac{1}{b}} + \beta_{q,r}^\kappa) \leq 7\ell^2 \sigma R c_{r,l}$$

where $\phi_r^{1-\frac{1}{b}} = (\phi_r)^{1-\frac{1}{b}}$ and

$$c_{r,l} := \phi_r^{1-\frac{1}{b}} + \beta_{q,r}^\kappa.$$

Notice that $\mathbb{E}F(\xi_n^{(1)}, \xi_{2n}^{(2)}, ..., \xi_{\ell n}^{(\ell)}) = \bar{F} = 0$ and therefore

$$|\mathbb{E}F(\Xi_{n,r})| \leq 7\sigma \ell^2 R c_{r,l} \qquad (1.3.45)$$

when $n > l$. When $n \leq l$ we deduce from (1.3.6) and the contraction of conditional expectations that

$$\|F(\Xi_{n,r})\|_{L^b} \leq K(1 + \ell \gamma_{\iota b}^\iota) \leq \ell \sigma R \qquad (1.3.46)$$

where we used that $m > \iota b$, which implies that $\gamma_{\iota b} \leq \gamma_m$. Since $|\mathbb{E}F(\Xi_{n,r})| \leq \|F(\Xi_{n,r})\|_{L^b}$, it follows from (1.3.44) and the above estimates that

$$\|Z_{N,r} - \bar{Z}_{N,r}\|_{L^\infty} \leq \ell \sigma R N^{-\frac{1}{2}} l + 7\sigma \ell^2 R N^{\frac{1}{2}} c_{r,l} \leq 7\sigma \ell^2 R s_{r,l} \qquad (1.3.47)$$

where

$$s_{r,l} := N^{-\frac{1}{2}} l + N^{\frac{1}{2}} \beta_{q,r}^\kappa + N^{\frac{1}{2}} \phi_r^{1-\frac{1}{b}}. \qquad (1.3.48)$$

Applying Lemma 1.1.1 we obtain from (1.3.43) and (1.3.47) that

$$d_K(\mathcal{L}(\sigma^{-1}Z_N), \mathcal{N}) \leq 3d_K(\mathcal{L}(\sigma^{-1}Z_{N,r}), \mathcal{N}) \qquad (1.3.49)$$
$$+ 4\left(\ell^2 R N^{\frac{1}{2}} \beta_{q,r}^\kappa\right)^{1-\frac{1}{b+1}} \leq 9d_K(\mathcal{L}(\sigma^{-1}\bar{Z}_{N,r}), \mathcal{N})$$
$$+ 4\ell^2 R^{1-\frac{1}{b+1}} a_{r,b} + 12 \cdot 7\ell^2 R s_{r,l}$$

where

$$a_{r,b} := \left(N^{\frac{1}{2}} \beta_{q,r}^{\kappa}\right)^{1 - \frac{1}{b+1}}. \tag{1.3.50}$$

Moreover, for any two random variables X and Y and a Lipschitz function h with the constant 1 we have $|\mathbb{E}h(X) - \mathbb{E}h(Y)| \leq \|X - Y\|_{L^1}$. Thus by (1.3.43) and (1.3.47),

$$d_W(\mathcal{L}(\sigma^{-1}Z_N), \mathcal{N}) \leq \sigma^{-1}\|Z_N - \bar{Z}_{N,r}\|_{L^1} \tag{1.3.51}$$

$$+ d_W(\mathcal{L}(\sigma^{-1}\bar{Z}_{N,r}), \mathcal{N}) \leq \ell^2 R\left(N^{\frac{1}{2}}\beta_{q,r}^{\kappa} + 7s_{r,l}\right) + d_W(\mathcal{L}(\sigma^{-1}\bar{Z}_{N,r}), \mathcal{N}).$$

1.3.7 *Back to graphical indexation*

Consider the graph $G(N, l)$ defined in Section 1.3.4. Set

$$X_v = X_{v,r,N} = N^{-\frac{1}{2}}\sigma^{-1}\left(F(\Xi_{v,r}) - \mathbb{E}F(\Xi_{v,r})\right), \; v \in V = \{1, 2, ..., N\} \tag{1.3.52}$$

and consider their sum

$$W = \sum_{v \in V} X_v$$

which satisfies $\sigma W = \bar{Z}_{N,r}$. In what follows we will estimate the remaining terms in Theorems 1.2.1 and 1.2.2. First, by (1.3.46) for any v we have

$$\|X_v\|_{L^b} \leq 2N^{-\frac{1}{2}}\sigma^{-1}\|F(\Xi_{v,r})\|_{L^b} \leq 2N^{-\frac{1}{2}}\ell R. \tag{1.3.53}$$

Next, we will estimate $|\mathbb{E}W^2 - 1|$. First, for any two random variables X and Y defined on the same probability space,

$$|\mathbb{E}X^2 - \mathbb{E}Y^2| \leq \|X - Y\|_{L^2}\|X + Y\|_{L^2} \leq \|X - Y\|_{L^2}(\|X - Y\|_{L^2} + 2\|Y\|_{L^2}).$$

Therefore, by (1.3.12),

$$\sigma^2|\mathbb{E}W^2 - 1| = |\mathbb{E}(\bar{Z}_{N,r})^2 - \sigma^2| \leq c_\ell C_0 N^{-\frac{1}{2}} + d_r(d_r + 2\|Z_N\|_{L^2}) \tag{1.3.54}$$

where $d_r = \|\bar{Z}_{N,r} - Z_N\|_{L^2}$. The right-hand side of (1.3.43) does not exceed $7\sigma\ell^2 R s_{r,l}$ and so it follows from (1.3.43) and (1.3.47) that

$$d_r \leq 14R\ell^2\sigma s_{r,l}.$$

In order to estimate $\|Z_N\|_{L^2}$ we first use (1.3.12) in order to deduce that $\|Z_N\|_{L^2} \leq c_\ell^{\frac{1}{2}} C_0^{\frac{1}{2}} + \sigma$. Notice that, in fact, $C_0 = R^2\sigma^2\Theta$, where $\Theta = \Theta(b, q, \kappa)$ is defined in (1.3.11), and it follows that $\|Z_N\|_{L^2} \leq \sigma(c_\ell^{\frac{1}{2}} \Theta^{\frac{1}{2}} R + 1)$. Finally, by the definition of $s_{r,l}$ and the inequality $(a+b+c)^2 \leq 3(a^2+b^2+c^2)$ we obtain that

$$s_{r,l}^2 \leq 3(N^{-1}l^2 + N\phi_r^{2-\frac{2}{b}} + N\beta_{q,r}^{2\kappa}).$$

We conclude from (1.3.54) and the above estimates that there exists a constant $a_\ell > 0$ which depends only on ℓ such that

$$|\mathbb{E}W^2 - 1| = \sigma^{-2}|\mathbb{E}(\bar{Z}_{N,r})^2 - \sigma^2| \qquad (1.3.55)$$

$$\leq a_\ell \max(R, R^2)\left(A_1 N^{-\frac{1}{2}} + A_2 N^{\frac{1}{2}}\beta_{q,r}^\kappa + A_3 N^{\frac{1}{2}}\phi_r^{1-\frac{1}{b}}\right)$$

where

$$A_1 = \Theta + l + l^2 N^{-\frac{1}{2}} + \Theta^{\frac{1}{2}},$$

$$A_2 = N^{\frac{1}{2}}\beta_{q,r}^\kappa + 1 \text{ and } A_3 = N^{\frac{1}{2}}\phi_r^{1-\frac{1}{b}} + 1.$$

Next, the following result is the first step towards estimating the δ_i's appearing in Theorems 1.2.1 and 1.2.2.

Lemma 1.3.15. *There exists a constant $c > 0$ which depends only on ℓ such that*

$$\max\left(\alpha(\mathcal{G}_{\{v\}}, \mathcal{G}_{\{u\}}), \alpha(\mathcal{G}_{\{v\}}, \mathcal{G}_{N_v^c}), \alpha(\mathcal{G}_{\{v_1, u_1\}}, \mathcal{G}_{\{v_2, u_2\}}), \qquad (1.3.56)\right.$$

$$\left.\alpha(\mathcal{G}_{\{v_1, u_1\}}, \mathcal{G}_{N_{v_1}^c \cap N_{u_1}^c}), \mathcal{E}_\infty(N_{v_1}, N_{v_2}), \alpha(V \setminus B(v, 2), N_v)\right) \leq c\phi_r$$

for any $v, u, v_1, u_1, v_2, u_2 \in V$ such that $d(v, u) > 1$, $d(v_1, v_2) > 3$ and $d(v_1, u_1) = d(v_2, u_2) = 1$, where $d = d_G$ is the shortest path distance associated with the graph $G = G_\ell(N, l)$ defined by (1.3.17).

Proof. In the course of the proof we will use the following notations and abbreviations. For any $B_1, ..., B_L \subset \mathbb{R}$ we will write

$$B_1 < B_2 < ... < B_L$$

when $b_1 < b_2 < ... < b_L$ for any $b_i \in B_i$, $i = 1, ..., L$. Next, for any $v, u \in V$ we abbreviate

$$\mathcal{G}_{\{v\}} = \mathcal{G}_v, \ \Gamma_{\{v\}} = \Gamma_v, \ \mathcal{G}_{\{v, u\}} = \mathcal{G}_{v, u} \text{ and } \Gamma_{\{v, u\}} = \Gamma_{v, u}$$

where for any $A \subset V$ the set Γ_A is defined by (1.3.15) and the σ-algebra $\mathcal{G}_A$ is defined in the beginning of Section 1.2.4.

In order to estimate the first two expressions on the left-hand side in (1.3.56), let $v \in V$ and $u \in N_v^c = V \setminus N_v$. Since $\mathcal{G}_u \subset \mathcal{G}_{N_v^c}$ we have

$$\alpha(\mathcal{G}_v, \mathcal{G}_u) \leq \alpha(\mathcal{G}_v, \mathcal{G}_{N_v^c}).$$

In order to estimate the above right-hand side, consider the set $\Gamma_{N_v^c}$. Then by (1.3.16) and the definition of the graph G,

$$dist(\Gamma_v, \Gamma_{N_v^c}) = d_\ell(\{v\}, N_v^c) \geq l.$$

Considering the points in $\Gamma_{N_v^c}$ which lie in the intervals $(iv, (i+1)v)$, $i = 0, ..., \ell - 1$ and $(\ell v, \infty)$, we can write

$$\Gamma_v \cup \Gamma_{N_v^c} = \bigcup_{i=1}^{L} B_i, \quad L \leq 2\ell + 1$$

where each B_i is a subset of either Γ_v or $\Gamma_{N_v^c}$,

$$B_1 < B_2 < ... < B_L \quad \text{and} \quad dist(B_{i+1}, B_i) \geq l, \ i = 1, 2, ..., L - 1. \quad (1.3.57)$$

Consider the random vectors

$$U_i = \{\xi_{b,r} : b \in B_i\}, \ i = 1, ..., L. \quad (1.3.58)$$

Since $l \geq 3r$, there exist numbers $m_i \leq n_i$ such that U_i is $\mathcal{F}_{m_i, n_i}$-measurable and

$$n_i \leq m_{i+1} - (l - 2r) \leq m_{i+1} - \frac{l}{3} \leq m_{i+1} - r, \ i = 1, ..., L - 1. \quad (1.3.59)$$

Since X_v is a function of $\{U_i : B_i \subset \Gamma_v\}$ and $X_{N_v^c}$ is a function of $\{U_i : B_i \subset \Gamma_{N_v^c}\}$ we obtain from Corollary 1.3.11 that

$$\alpha(\mathcal{G}_v, \mathcal{G}_{N_v^c}) \leq 4(2\ell + 1)\phi_r.$$

Next, in order to estimate the third and fourth expressions on the left hand side in (1.3.56), let $v_1, u_1 \in V$ be such that $d(v_1, u_1) = 1$. First notice that

$$\alpha(\mathcal{G}_{v_1, u_1}, \mathcal{G}_{v_2, u_2}) \leq \alpha(\mathcal{G}_{v_1, u_1}, \mathcal{G}_{N_{v_1}^c \cap N_{u_1}^c})$$

for any $v_2, u_2 \in V$ such that $d(v_1, v_2) > 3$ and $d(v_2, u_2) = 1$, and therefore it is sufficient to estimate the fourth expression. Set $A_{v_1, u_1} = N_{v_1}^c \cap N_{u_1}^c$. There exists no single edge connecting $\{v_1, u_1\}$ and A_{v_1, u_1} and therefore $d_\ell(\{v_1, u_1\}, A_{v_1, u_1}) \geq l$. Hence by (1.3.16),

$$dist(\Gamma_{v_1, u_1}, \Gamma_{A_{v_1, u_1}}) = d_\ell(\{v_1, u_1\}, A_{v_1, u_1}) \geq l.$$

Thus, by considering the points in A_{v_1, u_1} which lie either between two consecutive points of Γ_{v_1, u_1} or in one of the intervals $(0, \min(v_1, u_1))$ and $(\ell \max(v_1, u_1), \infty)$, we can write

$$\Gamma_{v_1, u_1} \cup \Gamma_{A_{v_1, u_1}} = \bigcup_{i=1}^{L} B_i, \quad L \leq 4\ell + 1$$

where each B_i is a subset of either Γ_{v_1, u_1} or $\Gamma_{A_{v_1, u_1}}$ and B_i's satisfy (1.3.57). Consider the random vectors $U_i, i = 1, ..., L$ given by (1.3.58) with the above B_i's. Since $l \geq 3r$, there exist numbers $n_i \leq m_i, i = 1, ..., L$ satisfying (1.3.59) so that each U_i is $\mathcal{F}_{m_i, n_i}$-measurable. Since (X_{v_1}, X_{u_1}) is a function of $\{U_i : B_i \subset \Gamma_{v_1,u_1}\}$ and $X_{A_{v_1,u_1}}$ is a function of $\{U_i : B_i \subset \Gamma_{A_{v_1,u_1}}\}$, we deduce from Corollary 1.3.11 that

$$\alpha(\mathcal{G}_{v_1,u_1}, \mathcal{G}_{N_{v_1}^c \cap N_{u_1}^c}) \leq 4(4\ell + 1)\phi_r.$$

Next, let v_1 and v_2 be such that $d(v_1, v_2) > 3$ and consider the sets N_{v_1} and N_{v_2}. In order to estimate $\mathcal{E}_\infty(N_{v_1}, N_{v_2})$, let $G = G(z_1, z_2)$ be a bounded Borel function, where $z_i = \{x_s : s \in N_{v_i}\}$, $i = 1, 2$. Consider the sets $M_i = \Gamma_{N_{v_i}}$, $i = 1, 2$. Since N_{v_1} and N_{v_2} are not connected by a single edge we have $d_\ell(N_{v_1}, N_{v_2}) \geq l$. Thus, by (1.3.16),

$$dist(M_1, M_2) = d_\ell(N_{v_1}, N_{v_2}) \geq l.$$

By (1.3.18) and the definition of the sets Γ_A's,

$$M_s = \{1, 2, ..., N\ell\} \cap \bigcup_{1 \leq i,j,k \leq \ell} I_{i,j,k}(v_s), \quad s = 1, 2$$

where

$$I_{i,j,k}(v) = (\frac{ki}{j}v - \frac{k}{j}l, \frac{ki}{j}v - \frac{k}{j}l).$$

As a consequence, since $dist(M_1, M_2) \geq l$ we can write

$$M_1 \cup M_2 = \bigcup_{i=1}^{L} B_i, \quad L \leq 2\ell^3 + 1$$

where each B_i is contained in either M_1 or M_2, and B_i's satisfy (1.3.57). Consider the random vectors U_i, $i = 1, ..., L$ given by (1.3.58) with the above B_i's. Since $l \geq 3r$, there exist numbers $n_i \leq m_i$, $i = 1, ..., L$ satisfying (1.3.59) so that each U_i is $\mathcal{F}_{m_i,n_i}$-measurable. Since $X_{N_{v_i}}$ is a function of $\{U_i : B_i \subset M_i\}$, $i = 1, 2$ it follows from Corollary 1.3.11 that

$$|\mathbb{E}G(X_{N_{v_1}}, X_{N_{v_2}}) - \mathbb{E}G(X_{N_{v_1}}, X_{N_{v_2}}^*)| \leq 4(2\ell^3 + 1)\sup|G|\phi_r$$

where $X_{N_{v_2}}^*$ is a copy of $X_{N_{v_2}}$ which is independent of $X_{N_{v_1}}$. Since the above estimate holds true for an arbitrary bounded Borel function G, we deduce that $\mathcal{E}_\infty(N_{v_1}, N_{v_2}) \leq 4(2\ell^3 + 1)\phi_r$.

Finally, let $v \in V$ and set $\mathcal{B}_v = V \setminus B(v, 2)$. Then $d_\ell(\mathcal{B}_v, N_v) \geq l$ and hence

$$dist(\Gamma_{\mathcal{B}_v}, \Gamma_{N_v}) = d_\ell(\mathcal{B}_v, N_v) \geq l.$$

Considering the points in $\Gamma_{\mathcal{B}_v}$ which lie either between two (non overlapping) intervals of the form $I_{i,j,k}(v)$ or lie to the right (left) from the maximal (minimal) point in Γ_{N_v}, we see that the union $\Gamma_{\mathcal{B}_v} \cup \Gamma_{N_v}$ can be written as a union of at most $2\ell^3 + 1$ sets $B_1, ..., B_L$ satisfying (1.3.57) such that each B_i is either a subset of $\Gamma_{\mathcal{B}_v}$ or a subset of Γ_{N_v}. Similarly to the above, we deduce that

$$\alpha(\mathcal{B}_v, N_v) \leq 4(2\ell^3 + 1)\phi_r$$

and the proof of the lemma is complete. $\qquad\square$

The following corollary follows from Lemmas 1.2.3, 1.2.4, 1.3.15, (1.3.19) and (1.3.53).

Corollary 1.3.16. *With* $R = \frac{K(1+\gamma_m^\ell)}{\sigma}$, *there exists* $c_1 \geq 1$ *which depends only on* ℓ *such that*

$$\delta_2, \delta_5, \delta_6 \leq c_1 R^2 N \phi_r^{1-\frac{2}{b}}, \quad \delta_4 \leq c_1 \Lambda \max(R, R^2) \phi_r^{\frac{1}{2}-\frac{1}{b}},$$

$$\delta_1 \leq c_1 R N^{\frac{1}{2}} \phi_r^{1-\frac{1}{b}} \quad and \quad (\delta_3)^2 \leq c_1 l^2 R^4 \phi_r^{1-\frac{4}{b}}$$

where $\phi_r^s = (\phi_r)^s$ *for any* $s > 0$ *and* $\Lambda = \|W\|_{L^2} N^{\frac{1}{2}} + D^2$.

Proof of Theorem 1.3.7 *(i)*. Suppose that Assumptions 1.3.1 and 1.3.2 hold true with $b = 5$ and set $c_5 = c^{1-\frac{1}{5}} = c^{\frac{4}{5}}$. Then

$$\Theta(5, q, \kappa) \leq \frac{2d}{(1 - c_5)^2} \tag{1.3.60}$$

since $\sum_{n=0}^{\infty}(n+1)x^n = \left(\frac{x}{1-x}\right)'$ for any $x \in (0,1)$, where $\Theta(5, q, \kappa)$ is defined in (1.3.11). Next, applying Corollary 1.3.16 with $b = 5$ and then using (1.3.9) yields

$$\delta_1, \delta_2, \delta_3, \delta_4, \delta_5, \delta_6^{\frac{1}{2}} \leq c_1(1 + \|W\|_{L^2})d \max(R, R^2) \max(l^2, N)c^{\frac{r}{10}}.$$

Taking $l = A \ln(N+1) + 3$ where $A = \frac{45}{\ln(c^{-1})} + 1$ and recalling that $r = [\frac{l}{3}]$ we have,

$$c^{\frac{r}{10}} \leq c^{\frac{l-3}{30}} \leq N^{-\frac{3}{2}} \tag{1.3.61}$$

and since $\ln(x + 1) \leq x$ for any $x \geq 1$ we deduce that

$$\delta_1, ..., \delta_5, \delta_6^{\frac{1}{2}} \leq 4c_1 d(1 + \|W\|_{L^2}) \max(R, R^2) A^2 N^{-\frac{1}{2}}. \tag{1.3.62}$$

Next, using (1.3.55) with $b = 5$ and then (1.3.9) we derive that there exists a constant g_ℓ which depends only on ℓ such that

$$|\mathbb{E}W^2 - 1| \leq g_\ell \max(R, R^2) C N^{-\frac{1}{2}} \ln(N + 1) \tag{1.3.63}$$

where with $\Theta = \Theta(5, q, \kappa)$,

$$C = 1 + \Theta + \Theta^{\frac{1}{2}} + A^2 + d^2$$

and we used that $\frac{1}{2} \leq \ln(N + 1) \leq N^{\frac{1}{2}}$ for any $N \geq 1$. In particular, we derive that

$$\|W\|_{L^2} \leq 1 + \left(g_\ell \max(R, R^2) C\right)^{\frac{1}{2}}. \tag{1.3.64}$$

Applying Theorems 1.2.1 and 1.2.2 with the X_v's and with $\rho = 1$, taking into account that $|V| = N$ and the definition (1.3.19) of D, and then using

Lemma 1.3.9, (1.3.53), (1.3.62), (1.3.63) and (1.3.64) we deduce that there exists a constant $z > 0$ which depends only on ℓ such that

$$d_W(\mathcal{L}(\sigma^{-1}\bar{Z}_{N,r}), \mathcal{N}), \ d_K(\mathcal{L}(\sigma^{-1}\bar{Z}_{N,r}), \mathcal{N}) \qquad (1.3.65)$$
$$\leq z(1+R)\max(R, R^2)C^{\frac{3}{2}}N^{-\frac{1}{2}}\ln^2(N+1) := \mathcal{D}$$

where we used that $1 + \max(R^{\frac{1}{2}}, R) \leq 2(1+R)$ and $\max(R^2, R^3) \leq (1+R)\max(R, R^2)$ for any $R \geq 0$ and that $\ln(N+1) \geq \frac{1}{2}$ for any $N \geq 1$, which implies that $l \leq 4A\ln(N+1)$.

Finally, consider the terms $s_{r,l}$ and $a_{r,b}$ in (1.3.48) and (1.3.50). Then by (1.3.9), taking into account that $l = A\ln(N+1)$ and $b = 5$ we obtain

$$s_{r,l} \leq 7AN^{-\frac{1}{2}}\ln(N+1) + dN^{-\frac{1}{2}} \quad \text{and} \quad a_{r,5} \leq dN^{-\frac{1}{2}} \qquad (1.3.66)$$

where in the first inequality we used that $l \leq 4A\ln(N+1)$. We conclude from (1.3.65), (1.3.51), (1.3.49) and (1.3.66) that there exists a constant $z' > 0$ which depends only on ℓ such that

$$d_W(\mathcal{L}(\sigma^{-1}Z_N), \mathcal{N}), \ d_K(\mathcal{L}(\sigma^{-1}Z_N), \mathcal{N})$$
$$\leq \mathcal{D} + \max(R^{\frac{5}{6}}, R)z'\left(dN^{-\frac{1}{2}} + AN^{-\frac{1}{2}}\ln(N+1)\right)$$

which completes the proof of Theorem 1.3.7, taking into account (1.3.60) and that $C \leq 2 + 2\Theta + A^2 + d^2$ and $A^2 \leq 2 + \frac{2 \cdot 45^2}{\ln^2(c-1)}$, which follows from the inequalities $x + x^2 \leq 1 + 2x^2$ and $(x+y)^2 \leq 2(x^2+y^2)$, $x, y \geq 0$. $\quad\square$

Proof of Theorem 1.3.7 (ii). Suppose that Assumptions 1.3.1 and 1.3.3 hold true, with θ and b such that $\theta > \frac{2b}{b-4}$. Consider l's of the form $l = 3N^{\zeta} + 1$, where ζ is yet to be determined. For such l's we have

$$\phi_r, \beta_{q,r}^{\kappa} \leq dr^{-\theta} \leq dN^{-\zeta\theta}. \qquad (1.3.67)$$

We assume now that $\zeta\theta \geq \frac{1}{2(1-\frac{4}{b})}$ and $\zeta \leq \frac{1}{2}$, which guarantees that $A_2, A_3 \leq d+1$ and $l^2 \leq N$, with A_2 and A_3 defined right after (1.3.55). From this together with (1.3.67), (1.3.48), (1.3.55) and Corollary 1.3.16 we obtain the following estimates:

$$\delta_1, ..., \delta_5, \delta_6^{\frac{1}{2}} \leq (1 + \|W\|_{L^2})\max(R, R^2)N^{1-(\frac{1}{2}-\frac{2}{b})\zeta\theta}, \qquad (1.3.68)$$
$$s_{l,r} \leq 4N^{-\frac{1}{2}+\zeta} + 2dN^{\frac{1}{2}-\zeta\theta(1-\frac{1}{b})},$$
$$|\mathbb{E}W^2 - 1| \leq 4\max(R, R^2)\left(dN^{\frac{1}{2}-\zeta\theta(1-\frac{1}{b})} + QN^{-\frac{1}{2}+\zeta}\right),$$
$$l^2N^{-\frac{1}{2}} \leq 3N^{2\zeta-\frac{1}{2}} \quad \text{and} \quad a_{r,b} \leq N^{(\frac{1}{2}-\zeta\theta)(1-\frac{1}{b+1})}$$

where $Q = 4 + \Theta(b, q, \kappa) + d$. The powers of N appearing in (1.3.68) do not exceed $1 - (\frac{1}{2} - \frac{2}{b})\zeta\theta$ and therefore the sum of the expressions in (1.3.68) does not exceed $K_0(N^{1-(\frac{1}{2}-\frac{2}{b})\zeta\theta} + N^{2\zeta-\frac{1}{2}})$, where

$$K_0 = 4(1 + \|W\|_{L^2} + d + Q)\max(R, R^2) + 10d.$$

Let $\zeta = 3\big(\theta(1 - \frac{4}{b}) + 4\big)^{-1}$ be the solution of the equation

$$2\zeta - \frac{1}{2} = 1 - (\frac{1}{2} - \frac{2}{b})\zeta\theta.$$

This ζ indeed satisfies $\zeta\theta \geq \frac{1}{2(1-\frac{1}{b})}$ and $\zeta \leq \frac{1}{2}$ since we have assumed that $\theta \geq \frac{2b}{b-4}$, and so the sum of the expressions in (1.3.68) does not exceed $2K_0 N^{2\zeta - \frac{1}{2}}$.

Observe next that

$$\Theta(b, q, \kappa) \leq 1 + \beta_{q,0}^\kappa + 4d \sum_{n=1}^{\infty} n^{-\theta_b + 1} \leq 1\beta_{q,0}^\kappa + 4d + 4d(\theta_b - 2)^{-1} \quad (1.3.69)$$

where $\theta_b = \theta(1 - \frac{1}{b})$ and we used that $\sum_{n=1}^{\infty} n^{-p} \leq 1 + \int_1^{\infty} x^{-p} dx = 1 + \frac{1}{p-1}$ for any $p > 1$. Applying Theorems 1.2.1 and 1.2.2 we deduce from (1.3.51), (1.3.49), (1.3.53) and (1.3.68) that

$$d_W(\mathcal{L}(\sigma^{-1} Z_N), \mathcal{N}), \ d_K(\mathcal{L}(\sigma^{-1} Z_N)), \mathcal{N})$$
$$\leq c(1 + \|W\|_2) \max(R^{1 - \frac{1}{b+1}}, R^2)(Q + d^2) N^{-(\frac{1}{2} - 2\zeta)}$$

where c depends only on ℓ. Now Theorem 1.3.7 *(ii)* follows, taking into account (1.3.69) and the inequality $\|W\|_2^2 \leq 1 + d + Q$ which follows from (1.3.68). □

1.4 General Stein's estimates: proofs

1.4.1 *Proof of Theorem 1.2.1*

In view of (1.2.2), in order to obtain an upper bound of $d_W(\mathcal{L}(W), \mathcal{N})$, it is sufficient to estimate expressions of the form $|E[f'(W) - Wf(W)]|$ for twice differentiable functions $f : \mathbb{R} \to \mathbb{R}$ which satisfy

$$\max \big(\sup |f|, \sup |f'|, \sup |f''| \big) \leq 1. \quad (1.4.1)$$

We first show that for any such f,

$$|\mathbb{E}[f'(W) - Wf(W)]| \leq \delta_1 + \delta_2 + r_1 + 2r_2 + r_3 + |\mathbb{E}W^2 - 1| \quad (1.4.2)$$

where with $Y_v = \sum_{u \in N_v} X_u$, $v \in V$,

$$r_1 = \mathbb{E}|\sum_{v \in V}(X_v Y_v - \mathbb{E}X_v Y_v)|, \ r_2 = \sum_{v \in V} \mathbb{E}|X_v Y_v| \mathbb{I}_{\{|Y_v| > 1\}} \quad (1.4.3)$$

$$\text{and} \ \ r_3 = \sum_{v \in V} \mathbb{E}|X_v| \min(Y_v^2, 1)$$

and for any $B \in \mathcal{F}$, $\mathbb{I}_B$ denotes the random variable which equals 1 on B and 0 outside B. Before proving (1.4.2) we assume that it holds true and complete the proof of Theorem 1.2.1. First, by the definition of r_3 and then by the Cauchy-Schwarz inequality,

$$r_3 \leq \sum_{v \in V} \mathbb{E}|X_v|Y_v^2 \leq \sum_{v \in V} \|X_v\|_{L^2} \|Y_v\|_{L^4}^2 \leq |V| D^2 \gamma_4^3 \qquad (1.4.4)$$

and by the Cauchy-Schwarz and Markov inequalities,

$$r_2 \leq \sum_{v \in V} \|X_v\|_{L^2} \|Y_v \mathbb{I}_{\{|Y_v|>1\}}\|_{L^2} \qquad (1.4.5)$$

$$\leq \sum_{v \in V} \|X_v\|_{L^4} \|Y_v\|_{L^4} \|\mathbb{I}_{\{|Y_v|>1\}}\|_{L^4} \leq \sum_{v \in v} \|X_v\|_{L^4} \|Y_v\|_{L^4}^2 \leq |V| \gamma_4^3 D^2$$

where we used that $\|X_v\|_{L^4} \leq \gamma_4^4$ and $|N_v| \leq D$, for any $v \in V$.

In order to bound r_1, set $\zeta_{v,u} = X_v X_u - \mathbb{E} X_v X_u$, $v, u \in V$. Then by the definition of Y_v's,

$$r_1 = \mathbb{E}\left| \sum_{v \in V} \sum_{u \in N_v} \zeta_{v,u} \right|.$$

By the Cauchy-Schwarz inequality,

$$r_1^2 \leq \mathbb{E}\left| \sum_{v \in V} \sum_{u \in N_v} \zeta_{v,u} \right|^2 = \sum_{v_1 \in V} \sum_{u_1 \in N_{v_1}} (A_{v_1,u_1} + B_{v_1,u_1})$$

where

$$A_{u_1,v_1} = \sum_{v_2 \in V : d(v_1,v_2) > 3} \sum_{u_2 \in N_{v_2}} \mathrm{Cov}(\zeta_{v_1,u_1}, \zeta_{v_2,u_2})$$

and

$$B_{v_1,u_1} = \sum_{v_2 \in V : d(v_1,v_2) \leq 3} \sum_{u_2 \in N_{v_2}} \mathrm{Cov}(\zeta_{v_1,u_1}, \zeta_{v_2,u_2}).$$

Notice that

$$\left| \sum_{v_1 \in V} \sum_{u_1 \in N_{v_1}} A_{u_1,v_1} \right| = \delta_3^2$$

where δ_3 is defined in the statement of Theorem 1.2.1. On the other hand, let $v_1, u_1, v_2, u_2 \in V$. By the Hölder inequality we have $|\mathrm{Cov}(\zeta_{v_1,u_1}, \zeta_{v_2,u_2})| \leq 2\gamma_4^4$ and therefore $B_{v_1,u_1} \leq 2c_0 D^{\rho+1} \gamma_4^4$, where we used that $|B(v_1, 3)| \leq c_0 D^\rho$. Hence,

$$r_1^2 \leq \delta_3^2 + 2c_0 |V| D^{\rho+2} \gamma_4^4 \qquad (1.4.6)$$

and Theorem 1.2.1 follows now from (1.2.2), (1.4.2), (1.4.5), (1.4.4) and (1.4.6).

For the purpose of proving (1.4.2) we will need the following notations from [15] which will be used in the proof of Theorem 1.2.2, as well. For any $v \in V$ and $t \in \mathbb{R}$ set

$$\hat{K}_v(t) = X_v\big(\mathbb{I}_{\{-Y_v \leq t < 0\}} - \mathbb{I}_{\{0 \leq t \leq -Y_v)\}}\big), \; K_v(t) = \mathbb{E}\hat{K}_v(t), \qquad (1.4.7)$$

$$\hat{K}(t) = \sum_{v \in \mathcal{I}} \hat{K}_v(t) \; \text{ and } \; K(t) = \mathbb{E}\hat{K}(t) = \sum_{v \in \mathcal{I}} K_v(t).$$

Notice that

$$\int_{-\infty}^{\infty} K(t)dt = \sum_{v \in V} \mathbb{E}X_v Y_v = \mathbb{E}W^2 - \Delta_2 = 1 - (\Delta_2 + 1 - \mathbb{E}W^2) \quad (1.4.8)$$

where $\Delta_2 = \sum_{v \in V} \sum_{u \in N_v^c} \mathbb{E}X_v X_u$ which satisfies $|\Delta_2| = \delta_2$.

In order to bound $|\mathbb{E}f'(W) - \mathbb{E}Wf(W)|$ we will need first the following Stein's identity from [15], which for readers' convenience is formulated in the following lemma.

Lemma 1.4.1. *Let f be a bounded differentiable function with bounded derivative. Then*

$$\mathbb{E}f'(W) - \mathbb{E}Wf(W) = R_1 + R_2 + R_3 + R_4 + (\Delta_2 + r)\mathbb{E}f'(W) - \delta_1(f) \quad (1.4.9)$$

where $\delta_1(f) = \sum_{v \in V} \mathbb{E}X_v f(W - Y_v)$, $r = 1 - \mathbb{E}W^2$,

$$R_1 = \mathbb{E}f'(W)\int_{-\infty}^{\infty}(K(t) - \hat{K}(t))dt,$$

$$R_2 = \mathbb{E}\int_{|t|>1}\big(f'(W) - f'(W+t)\big)\hat{K}(t)dt,$$

$$R_3 = \mathbb{E}\int_{|t|\leq 1}\big(f'(W) - f'(W+t)\big)\big(\hat{K}(t) - K(t)\big)dt \; \text{ and}$$

$$R_4 = \mathbb{E}\int_{|t|\leq 1}\big(f'(W) - f'(W+t)\big)K(t)dt.$$

Proof. First we write

$$\mathbb{E}Wf(W) = \sum_{v \in V}\mathbb{E}X_v\big(f(W) - f(W - Y_v)\big) + \delta_1(f).$$

The first expression in the above right-hand side can be written as

$$\sum_{v \in V}\mathbb{E}X_v\big(f(W) - f(W - Y_v)\big) = \sum_{v \in V}\mathbb{E}X_v\int_{-Y_v}^{0}f'(W+t)dt$$

$$= \sum_{v \in V}\mathbb{E}\int_{-\infty}^{\infty}f'(W+t)\hat{K}_v(t)dt = \mathbb{E}\int_{-\infty}^{\infty}f'(W+t)\hat{K}(t)dt.$$

Next, by (1.4.8),

$$\mathbb{E}f'(W) = \mathbb{E}f'(W)\left(\Delta_2 + r + \int_{-\infty}^{\infty} K(t)dt\right)$$

$$= \mathbb{E}\int_{-\infty}^{\infty} f'(W)K(t)dt + \left(\Delta_2 + r\right)\mathbb{E}f'(W)$$

and therefore

$$\mathbb{E}[f'(W) - Wf(W)] = \mathbb{E}\int_{-\infty}^{\infty} f'(W)K(t)dt$$

$$-\mathbb{E}\int_{-\infty}^{\infty} f'(W+t)\hat{K}(t)dt + (\Delta_2 + r)\mathbb{E}f'(W) - \delta_1(f)$$

and (1.4.9) follows by the definition of the R_i's. $\qquad\square$

We will also use the following relations:

$$r_1 = \mathbb{E}\left|\int_{-\infty}^{\infty} \left(\hat{K}(t) - K(t)\right)dt\right|, \qquad (1.4.10)$$

$$\sum_{v \in V} \mathbb{E}\int_{|t|>1} |\hat{K}_v(t)|dt = \sum_{v \in V} \mathbb{E}|X_v|(|Y_v| - 1)\mathbb{I}_{\{|Y_v|>1\}} \le r_2 \qquad (1.4.11)$$

$$\text{and } r_3 = 2\sum_{v \in V} \mathbb{E}\int_{|t|\le 1} |t\hat{K}_v(t)|dt \qquad (1.4.12)$$

where r_1, r_2 and r_3 are defined in (1.4.3).

In order to prove (1.4.2), let $f : \mathbb{R} \to \mathbb{R}$ be a twice differentiable function satisfying (1.4.1). We will estimate now the terms on the right-hand side of (1.4.9). First, by the definitions of δ_1, δ_2 and r and since both $\sup|f|$ and $\sup|f'|$ do not exceed 1,

$$|\delta_1(f)| \le \sup|f|\delta_1 \le \delta_1 \text{ and} \qquad (1.4.13)$$

$$|(\Delta_2 + r)\mathbb{E}f'(W)| \le \sup|f'|(\delta_2 + |\mathbb{E}W^2 - 1|) \le \delta_2 + |\mathbb{E}W^2 - 1|.$$

Next, since $\sup|f'| \le 1$ we deduce from (1.4.10) that

$$|R_1| \le \sup|f'|r_1 \le r_1. \qquad (1.4.14)$$

By the mean value theorem together with the inequality $\sup|f'| \le 1$ we have

$$|f'(W) - f'(W+t)| \le 2\sup|f'| \le 2$$

for any $t \in \mathbb{R}$. Therefore,

$$|R_2| \le 2\mathbb{E}\int_{|t|>1} |\hat{K}(t)|dt \le 2\sum_{v \in V} \mathbb{E}\int_{|t|>1} |\hat{K}_v(t)|dt \le 2r_2 \qquad (1.4.15)$$

where the last inequality follows from (1.4.11). Finally, set

$$S_3 = R_3 + R_4 = \mathbb{E} \int_{|t| \leq 1} \left(f'(W) - f'(W+t) \right) \hat{K}(t) dt.$$

Since $\sup |f''| \leq 1$ by the mean value theorem we have

$$|f'(W) - f'(W+t)| \leq |t|$$

and therefore

$$|S_3| \leq \mathbb{E} \int_{|t| \leq 1} |f'(W) - f'(W+t)| |\hat{K}(t)| dt \qquad (1.4.16)$$

$$\leq \mathbb{E} \int_{|t| \leq 1} |t| |\hat{K}(t)| dt \leq \sum_{v \in V} \mathbb{E} \int_{|t| \leq 1} |t| |\hat{K}_v(t)| dt = \frac{1}{2} r_3$$

where the last equality follows from (1.4.12). Inequality (1.4.2) follows now from (1.4.9) and (1.4.13)-(1.4.16), and the proof of Theorem 1.2.1 is complete. □

1.4.2　Proof of Theorem 1.2.2

The following proof is based on the proof of Theorem 2.1 in [15]. Let $z \in \mathbb{R}$ and $\alpha > 0$. Consider the function $h_{z,\alpha} : \mathbb{R} \to \mathbb{R}$ given by the formula

$$h_{z,\alpha}(w) = \begin{cases} 1 & \text{if } w \leq z \\ 1 + \frac{z-w}{\alpha} & \text{if } z < w \leq z + \alpha \\ 0 & \text{if } w > z + \alpha \end{cases} \qquad (1.4.17)$$

and set $\mathbf{N} h_{z,\alpha} = (2\pi)^{-\frac{1}{2}} \int_{-\infty}^{\infty} h_{z,\alpha}(x) e^{-\frac{1}{2}x^2} dx$. Let $f = f_{z,\alpha}$ be the unique bounded solution of the ordinary differential equation

$$\mathcal{A}f(w) = f'(w) - w f(w) = h_{z,\alpha}(w) - \mathbf{N} h_{z,\alpha} \qquad (1.4.18)$$

which is given by the formula

$$f(w) = e^{\frac{1}{2}w^2} \int_w^{\infty} e^{-\frac{1}{2}x^2} \left(h(x) - \mathbf{N} h_{z,\alpha} \right) dx.$$

Since

$$\mathbb{E} h_{z-\alpha,\alpha}(W) \leq P(W \leq z) \leq \mathbb{E} h_{z,\alpha}(W) \text{ and}$$

$$|\Phi(z) - \mathbf{N} h_{z,\alpha}|, \ |\Phi(z+\alpha) - \mathbf{N} h_{z,\alpha}| \leq \Phi(z+\alpha) - \Phi(z) \leq (2\pi)^{-1/2} \alpha$$

for any $z \in \mathbb{R}$, where Φ is the standard normal distribution function, we deduce that

$$\sup_z |P(W \leq z) - \Phi(z)| \leq \sup_z |\mathbb{E} h_{z,\alpha}(W) - \mathbf{N} h_{z,\alpha}| + 0.5\alpha \qquad (1.4.19)$$

$$= \sup_z |\mathbb{E} \mathcal{A} f_{z,\alpha}(W)| + 0.5\alpha.$$

As in (4.5)-(4.8) in [15] the function $f = f_{z,\alpha}$ satisfies the following properties (in fact, these estimates come from [62]):

$$0 \leq f(w) \leq 1, \ |f'(w)| \leq 1, \ |f'(w) - f'(v)| \leq 1 \ \text{ and } \tag{1.4.20}$$

$$|f'(w+s) - f'(w+t)| \tag{1.4.21}$$

$$\leq (|w| + 1)\min(|s| + |t|, 1) + \alpha^{-1}|\int_s^t \mathbb{I}_{\{z \leq w + u \leq z + \alpha\}} du|$$

$$\leq (|w| + 1)\min(|s| + |t|, 1) + \mathbb{I}_{\{z - \max(s,t) \leq w \leq z - \min(s,t) + \alpha\}}$$

for any $w, v, s, t \in \mathbb{R}$, where $\mathbb{I}_{\{a \leq b \leq c\}} = 1$ when $a \leq b \leq c$ and otherwise $= 0$, for any $a, b, c \in \mathbb{R}$. In view of (1.4.19), our goal now is to bound $|\mathbb{E}f'(W) - \mathbb{E}Wf(W)|$ from above for functions f satisfying (1.4.20) and (1.4.21) where $z \in \mathbb{R}$ and $\alpha > 0$ are fixed. First, by Lemma 1.4.1,

$$|\mathbb{E}f'(W) - \mathbb{E}Wf(W)| \leq \sum_{i=1}^4 |R_i| + |\delta_1(f)| + |\mathbb{E}f'(W)|(\delta_2 + |\mathbb{E}W^2 - 1|).$$

$$\tag{1.4.22}$$

Since $\max(\sup|f|, \sup|f'|) \leq 1$,

$$|\delta_1(f)| \leq \delta_1 \ \text{ and } \ |\mathbb{E}f'(W)|(\delta_2 + |\mathbb{E}W^2 - 1|) \leq \delta_2 + |\mathbb{E}W^2 - 1| \tag{1.4.23}$$

and since $\sup|f'| \leq 1$ and $|f'(w) - f'(v)| \leq 1$ for any $w, v \in \mathbb{R}$, similarly to (1.4.14) and (1.4.15) we obtain that

$$|R_1| \leq r_1 \ \text{ and } \ |R_2| \leq r_2. \tag{1.4.24}$$

Estimating R_3 and R_4 is a more complicated task and it requires the following concentration inequality. Set

$$r_4 = \sum_{v \in V} \mathbb{E}|WX_v \min(Y_v^2, 1)|, \tag{1.4.25}$$

$$r_5 = |\sum_{v,u \in V} \mathbb{E}X_v X_u g_5(Y_v, Y_u) - \mathbb{E}X_v X_u^* g_5(Y_v, Y_u^*)| \ \text{ and } \tag{1.4.26}$$

$$r_6 = |\frac{1}{2} \sum_{v,u \in V} \mathbb{E}X_v X_u g_6(Y_v, Y_u) - \mathbb{E}X_v X_u^* g_6(Y_v, Y_u^*)|^{\frac{1}{2}} \tag{1.4.27}$$

where the functions g_5 and g_6 are defined in Theorem 1.2.2, and recall that (X_u^*, Y_u^*) is a copy of (X_u, Y_u) which is independent of (X_v, Y_v).

Proposition 1.4.2. *For any $a < b$ we have*

$$P(a \leq W \leq b) \leq (0.125 + 0.5\|W\|_{L^2} + 0.5\delta_1)(b - a) + 2\delta_2$$

$$+(0.625 + 1.5\|W\|_{L^2} + 1.5\delta_1)r_3 + 4r_2 + 4r_5 + 2|\mathbb{E}W^2 - 1|.$$

Before proving Proposition 1.4.2 we assume that it holds true and complete the proof of Theorem 1.2.2. Together with (1.4.8), (1.4.10), (1.4.11) and (1.4.12) we will use the following identities:

$$r_4 = 2 \sum_{v \in V} \mathbb{E}|W| \int_{|t| \leq 1} |t \hat{K}_v(t)| dt \quad (1.4.28)$$

$$r_5 = \int_{|t| \leq 1} \mathrm{Var}\big(\hat{K}(t)\big) dt \quad \text{and} \quad r_6^2 = \int_{|t| \leq 1} |t| \mathrm{Var}\big(\hat{K}(t)\big) dt \quad (1.4.29)$$

(see the proof of (1.4.29) after the statement of Theorem 2.1 in [15]). We estimate next R_3 as in (4.11) in [15]. It follows from the definition of R_3 and from (1.4.21), (1.4.12) and (1.4.28) that

$$|R_3| \leq r_3 + r_4 + R_{3,1} + R_{3,2} \quad (1.4.30)$$

where

$$R_{3,1} = \mathbb{E} \int_0^1 \mathbb{I}_{\{z-t \leq W \leq z+\alpha\}} |\hat{K}(t) - K(t)| dt \quad \text{and}$$

$$R_{3,2} = \mathbb{E} \int_{-1}^0 \mathbb{I}_{\{z \leq W \leq z-t+\alpha\}} |\hat{K}(t) - K(t)| dt.$$

Set

$$\delta' = \mathcal{D}_1 \alpha + 2\delta_2 + \Big(0.625 + \frac{3}{2}\|W\|_{L^2} + \frac{3}{2}\delta_1\Big) r_3 + 4r_2 + 4r_5 + 2|\mathbb{E}W^2 - 1|$$

where

$$\mathcal{D}_1 = 0.125 + \frac{1}{2}\|W\|_{L^2} + \frac{1}{2}\delta_1.$$

Then by Proposition 1.4.2 for any $t \geq 0$,

$$P(z - t \leq W \leq z + \alpha) \leq \delta' + \mathcal{D}_1 t. \quad (1.4.31)$$

If either $\delta_1 \geq 1$, $\delta_2 \geq 1$ or $\|W\|_{L^2} \geq 2$ then the inequality in the statement of Theorem 1.2.2 clearly holds true, and so we can assume without loss of generality that $\max(\delta_1, \delta_2, \frac{1}{2}\|W\|_{L^2}) \leq 1$ and in this case

$$\delta' \leq 2\alpha + 2\delta_2 + 6r_3 + 4r_2 + 4r_5 + 2|\mathbb{E}W^2 - 1| := \delta''. \quad (1.4.32)$$

In order to estimate $R_{3,1}$, we first notice that for any $a, b \in \mathbb{R}$ and $c > 0$,

$$2ab \leq ca^2 + c^{-1}b^2.$$

Fix some $t \in [0,1]$. Similarly to (4.12) in [15], applying that inequality with $a = \mathbb{I}_{\{z-t \leq W \leq z+\alpha\}}$, $b = |\hat{K}(t) - K(t)|$ and $c = c(t) = \alpha(\delta'' + \mathcal{D}_1 t)^{-1}$ we deduce from (1.4.31) and (1.4.32) that

$$R_{3,1} \leq \frac{1}{2}\mathbb{E}\int_0^1 c(t)\mathbb{I}_{\{z-t \leq W \leq z+\alpha\}}dt$$

$$+\frac{1}{2}\mathbb{E}\int_0^1 \left(c(t)\right)^{-1}|\hat{K}(t) - K(t)|^2 dt$$

$$\leq \frac{1}{2}\alpha + \frac{1}{2}\alpha^{-1}\delta''\int_0^1 \mathrm{Var}\left(\hat{K}(t)\right)dt + \frac{1}{2}\alpha^{-1}\mathcal{D}_1\int_0^1 |t|\mathrm{Var}\left(\hat{K}(t)\right)dt$$

and a similar inequality holds true with $R_{3,2}$. Relying on (1.4.29), we deduce that

$$R_{3,1} + R_{3,2} \leq \alpha + \frac{1}{2}\alpha^{-1}\left(\delta''r_5 + \mathcal{D}_1 r_6^2\right)$$

$$\leq \alpha + \frac{1}{2}\alpha^{-1}\left(\delta''r_5 + 2r_6^2\right)$$

where in the last inequality we used our assumption that

$$\max(\delta_1, \delta_2, \frac{1}{2}\|W\|_{L^2}) \leq 1.$$

Then by (1.4.30),

$$|R_3| \leq \alpha + \frac{1}{2}\alpha^{-1}\left(\delta''r_5 + r_6^2\right) + r_3 + r_4. \tag{1.4.33}$$

Next, similarly to (4.13) in [15] by (1.4.21) and Proposition 1.4.2 we obtain

$$|R_4| \leq \mathbb{E}\int_{-1}^1 (|W| + 1)|tK(t)|dt \tag{1.4.34}$$

$$+\alpha^{-1}\int_{-1}^1 \left|\int_0^t P(z \leq W + u \leq z+\alpha)du\right||K(t)|dt$$

$$\leq (1 + \|W\|_{L^2})r_3 + \alpha^{-1}\delta''r_3 \leq 3r_3 + \alpha^{-1}\delta''r_3$$

where in the first inequality we used that $|K_v(t)| = |\mathbb{E}\hat{K}_v(t)| \leq \mathbb{E}|\hat{K}_v(t)|$ and the identity (1.4.11), while in the second inequality we used our assumption that $\|W\|_{L^2} \leq 2$. It follows from (1.4.9), (1.4.23), (1.4.24), (1.4.33) and (1.4.34) that for any real z,

$$|\mathbb{E}h_{z,\alpha}(W) - \mathbf{N}h_{z,\alpha}| = |\mathbb{E}\mathcal{A}f_{z,\alpha}(W)| \leq A + \alpha + \alpha^{-1}B \tag{1.4.35}$$

where

$$A = \delta_1 + \delta_2 + |\mathbb{E}W^2 - 1| + r_1 + r_2 + 6r_3 + r_4 + 2r_5 \quad \text{and}$$
$$B = (r_5 + r_3)(\delta_2 + |\mathbb{E}W^2 - 1| + 6r_3 + 4r_2 + 4r_5) + 2r_6^2.$$

Therefore, by (1.4.19) we have

$$d_K(\mathcal{L}(W), \mathcal{N}) \le A + \frac{3}{2}\alpha + \alpha^{-1}B. \tag{1.4.36}$$

Taking $\alpha = \sqrt{2/3}B^{1/2}$ we conclude that

$$d_K(\mathcal{L}(W), \mathcal{N}) \le \frac{3}{2}\alpha + A + \sqrt{3/2}B^{1/2} = A + \sqrt{6}B^{1/2}. \tag{1.4.37}$$

Note that when B vanishes then we obtain (1.4.37) directly by letting $\alpha \to 0$. Relying on the inequalities $\sqrt{a+b} \le \sqrt{a} + \sqrt{b}$ and $2ab \le a^2 + b^2$, $a, b \ge 0$ and the definitions of A and B it follows that

$$d_K(\mathcal{L}(W), \mathcal{N}) \le r_1 + 3r_2 + 13r_3 + r_4 + 6r_5 + 4r_6 + \delta_1 + 4\delta_2 + 4|\mathbb{E}W^2 - 1|. \tag{1.4.38}$$

We will prove now Theorem 1.2.2 relying on (1.4.38). Assume without loss of generality that $|\mathbb{E}W^2 - 1| \le 1$ since otherwise Theorem 1.2.2 trivially holds true. In order to estimate r_4 we first have,

$$r_4 \le \sum_{v \in V} \mathbb{E}|W - Z_v||X_v \min(Y_v^2, 1)| + \sum_{v \in V} \mathbb{E}|Z_v X_v \min(Y_v^2, 1)| := I_1 + I_2.$$

By the Hölder inequality, $\mathbb{E}|X_v Z_v Y_v^2| \le \|X_v\|_{L^4}\|Z_v\|_{L^2}\|Y_v\|_{L^4}^2$ for any $v \in V$ and therefore by the definitions of Y_v and Z_v,

$$I_2 \le D^4|V|\gamma_4^4.$$

Next, notice that

$$I_1 = \delta_4 + \sum_{v \in V} \mathbb{E}|W - Z_v|\mathbb{E}|X_v \min(Y_v^2, 1)|$$

where δ_4 is defined in Theorem 1.2.2. Let $v \in V$. Then by the Cauchy-Schwarz inequality and the definition of Z_v,

$$\mathbb{E}|W - Z_v| \le \|W - Z_v\|_{L^2} \le \|W\|_{L^2} + \|Z_v\|_{L^2}$$
$$\le 1 + \mathbb{E}W^2 + D^2\gamma_4 \le 2 + |\mathbb{E}W^2 - 1| + D^2\gamma_4$$

and by the Hölder inequality, $\mathbb{E}|X_v \min(Y_v^2, 1)| \le D^2\gamma_4^3$. We conclude that

$$r_4 \le D^4|V|\gamma_4^4 + \delta_4 + D^2|V|\gamma_4^3(2 + |\mathbb{E}W^2 - 1| + D^2\gamma_4) \tag{1.4.39}$$
$$\le D^4|V|\gamma_4^4 + \delta_4 + D^2|V|\gamma_4^3(3 + D^2\gamma_4) = \delta_4 + 2D^4|V|\gamma_4^4 + 3D^2|V|\gamma_4^3$$

where in the last inequality our assumption that $|\mathbb{E}W^2 - 1| \le 1$ is used.

Next, we will estimate r_5. Let $v, u \in V$. Then by the Hölder inequality and since $g_5(Y_v, Y_u) \le |Y_v|$,

$$\mathbb{E}|X_v X_u g_5(Y_v, Y_u)| \le \|X_v\|_{L^3}\|X_u\|_{L^3}\|Y_v\|_{L^3} \le D\gamma_4^3$$

and similar inequality holds true when X_u and Y_u are replaced with X_u^* and Y_u^*, respectively. Thus, by the definition of δ_5 and our assumption that $|B(v,3)| \leq c_0 D^\rho$,

$$r_5 \leq \delta_5 + \sum_{v \in V} \sum_{u: \in B(v,3)} 2D\gamma_4^3 \leq \delta_5 + 2c_0 D^{1+\rho}|V|\gamma_4^3. \qquad (1.4.40)$$

Since $g_6(Y_v, Y_u) \leq Y_v^2$ and $\|X_v X_u Y_v^2\|_{L^1} \leq \|X_v\|_{L^4}\|X_u\|_{L^4}\|Y_v\|_{L^4}^2 \leq D^2\gamma_4^4$ it follows in a similar way that

$$r_6^2 \leq \delta_6 + D^4|V|\gamma_4^4. \qquad (1.4.41)$$

Theorem 1.2.2 is deduced now from (1.4.38) and the estimates (1.4.6), (1.4.5), (1.4.4), (1.4.39), (1.4.40) and (1.4.41) where we have taken into account that $D^4|V|\gamma_4^4 \leq D^2|V|^{\frac{1}{2}}\gamma_4^2$ unless $D^2|V|^{\frac{1}{2}}\gamma_4^2 > 1$, while in the latter case the inequality in the statement of Theorem 1.2.2 trivially holds true.

Proof of Proposition 1.4.2. First notice that $\mathbb{E}W^2 = 0$ when $r_3 = 0$ and in this situation Proposition 1.4.2 trivially holds true, and so we assume without loss of generality that $r_3 > 0$. Set $\alpha = r_3$ and let $f : \mathbb{R} \to \mathbb{R}$ be the function defined by

$$f(x) = \begin{cases} -\frac{b-a+\alpha}{2} & \text{if } x \leq a - \alpha \\ \frac{1}{2\alpha}(x - a + \alpha)^2 - \frac{b-a+\alpha}{2} & \text{if } a - \alpha < x \leq a \\ x - \frac{a+b}{2} & \text{if } a < x \leq b \\ -\frac{1}{2\alpha}(x - b - \alpha)^2 + \frac{b-a+\alpha}{2} & \text{if } b < x \leq b + \alpha \\ \frac{b-a+\alpha}{2} & \text{if } x > b + \alpha. \end{cases}$$

Then

$$\sup|f| = \frac{b - a + \alpha}{2} \qquad (1.4.42)$$

and f' is a continuous function given by

$$f'(x) = \begin{cases} 1 & \text{if } a \leq x \leq b \\ 0 & \text{if } x \leq a - \alpha \text{ or } x \geq b + \alpha \\ \text{linear} & \text{otherwise.} \end{cases} \qquad (1.4.43)$$

By (1.4.42) and the Cauchy-Schwarz inequality,

$$\|W\|_{L^2}(b - a + \alpha)/2 = \|W\|_{L^2}\sup|f| \geq \mathbb{E}Wf(W) \qquad (1.4.44)$$
$$= \sum_{v \in V} \mathbb{E}X_v\big(f(W) - f(W - Y_v)\big) + \delta_1(f)$$

where $\delta_1(f) = \sum_{v \in V} \mathbb{E} X_v f(W \dotdiv Y_v)$. By the definition of δ_1 we have

$$|\delta_1(f)| \leq \sup |f| \delta_1 = \frac{b - a + \alpha}{2} \delta_1. \qquad (1.4.45)$$

Next, with $K_v(t), \hat{K}_v(t), K(t)$ and $\hat{K}(t)$ defined in (1.4.7),

$$\sum_{v \in V} \mathbb{E} X_v \big(f(W) - f(W - Y_v)\big) = \sum_{v \in V} \mathbb{E} X_v \int_{-Y_v}^0 f'(W + t) dt$$

$$= \sum_{v \in V} \mathbb{E} \int_{-\infty}^{\infty} f'(W + t) \hat{K}_v(t) dt$$

$$= \mathbb{E} \int_{-\infty}^{\infty} f'(W + t) \hat{K}(t) dt := H_1 + H_2 + H_3 + H_4$$

where

$$H_1 = \mathbb{E} f'(W) \int_{|t| \leq 1} K(t) dt, \quad H_2 = \mathbb{E} \int_{|t| \leq 1} \big(f'(W + t) - f'(W)\big) K(t) dt,$$

$$H_3 = \mathbb{E} \int_{|t| > 1} f'(W + t) \hat{K}(t) dt \quad \text{and}$$

$$H_4 = \mathbb{E} \int_{|t| \leq 1} f'(W + t) \big(\hat{K}(t) - K(t)\big) dt.$$

The inequalities

$$|H_3| \leq r_2, \quad |H_4| \leq \frac{b - a + 2\alpha}{8} + 2r_5 \quad \text{and} \quad |H_2| \leq 0.5 L(\alpha) \qquad (1.4.46)$$

where

$$L(\alpha) = \sup_{x \in \mathbb{R}} P(x \leq W \leq x + \alpha)$$

are established exactly as in the proof of Proposition 3.1 in [15], relying only on (1.4.8), (1.4.10)-(1.4.12), (1.4.29) and the properties of f. To estimate H_1, set $r = 1 - \mathbb{E} W^2$. Then by (1.4.8) and (1.4.11),

$$H_1 = \mathbb{E} f'(W) \int_{|t| \leq 1} K(t) dt \qquad (1.4.47)$$

$$= \mathbb{E} f'(W) \Big(\int_{-\infty}^{\infty} K(t) dt - \int_{|t| > 1} K(t) dt \Big)$$

$$\geq \mathbb{E} f'(W)(1 - \Delta_2 - r - r_2) \geq P(a \leq W \leq b) - |\Delta_2| - r_2 - |r|$$

where the second inequality holds true since $\mathbb{I}_{\{a \leq x \leq b\}} \leq f'(x) \leq 1$ for any real x. We conclude from (1.4.44)-(1.4.47) that

$$P(a \leq W \leq b) \leq (0.125 + 0.5\|W\|_{L^2} + 0.5\delta_1)(b - a) \qquad (1.4.48)$$

$$+ (0.25 + 0.5\|W\|_{L^2} + 0.5\delta_1)\alpha + 2r_2 + 2r_5 + |\Delta_2| + |r| + 0.5 L(\alpha).$$

Substituting $a = x$ and $b = x + \alpha$ in (1.4.48), we obtain

$$L(\alpha) \leq (0.375 + \|W\|_{L^2} + \delta_1)\alpha + 2r_2 + 2r_5 + |\Delta_2| + |r| + 0.5L(\alpha)$$

and hence

$$L(\alpha) \leq (0.75 + 2\|W\|_{L^2} + 2\delta_1)\alpha + 4r_2 + 4r_5 + 2|\Delta_2| + 2|r|. \qquad (1.4.49)$$

Finally, combining (1.4.48) and (1.4.49) we obtain (1.4.28), recalling that $\alpha = r_3$. □

1.5 Stein's method for diffusion approximations

In the next section we will prove a functional central limit theorem for random functions of the form

$$\mathcal{Z}_N(t) = N^{-\frac{1}{2}} \sum_{n=1}^{[Nt]} F(\xi_n, \xi_{2n}, ..., \xi_{\ell n})$$

and for similar ones. Stein's method in the situation of a functional CLT requires some notations which will be introduced in the following section.

1.5.1 *A functional CLT via Stein's method*

Let $(\Omega, \mathcal{F}, P)$ be a probability space and let $D = D[0, 1]$ be the Skorokhod space of all functions $w : [0, 1] \to \mathbb{R}$ which are continuous from the right and have left limits. Each element w of D is a bounded function, and we denote by $\sup |w|$ the supremum of $|w|$. Then the space D endowed with the supremum norm is a Banach space. Recall that a function $f : D \to \mathbb{R}$ is twice Fréchet differentiable if for any $w \in D$ there exist a continuous linear functional $Df(w)[\cdot] : D \to \mathbb{R}$ and a continuous bilinear form $D^2 f(w)[\cdot, \cdot] : D \times D \to \mathbb{R}$ such that

$$\lim_{h \to 0} \frac{|f(w + h) - f(w) - Df(w)[h] - \frac{1}{2}D^2 f(w)[h, h]|}{(\sup |h|)^2} = 0.$$

For any linear functional $\gamma[\cdot] : D \to \mathbb{R}$ and a bilinear form $\Gamma[\cdot, \cdot] : D \times D \to \mathbb{R}$ let

$$\|\gamma\| = \sup_{h \in D: \, \sup |h| = 1} |\gamma[h]|$$

and

$$\|\Gamma\| = \sup_{h \in D: \, \sup |h| = 1} |\Gamma[h, h]|$$

be their (operator) norms with respect to the supremum norm.

As in [1], let M be the space of all twice Fréchet differentiable functions $f : D \to \mathbb{R}$ such that

$$\|f\|_M := \sup_{w \in D}(1 + \sup|w|^3)^{-1}|f(w)| + \sup_{w \in D}(1 + \sup|w|^2)^{-1}\|Df(w)\|$$

$$+ \sup_{w \in D}(1 + \sup|w|)^{-1}\|D^2 f(w)\|$$

$$+ \sup_{w,h \in D, h \neq 0}(\sup|h|)^{-1}\|D^2 f(w+h) - D^2 f(w)\| < \infty.$$

Then $\| \cdot \|_M$ is a norm on M. Let M_0 be the space of all functions $f \in M$ such that

$$\|f\|_{M_0} := \|f\|_M + \sup_{w \in D}|f(w)| + \sup_{w \in D}\|Df(w)\| + \sup_{w \in D}\|D^2 f(w)\| < \infty.$$

$$(1.5.1)$$

Following [1], for any two probability measures $\mathbf{F}$ and $\mathbf{G}$ on D set

$$d_{M_0}(\mathbf{F}, \mathbf{G}) = \sup_{f \in M_0 : \|f\|_{M_0} \leq 1} \left| \int f d\mathbf{F} - \int f d\mathbf{G} \right|.$$

Then as in [1], $d_{M_0}(\cdot, \cdot)$ is a probability metric on the set of probability measures on D.

Next, let $N \in \mathbb{N}$ and let $G = (V, E)$ be a graph on $V = \{1, ..., N\}$ (where E is the set of edges and V is the set of vertices). Denote by d_G the shortest path distance on the graph and for any $n \in V$ set

$$N_n = \{m \in V : d_G(n, m) \leq 1\}.$$

Let $\{X_{1,N}, X_{2,N}, ..., X_{N,N}\}$, $N \in \mathbb{N}$ be a triangular array of centered random variables defined on the probability space $(\Omega, \mathcal{F}, P)$ and consider the random function $\mathcal{W}_N : [0, 1] \to \mathbb{R}$ given by

$$\mathcal{W}_N(t) = \sum_{n=1}^{[Nt]} X_{n,N}$$

which is a random element of D. We denote by $c_N(\cdot, \cdot)$ the covariance function of $\mathcal{W}_N(\cdot)$, namely, the function given by the formula $c_N(t, s) = \text{Cov}(\mathcal{W}_N(t), \mathcal{W}_N(s))$.

Theorem 1.5.1. *(i) Let $G_N(\cdot)$ be a Gaussian process which has the covariance function $c_N(\cdot, \cdot)$ and let $p, q \geq 1$. Then there exists an absolute constant C such that*

$$d_{M_0}(\mathcal{L}(\mathcal{W}_N), \mathcal{L}(G_N)) \leq C(d_1 + d_2 + d_3 + d_4) := C\tau_N \qquad (1.5.2)$$

where with $X_n = X_{n,N}$ and $\sigma_{n,m} = \mathbb{E}X_n X_m$,

$$d_1 = \sum_{n=1}^{N} \|X_n\|_{L^p} \varepsilon_{p,\infty}(\{n\}, N_n^c),$$

$$d_2 = \sum_{n=1}^{N} \sum_{m \in N_n} \|X_n X_m - \sigma_{n,m}\|_{L^q} \varepsilon_{q,\infty}(\{n, m\}, N_n^c \cap N_m^c),$$

$$d_3 = \sum_{n=1}^{N} \sum_{m,k \in N_n} \left(\mathbb{E}|X_n X_m X_k| + \mathbb{E}|X_n X_m| \mathbb{E}|X_k| \right),$$

$$d_4 = \sum_{n=1}^{N} \sum_{m \in N_n^c} \|X_n\|_{L^p} \|X_m\|_{L^p} \varepsilon_{p,p}(\{n\}, \{m\})$$

and the (mixing) coefficients $\varepsilon_{a,b}(\cdot, \cdot)$ are defined in (1.2.6).

(ii) Suppose that there exists $\Gamma > 0$ such that

$$\|\mathcal{W}_N(s) - \mathcal{W}_N(t)\|_{L^2} \leq \Gamma \frac{[Ns] - [Nt]}{N} \qquad (1.5.3)$$

for any $N \in \mathbb{N}$ and $0 \leq t \leq s \leq 1$. Further assume that the limits $\lim_{N \to \infty} c_N(t, s) = c(t, s), s, t \in [0, 1]$ exist and that

$$\lim_{N \to \infty} \tau_N \ln^2 N = 0.$$

Then $\mathcal{W}_N$ and G_N weakly converge in the Skorokhod space D to a continuous centered Gaussian process G with covariance function $c(\cdot, \cdot)$.

Theorem 1.5.1 *(i)* follows from the arguments in the proof of Lemma 3.1 in [1] (see the remark proceeding it). Not all the details are given in [1], and in the proof below we will provide the missing details and refer to [1] when the corresponding arguments go exactly as in [1]. The proof of Theorem 1.5.1 *(ii)* is a consequence of Proposition 3.1 in [2] and a standard tightness criterion from [6], and it is given here for readers' convenience, as well.

Proof of Theorem 1.5.1. Let $B = B(\cdot)$ be a standard Brownian motion on $[0, \infty)$ and let Z be a standard normal random variable which is independent of B. Let Y be the Ornstein-Uhlenbeck process with standard normal equilibrium distribution. Namely, Y is the unique strong solution of the stochastic differential equation

$$dY(t) = -Y(t)dt + \sqrt{2}dB(t), \quad Y(0) = Z$$

which is given by

$$Y(t) = e^{-t}Z + \sqrt{2} \int_0^t e^{-(t-s)} dB(s). \qquad (1.5.4)$$

By (1.5.4) and the Itô isometry, for any $u, v \geq 0$ we have

$$Y(u+v) - e^{-v}Y(u) \stackrel{d}{=} \sigma(v)Z \qquad (1.5.5)$$

where $\sigma(v) = \sqrt{1 - e^{-2v}}$ and $\mathcal{X} \stackrel{d}{=} \mathcal{Y}$ means that two random variables $\mathcal{X}$ and $\mathcal{Y}$ have the same distribution. Let $Y_1, ..., Y_N$ be N independent copies of Y and consider the random vector $\bar{Y} = \bar{Y}_N = \{Y_1, ..., Y_N\}$. Let $\Sigma = \Sigma_N$ be the covariance matrix of the random vector $\{X_{n,N} : 1 \leq n \leq N\}$. Then we can write $\Sigma = AA^*$ for some matrix $A = A_N$, where A^* stands for the transpose of A. Let $\tilde{Y} = \tilde{Y}_N$ be the random vector given by

$$\tilde{Y} = A\bar{Y} := \{\tilde{Y}_n : 1 \leq n \leq N\}.$$

Then by (1.5.5), for any $u, v \geq 0$,

$$\tilde{Y}(u+v) - e^{-v}\tilde{Y}(u) \stackrel{d}{=} \sigma(v)Z_\Sigma \qquad (1.5.6)$$

where $Z_\Sigma = \{Z_1, ..., Z_N\}$ is a centered Gaussian vector whose covariance matrix is Σ. Consider the D-valued Markov process $\{W(\cdot, u) : u \geq 0\}$ given by

$$W(t, u) = \sum_{n=1}^{N} \tilde{Y}_n(u) J_{\frac{n}{N}}(t) = \sum_{n=1}^{[Nt]} \tilde{Y}_n(u), \ t \in [0, 1]$$

where $J_a(t) = 1$ if $a \leq t$ and equals 0 otherwise. Then by (1.5.6), for any $u, v \geq 0$,

$$W(\cdot, u+v) - e^{-v}W(\cdot, u) \stackrel{d}{=} \sigma(v)G_N(\cdot). \qquad (1.5.7)$$

Here $G_N(\cdot)$ is a centered Gaussian process with the covariance function $c_N(\cdot, \cdot)$ which can be represented in the form

$$G_N(t) \stackrel{d}{=} \sum_{n=1}^{N} Z_n J_{\frac{n}{N}}(t) = \sum_{n=1}^{[Nt]} Z_n. \qquad (1.5.8)$$

Consider the semigroup $\{T_u : u \geq 0\}$ associated with $\{W(\cdot, u) : u \geq 0\}$ which acts on functions $f : D \to \mathbb{R}$ and is given by $T_u f(w) = \mathbb{E}[f(W(\cdot, u))|W(\cdot, 0) = w]$. Since $Y(0)$ and $Y(u) - e^{-u}Y(0)$ are independent, it follows from (1.5.7) that

$$T_u f(w) = \mathbb{E}f(we^{-u} + \sigma(u)G_N). \qquad (1.5.9)$$

Relying on (1.5.9), the proof of (2.9) in [1] is carried out here in the same way with the Gaussian processes $G_N(\cdot)$ in place of the Brownian motion

$B(\cdot)$ (denoted there by Z), and it follows that the infinitesimal generator $\mathcal{A}$ of $\{W(\cdot, u) : u \geq 0\}$ is given for any $f \in M$ by

$$\mathcal{A}f(w) := \lim_{u \to 0} \frac{T_u f(w) - f(w)}{u} = -Df(w)[w] \quad (1.5.10)$$

$$+\mathbb{E}D^2 f(w)[G_N, G_N] = Df(w)[w] + \sum_{n,m=1}^{N} \mathbb{E}X_{n,N} X_{m,N} D^2 f(w)[J_{\frac{n}{N}}, J_{\frac{m}{N}}]$$

where in the second equality we used (1.5.8) and the definition of Σ. Let $g \in M_0$ satisfy $\mathbb{E}g(G_N) = 0$ and consider the (differential) equation

$$\mathcal{A}f = g. \quad (1.5.11)$$

Then the function $f = f_g : D \to \mathbb{R}$ given by

$$f_g(w) = -\int_0^\infty T_u g(w) du$$

is well defined, in the domain of $\mathcal{A}$, solves (1.5.11) and satisfies $\|f_g\|_{M_0} \leq C_0 \|g\|_{M_0}$ for some absolute constant C_0. Indeed, the first three properties follow relying on (1.5.9) in a similar way to Section 4 of [17]. We remark that these properties were originally proved in [1], but there was a mistake in the proof, and this issue was sorted out in [17]. The upper bound $\|f_g\|_{M_0} \leq C_0 \|g\|_{M_0}$ is obtained similarly to [1]. Note that in [1] a similar upper bound was obtained only for the norm $\|\cdot\|_M$ relying on the dominated convergence theorem, but the same arguments yield the upper bound for the norm $\|\cdot\|_{M_0}$.

Finally, similarly to Lemma 3.1 in [1] and the approximations preceding Theorem 3 from there, for any function $f \in M_0$ we have

$$\left| \mathbb{E}Df(\mathcal{W}_N)[\mathcal{W}_N] - \sum_{1 \leq n,m \leq N} \mathbb{E}X_{n,N} X_{m,N} D^2 f(\mathcal{W}_N)[J_{\frac{n}{N}}, J_{\frac{m}{N}}] \right| \quad (1.5.12)$$

$$\leq 16(d_1 + d_2 + d_3 + d_4)\|f\|_{M_0}$$

where the definition of the mixing coefficients $\varepsilon_{s,t}$ is used to bound from above the expressions from (3.4) in [1]. Taking $f = f_g$, for which

$$\mathbb{E}\mathcal{A}f_g(\mathcal{W}_N) = \mathbb{E}g(\mathcal{W}_N) = \mathbb{E}g(\mathcal{W}_N) - \mathbb{E}g(G_N)$$

and then using (1.5.10) and (1.5.12) we complete the proof of Theorem 1.5.1 *(i)*.

For the proof of Theorem 1.5.1 *(ii)*, let $G = G(t)$ be a centered Gaussian process on $[0, 1]$ with a covariance function $c(\cdot, \cdot)$. We will prove next that G_N converges in distribution to G and that G has a continuous modification. This is indeed sufficient in order to derive Theorem 1.5.1 *(ii)* by virtue of Proposition 3.1 in [2], Theorem 1.5.1 *(i)* and our assumption that $\lim_{N\to\infty} \tau_N \ln^2 N = 0$. Since the covariance function of G_N converges to the covariance function of G, any finite dimensional distribution of G_N weakly converges to the corresponding finite dimensional distribution of G. We show now that $\{G_N : N \geq 1\}$ forms a tight sequence of D-valued random variables as $N \to \infty$. Indeed, by (1.5.3),

$$\mathbb{E}|G_N(s) - G_N(t)|^2 = c_N(s,s) - 2c_N(t,s) + c_N(t,t) \qquad (1.5.13)$$

$$= \mathbb{E}|\mathcal{W}_N(s) - \mathcal{W}_N(t)|^2 \leq \Gamma \frac{[Ns] - [Nt]}{N}$$

for any $N \in \mathbb{N}$ and $0 \leq t \leq s \leq 1$. Since G_N is Gaussian we have

$$\mathbb{E}|G_N(s) - G_N(t)|^4 \qquad (1.5.14)$$

$$= 3\big(\mathbb{E}|G_N(s) - G_N(t)|^2\big)^2 \leq 3\Gamma^2\Big(\frac{[Ns] - [Nt]}{N}\Big)^2.$$

We conclude from the Cauchy-Schwarz inequality that for any $0 \leq t \leq s \leq u \leq 1$,

$$\mathbb{E}|G_N(s) - G_N(t)|^2|G_N(u) - G_N(s)|^2 \qquad (1.5.15)$$

$$\leq \sqrt{\mathbb{E}|G_N(s) - G_N(t)|^4 \mathbb{E}|G_N(u) - G_N(s)|^4} \leq 3\Gamma^2\Big(\frac{[Ns] - [Nt]}{N}\Big)^2$$

and therefore by Ch. 15 of [6], the family of distributions $\{\mathcal{L}(G_N) : N \geq 1\}$ is tight and it follows that G_N converges in distribution to G. In order to show that G has a continuous modification, taking into account that G is Gaussian, for any $0 \leq t, s \leq 1$ we have

$$\mathbb{E}|G(s) - G(t)|^4 = 3\big(E|G(s) - G(t)|^2\big)^2.$$

Since $c(t,s) = \lim_{N\to\infty} c_N(t,s)$, the above right-hand side is the limit as $N \to \infty$ of $3\big(E|G_N(s) - G_N(t)|^2\big)^2$ and therefore by (1.5.14),

$$\mathbb{E}|G(s) - G(t)|^4 \leq 3\Gamma^2|s - t|^2$$

and G indeed has a continuous modification by the Kolmogorov continuity theorem, which completes the proof of Theorem 1.5.1 *(ii)*.   □

1.5.2   *Finite dimensional convergence rate*

In some situations we can consider (versions of) G and G_N on one probability space and obtain that $\mathbb{E}\sup_{0\leq t\leq 1}|G_N(t) - G(t)|$ converges to 0 when $N \to \infty$ with some convergence rate. This type of estimates together with Theorem 1.5.1 yield explicit estimates of $d_{M_0}(\mathcal{W}_N, G)$. One simple example is the case when $X_{n,N}$'s are uncorrelated with variance 1. In this situation we have $\mathcal{A}f(w) = -Df(w)[w] + \sum_{n=1}^{N} D^2 f(w)[J_{\frac{n}{N}}, J_{\frac{n}{N}}]$ which is the infinitesimal generator of the random function B_N given by $B_N(t) = B(\frac{[tN]}{N})$. Since (see Theorem 1 in [1]),

$$\mathbb{E} \sup_{0 \leq t \leq 1} |B(t) - B_N(t)| \leq CN^{-\frac{1}{2}} \ln N$$

for some absolute constant C, we obtain from Theorem 1.5.1 *(i)* an explicit convergence rate. The situation when $X_{n,N}$'s are "asymptotically correlated" as $N \to \infty$ can be considered, as well. See Section 4 in [2] for another example in which similar realization exists. In general, it is unclear whether two Gaussian processes whose covariance functions are "close" admit such a realization, and in our application to nonconventional sums the weak limit (in general) will not have independent increments. Still, the following estimates hold true.

Lemma 1.5.2. *Let* $d, N \in \mathbb{N}$ *and let*

$$k_t : \mathbb{R}^d \to \mathbb{R}, \; t = (t_1, ..., t_d) \in [0,1]^d$$

be a family of functions. Let μ *be a probability measure on* $[0,1]^d$ *and consider the function* $g = g_{k,\mu,d} : D \to \mathbb{R}$ *given by* $g(w) = \int k_t(w(t_1), ..., w(t_d)) d\mu(t)$. *Suppose that for any* $t \in \mathbb{R}^d$ *the function* k_t *and its Fourier transform are absolutely integrable. Then*

$$|\mathbb{E}g(G_N) - \mathbb{E}g(G)| \leq \sup_{t \in supp(\mu)} C_{N,t,d} \int |\hat{k}_t(x)| \|x\|_1^2 dx$$

where $\|x\|_1 = \sum_{i=1}^{d} |x_i|$, $C_{N,t,d} = \sup_{1 \leq i,j \leq d} |c_N(t_i, t_j) - c(t_i, t_j)|$ *and* $c(t, s)$ *are the limits defined in Theorem 1.5.1 (ii).*

Proof. By the Fubini theorem, it suffices to prove the lemma in the case when μ assigns unit mass to a single point $t = (t_1, ..., t_d) \in \mathbb{R}^d$. Let $h : \mathbb{R}^d \to \mathbb{R}$ be an absolutely integrable function with absolutely integrable Fourier transform. Let $\varphi_{t,d}$ and $\varphi_{t,d,N}$ be the characteristic functions of the random vectors $(G(t_1), ..., G(t_d))$ and $(G_N(t_1), ..., G_N(t_d))$, respectively, which are given by

$$\varphi_{t,d}(x) = \exp\left(-\frac{1}{2} \sum_{1 \leq i,j \leq d} c(t_i, t_j) x_i x_j\right) \text{ and}$$

$$\varphi_{t,d,N}(x) = \exp\left(-\frac{1}{2} \sum_{1 \leq i,j \leq d} c_N(t_i, t_j) x_i x_j\right).$$

By the mean value theorem,

$$|\varphi_{t,d,N}(x) - \varphi_{t,d}(x)| \leq C_{N,t,d} \|x\|_1^2 \tag{1.5.16}$$

and we conclude from the Fourier inversion formula that

$$\big|\mathbb{E}h\big(G_N(t_1), ..., G_N(t_d)\big) - \mathbb{E}h\big(G(t_1), ..., G(t_d)\big)\big|$$

$$= \Big| \int_{\mathbb{R}^d} \hat{h}(x)\big(\varphi_{t,d,N}(x) - \varphi_{t,d}(x)\big)dx \Big| \le C_{N,t,d} \int_{\mathbb{R}^d} |\hat{h}(x)|\|x\|_1^2 dx$$

and the lemma follows.   □

Note that d may depend on N and so, when $\sup_{0 \le t,s \le 1} |c_N(t,s) - c(t,s)|$ converges to 0 as $N \to \infty$ then Lemma 1.5.2 provides estimates on the difference $|\mathbb{E}g(G_N) - \mathbb{E}g(G)|$ for functions g which can be approximated sufficiently well by functions of the form $g_{\kappa,\mu,d}$, $d \in \mathbb{N}$ such that the integrals $\int |\hat{k}_t(x)|\|x\|_1^2 dx$ have some regularity as a function of t and d.

Next let $h : \mathbb{R}^d \to \mathbb{R}$ be a twice differentiable function and let $v \in \mathbb{R}^d$. Denote by $H_h(v)[\cdot, \cdot]$ the Hessian form of h at v which is given by

$$H_h(v)[u, p] = \sum_{1 \le i,j \le d} u_i p_j \frac{\partial^2 h(v)}{\partial v_i \partial v_j}.$$

For any $u \in \mathbb{R}^2$ set $\|u\|_\infty = \max\{|u_i| : 1 \le i \le d\}$ and let

$$\|H_v\|_\infty = \sup\{|H_v[u, u]| : \|u\|_\infty = 1\}$$

and

$$L_\infty(H_h) = \sup_{v,u \in \mathbb{R}^d, u \ne 0} \frac{\|H_h(v + u) - H_h(v)\|_\infty}{\|u\|_\infty}$$

be the (operator) norm and Lipschitz constant of H_v with respect to the supremum norm, respectively. We consider the following two norms of h:

$$\|h\|_{M,d} = \rho_{\infty,3}(h) + \rho_{\infty,2}(\|\nabla h(\cdot)\|_1) + \rho_{\infty,1}(\|H_h(\cdot)\|_\infty) + L_\infty(H_h) \text{ and}$$

$$\|h\|_{M_0,d} = \|h\|_{M,d} + 2\rho_{\infty,0}(h) + \rho_{\infty,0}(\|\nabla h(\cdot)\|_1) + \rho_{\infty,0}(\|H_h(\cdot)\|_\infty)$$

where for any $k \ge 0$ and $f : \mathbb{R}^d \to \mathbb{R}$,

$$\rho_{\infty,k}(f) = \sup_{v \in \mathbb{R}^d} \frac{|f(v)|}{1 + (\|v\|_\infty)^k}.$$

The following result follows.

Corollary 1.5.3. *Let $g = g_{k,\mu,d} : D \to \mathbb{R}$ be as in Lemma 1.5.2. Suppose, in addition, that the functions $k_t(\cdot)$, $t \in supp(\mu)$ are twice differentiable. Then there exists an absolute constant c_0 such that*

$$|\mathbb{E}g(\mathcal{W}_N) - \mathbb{E}g(G)|$$

$$\le c_0\Big(\tau_N \sup_{t \in supp(\mu)} \|k_t(\cdot)\|_{M_0,d} + \sup_{t \in supp(\mu)} C_{N,t,d} \int |\hat{k}_t(x)|\|x\|_1^2 dx\Big).$$

Proof. By Theorem 1.5.1 *(i)* and Lemma 1.5.2 it is sufficient to show that

$$\|g_{k,\mu,d}\|_{M_0} \leq c_0 \sup_{t \in supp(\mu)} \|k_t(\cdot)\|_{M_0,d}. \tag{1.5.17}$$

By the dominated convergence theorem it is sufficient to consider the case when μ assigns unit mass to some point $t \in [0,1]^d$. Let $t \in [0,1]^d$ and consider the map $\pi_t : D \to \mathbb{R}^d$ given by $\pi_t(w) = (w(t_1), ..., w(t_d))$. Let $h : \mathbb{R}^d \to \mathbb{R}$ be a twice differentiable function such that $\|h\|_{M_0,d} < \infty$. In order to prove (1.5.17), it is sufficient to show that the lifted function h_t given by $h_t = h \circ \pi_t$ satisfies

$$\|h_t\|_M \leq \|h\|_{M,d} \text{ and } \|h_t\|_{M_0} \leq \|h\|_{M_0,d} \tag{1.5.18}$$

and this is a consequence of the equalities $Dh_t(w)[w_1] = \nabla h(\pi_t(w)) \cdot \pi_t(w_1)$ and $D^2 h_t(w)[w_1, w_2] = H_h(\pi_t(w))[\pi_t(w_1), \pi_t(w_2)]$ where $u \cdot v$ stands for the standard inner product of two vectors $u, v \in \mathbb{R}^d$. $\square$

Remark 1.5.4. Let $d \in \mathbb{N}$ and $t = (t_1, ..., t_d) \in [0,1]^d$. Then Corollary 1.5.3 provides, in particular, rates in the multidimensional CLT for the random vectors $\mathcal{W}_N(\bar{t}) := (\mathcal{W}_N(t_1), ..., \mathcal{W}_N(t_d))$, namely, upper bounds of expressions of the form

$$|\mathbb{E}h(\mathcal{W}_N(\bar{t})) - \mathbb{E}h(G(t_1), G(t_2), ..., G(t_d))|$$

for some family of twice differentiable functions h. It is customary to have explicit dependence on d in such rates, under some restrictions on the growth rates of some norms of the function h and its derivatives with respect to the Euclidean norm $\|\cdot\|_2$. Our estimates, of course, can be reformulated in this way since $\|v\|_\infty \leq \|v\|_2 \leq \sqrt{d}\|v\|_\infty$ for any $v \in \mathbb{R}^d$.

1.6 A nonconventional functional CLT

Let $(\Omega, \mathcal{F}, P)$, $\mathcal{F}_{n,m}$, $\{\xi_n : n \geq 0\}$ and F be as described in Section 1.3.1. Consider the random function $\mathcal{Z}_N$ given by

$$\mathcal{Z}_N(t) = N^{-\frac{1}{2}} \sum_{n=1}^{[Nt]} F(\xi_n, \xi_{2n}, ..., \xi_{\ell n}).$$

The main goal of this section is to prove a functional CLT for these random functions. We need first the following lemma.

Lemma 1.6.1. *Suppose that the assumptions of Theorem 1.3.4 hold true. Let $t, s > 0$ and $N \in \mathbb{N}$ and set $b_N(t,s) = E\mathcal{Z}_N(t)\mathcal{Z}_N(s)$. Then the limit $\lim_{N\to\infty} b_N(t,s) = b(t,s)$ exists and for any $N \in \mathbb{N}$,*

$$|b_N(t,u) - b(t,u)| \leq c_\ell C_0 N^{-\frac{1}{2}} \tag{1.6.1}$$

where C_0 is defined in Theorem 1.3.4 and c_ℓ is a constant which depends only on ℓ.

The proof of this lemma follows from the arguments in [44] similarly to the proof of (1.3.12) in Theorem 1.3.4 (see [45] for an explicit formula for the limits $b(t, s)$ and for other properties of these limits see [31]).

Now we can state our nonconventional functional CLT.

Theorem 1.6.2. *(i) Suppose that Assumptions 1.3.1 and 1.3.3 hold true with θ and $b > 0$ such that $\theta > \frac{4b}{b-2}$. Then the random function $\mathcal{Z}_N(\cdot)$ converges in distribution to a centered Gaussian process $\eta(\cdot)$ whose covariance function is $b(\cdot, \cdot)$.*

(ii) Moreover, let k, μ and d be as in Lemma 1.5.2 and consider the function $g = g_{k,\mu,d}$. Further assume that $\|\kappa_t(\cdot)\|_{M_0} \leq 1$ and $\int |\hat{k}_t(x)| \|x\|_1^2 dx \leq 1$ for any $t \in supp(\mu)$. Then

$$|\mathbb{E}g(\mathcal{Z}_N) - \mathbb{E}g(\eta)| \leq a_\ell C_2 \max(B, B^3) N^{-(\frac{1}{2} - 2\theta_1)} \qquad (1.6.2)$$

where $B = K(1 + \gamma_m^\iota)$, C_2 is defined in Theorem 1.3.7,

$$\theta_1 = \frac{3b}{4b + 2\theta(b-2)} < \frac{1}{4}$$

and a_ℓ is a constant which depends only on ℓ. When Assumption 1.3.2 holds true then the above right-hand side can be replaced with $a_\ell C_1 \max(B, B^3) N^{-\frac{1}{2}} \ln^2(N+1)$, where C_1 is defined in Theorem 1.3.7.

As in Theorem 1.3.7, when $\beta_{r_0} = 0$ for some r_0 then all the above results hold true when F is only a Borel function which satisfies (1.3.6). In particular, when our mixing assumptions hold true with the σ-algebras $\mathcal{F}_{m,n} = \sigma\{\xi_m, ..., \xi_n\}$ then we can take $r_0 = 0$ and we obtain a functional nonconventional CLT when F only satisfies (1.3.6).

Proof of Theorem 1.6.2. Let $l \geq 1$, set $r = [\frac{l}{3}]$ and consider the random functions $\mathcal{Z}_{N,r}(\cdot)$ and $\bar{\mathcal{Z}}_{N,r}(\cdot)$ given by

$$\mathcal{Z}_{N,r}(t) = N^{-\frac{1}{2}} \sum_{n=1}^{[Nt]} F(\xi_{n,r}, \xi_{2n,r}, ..., \xi_{\ell n,r}) \text{ and}$$

$$\bar{\mathcal{Z}}_{N,r}(t) = \mathcal{Z}_{N,r}(t) - \mathbb{E}\mathcal{Z}_{N,r}(t).$$

Let $g \in M_0$ be such that $\|g\|_{M_0} \leq 1$. We first estimate the difference

$$|\mathbb{E}g(\mathcal{Z}_N(\cdot)) - \mathbb{E}g(\bar{\mathcal{Z}}_{N,r}(\cdot))|.$$

By the definition of M_0, for any $w, h \in D$ we have

$$|g(w + h) - g(w)| \le \sup |h|$$

and therefore

$$|\mathbb{E}g(\mathcal{Z}_N(\cdot)) - \mathbb{E}g(\bar{\mathcal{Z}}_{N,r}(\cdot))| \le \overline{\mathbb{E}} \sup |\mathcal{Z}_N(\cdot) - \bar{\mathcal{Z}}_{N,r}(\cdot)|. \qquad (1.6.3)$$

In order to estimate the right-hand side of (1.6.3) we apply first (1.3.45) and (1.3.46) to obtain

$$\sup |\bar{\mathcal{Z}}_{N,r}(\cdot) - \mathcal{Z}_{N,r}(\cdot)|$$

$$\le N^{-\frac{1}{2}} \sum_{n=1}^{N} |\mathbb{E}F(\Xi_{n,r})| \le 7\sigma \ell^2 R s_{r,l} := R_1, \quad P\text{-a.s.}$$

where $s_{r,l}$ is defined in (1.3.48). Secondly, we apply (1.3.42) and since $m > \iota b$ we have

$$\left\| \sup |\mathcal{Z}_N(\cdot) - \mathcal{Z}_{N,r}(\cdot)| \right\|_{L^b} \le N^{-\frac{1}{2}} \sum_{n=1}^{N} \|F(\Xi_n) - F(\Xi_{n,r})\|_{L^b}$$

$$\le N^{\frac{1}{2}} \ell^2 K (1 + \gamma_m^\iota) \beta_{q,r}^\kappa := R_2$$

concluding from (1.6.3) that

$$|\mathbb{E}g(\mathcal{Z}_N(\cdot)) - \mathbb{E}g(\bar{\mathcal{Z}}_{N,r}(\cdot))| \qquad (1.6.4)$$

$$\le \left\| \sup |\mathcal{Z}_N(\cdot) - \bar{\mathcal{Z}}_{N,r}(\cdot)| \right\|_{L^b} \le R_1 + R_2 := R(N, r).$$

Next, consider l's of the form $l = 3N^{\zeta_1} + 3$, where ζ_1 is a parameter yet to be chosen. Consider the triangular array $\{X_{1,N}, X_{2,N}, ..., X_{N,N}\}$ where the $X_{n,N}$'s are given by (1.3.52). The purpose of the following arguments is to show that Theorem 1.5.1 can be applied with the random functions

$$\mathcal{W}_N = \mathcal{W}_N(\cdot) = \bar{\mathcal{Z}}_{N,r}(\cdot).$$

We begin with showing that condition (1.5.3) holds true when ζ_1 is chosen appropriately. We first claim that there exists $C > 0$ such that

$$\mathrm{Var}(\mathcal{Z}_N(s) - \mathcal{Z}_N(t)) \le C^2 \frac{[Ns] - [Nt]}{N} \qquad (1.6.5)$$

for any $N \in \mathbb{N}$ and $0 \le t \le s \le 1$. Indeed, by Corollary 1.3.14 there exists a constant $c_0 > 0$ such that for any $n, m \in \mathbb{N}$,

$$|\mathrm{Cov}(F(\Xi_n), F(\Xi_m))| \le c_0 \tau([\frac{1}{3} d_\ell(n, m)])$$

where $\tau(k) = \phi_k^{1-\frac{1}{b}} + \beta_{q,k}^{\kappa} \leq dk^{-(1-\frac{1}{b})\theta}$, $k \geq 0$ and $d_\ell(n,m)$ is given by (1.3.13). For any $k \geq 0$ and $n \in \mathbb{N}$ there exist at most $2\ell^2$ natural m's such that $d_\ell(n,m) = k$. Therefore since $\theta(1 - \frac{1}{b}) > 1$,

$$\sum_{m=1}^{\infty} |\text{Cov}(F(\Xi_n), F(\Xi_m))| \leq 6\ell^2 c_0 \sum_{s=0}^{\infty} \tau(s) := A < \infty$$

for any natural n, and so

$$\text{Var}(\mathcal{Z}_N(s) - \mathcal{Z}_N(t)) \leq N^{-1} \sum_{n=[Nt]}^{[Ns]} \sum_{m=[Nt]}^{[Ns]} |\text{Cov}(F(\Xi_n), F(\Xi_m))|$$

$$\leq A \frac{[Ns] - [Nt]}{N}$$

implying (1.6.5). Next, by (1.3.42),

$$\|\mathcal{Z}_{N,r}(s) - \mathcal{Z}_{N,r}(t) - (\mathcal{Z}_N(s) - \mathcal{Z}_N(t))\|_{L^2}$$

$$\leq N^{-\frac{1}{2}} \sum_{n=[Nt]}^{[Ns]} c_2 \beta_{q,r}^{\kappa} \leq c_2 c_3 \frac{[Ns] - [Nt]}{N} \leq c_2 c_3 \sqrt{\frac{[Ns] - [Nt]}{N}}$$

where $c_3 = K\ell(1 + \ell\gamma_m^L)$, assuming that $\beta_{q,r}^{\kappa} \leq c_2 N^{-\frac{1}{2}}$ for some $c_2 > 0$. Finally, we recall that $\beta_{q,r}^{\kappa} \leq dr^{-\theta}$ and $r = [\frac{l}{3}] \geq N^{\zeta_1}$. Therefore $\beta_{q,r}^{\kappa} \leq dN^{-\frac{1}{2}}$ when $\zeta_1 \geq \frac{1}{2\theta}$ and in this case we can take $c_2 = d$. It follows that (1.6.5) holds true with $\mathcal{Z}_{N,r}$ in place of $\mathcal{Z}_N$, possibly with a different constant C' in place of C. Since

$$E|\bar{\mathcal{Z}}_{N,r}(s) - \bar{\mathcal{Z}}_{N,r}(t)|^2 = \text{Var}(\mathcal{Z}_{N,r}(s) - \mathcal{Z}_{N,r}(t))$$

this completes the proof that condition (1.5.3) holds true for the random functions $\mathcal{W}_N = \bar{\mathcal{Z}}_{N,r}$, $N \geq 1$.

Next, let $\bar{b}_N(\cdot, \cdot)$ be the covariance function of $\mathcal{W}_N$. We claim that, under additional restrictions on ζ_1, the limiting covariances of $\bar{b}_N(\cdot, \cdot)$ are $b(\cdot, \cdot)$ appearing in Lemma 1.6.1. Indeed, for any $0 \leq t, s \leq 1$ we have

$$|\bar{b}_N(t, s) - b_N(t, s)| = |\mathbb{E}\bar{\mathcal{Z}}_{N,r}(t)\bar{\mathcal{Z}}_{N,r}(s) - \mathbb{E}\mathcal{Z}_N(t)\mathcal{Z}_N(s)|$$

$$\leq \|\mathcal{Z}_N(t)\|_{L^2}\|\mathcal{Z}_N(s) - \bar{\mathcal{Z}}_{N,r}(s)\|_{L^2} + \|\bar{\mathcal{Z}}_{N,r}(s)\|_{L^2}\|\mathcal{Z}_N(t) - \bar{\mathcal{Z}}_{N,r}(t)\|_{L^2}$$

$$\leq R(N,r)(2c_\ell^{\frac{1}{2}} C_0^{\frac{1}{2}} + \sigma + R(N,r)) := B(N,r).$$

In the first inequality we used the Cauchy-Schwarz inequality while in the second inequality we used (1.6.4) and (1.3.12) with $[Nt]$ and $[Ns]$ in place of N. Let $\zeta_1 \geq \frac{1}{2\theta}$ be such that $\tau_N \ln^2 N$ and $B(N,r)$ converge to 0 as $N \to \infty$, where τ_N is defined in Theorem 1.5.1 (for some choice of p and q).

Then by Theorem 1.5.1 the random function $\mathcal{W}_N$ converges in distribution on the Skorokhod space D to a centered continuous Gaussian process η with the covariance function $b(\cdot, \cdot)$. Finally, by Proposition 3.1 in [2] we deduce that $\mathcal{Z}_N(\cdot)$ weakly converges to η, assuming that $R(N, r) \ln^2 N$ converges to 0 as $N \to \infty$. By (1.2.17), Lemma 1.3.15 and similarly to (1.3.53) the d_i's defined in Theorem 1.5.1 *(i)* satisfy

$$d_1 \leq cN^{\frac{1}{2}}B\phi_r^{1-\frac{1}{b}}, \quad d_2 \leq cB^2 D\phi_r^{1-\frac{2}{b}}, d_3 \leq cD^2 N^{-\frac{1}{2}}B^3$$

$$\text{and} \quad d_4 \leq cNB^2\phi_r^{1-\frac{2}{b}}$$

where c depends only on ℓ, $B = K(1 + \gamma_m^\ell)$, D is given by (1.3.19) and we have chosen here $p = 2q = b$. Since $r = [\frac{l}{3}]$ we have

$$\phi_r \leq dr^{-\theta} \leq dN^{-\theta\zeta_1} \tag{1.6.6}$$

and, taking into account (1.3.19), we deduce that there exists a constant c_4 which depends only on ℓ such that

$$\tau_N \leq c_4 d \max(B, B^3) N^{-a_1} \tag{1.6.7}$$

where

$$a_1 = \min\left((1 - \frac{2}{b})\zeta_1\theta - 1, \frac{1}{2} - 2\zeta_1\right).$$

Taking $\left(\theta(1 - \frac{2}{b})\right)^{-1} < \zeta_1 < \frac{1}{4}$ we have $a_1 > 0$ and $\tau_N \ln^2 N$ converges to 0 as $N \to \infty$. Such choice of ζ_1 is possible by our assumption that $\theta > \frac{4b}{b-2}$ and note that such ζ_1 indeed satisfies $\zeta_1 \geq \frac{1}{2\theta}$. Finally, by (1.6.6) the definition of $s_{r,l}$ in (1.3.48), (1.3.19) and similarly to (1.3.68) we have

$$R(N, r) = R_1 + R_2 \leq c_2 B(s_{r,l} + N^{\frac{1}{2}}\beta_{q,r}^k) \leq c_5 dB N^{-a_2}$$

where c_5 is a constant which depends only on ℓ and

$$a_2 = \min\left(\frac{1}{2} - \zeta_1, (1 - \frac{1}{b})\theta\zeta_1 - \frac{1}{2}\right) \geq a_1.$$

Therefore, for the above choice of ζ_1 the sequences $R(N, r) \ln^2 N$ and $B(N, r)$ converge to 0 as $N \to \infty$, as well, and we have completed the proof of the first statement of Theorem 1.6.

In order to prove the second statement, we take $\zeta_1 = \frac{3b}{4b+2\theta(b-2)}$ which solves the equation

$$\zeta_1\theta(1 - \frac{2}{b}) - 1 = \frac{1}{2} - 2\zeta_1.$$

This choice of ζ_1 together with Corollary 1.5.3 and the previous estimates yield (1.6.2), taking into account (1.3.69). We note that $a_1 > 0$ for this ζ_1 in view of our assumption that $\theta > \frac{4b}{b-2}$. When Assumption 1.3.2 holds true then as in the proof of Theorem 1.3.7 *(i)* we take $l = A \ln(N + 1) + 3$, where $A = \frac{45}{-\ln(c)}$ and obtain the desired upper bound.

1.7   Extensions to nonlinear indexes

1.7.1   *Preliminaries*

Let $q_1, ..., q_\ell$ be strictly increasing functions which map $\mathbb{N}$ to $\mathbb{N}$ and are ordered so that

$$q_1(n) < q_2(n) < ... < q_\ell(n)$$

for any sufficiently large n. For any $N \in \mathbb{N}$ consider the random function

$$\mathcal{Z}_N(t) = N^{-\frac{1}{2}} \sum_{n=1}^{[Nt]} \left(F(\xi_{q_1(n)}, \xi_{q_2(n)}, ..., \xi_{q_\ell(n)}) - \bar{F} \right) \qquad (1.7.1)$$

where $\bar{F}$ is given by (1.3.7). We further assume that the difference $q_i(n) - q_{i-1}(n)$ tends to ∞ as $n \to \infty$ for any $i = 1, 2, ..., \ell$, where $q_0 \equiv 0$, though the situation when some of these differences are nonnegative constants can be considered, as well (see Section 3 in [31]). By disregarding some of the first summands in (1.7.1), we can also consider the case when q_i's are strictly increasing on some ray $[Q, \infty)$, and then we can consider polynomial q_i's with positive leading coefficients which map $\mathbb{N}$ to itself.

Next, for any $n, m \in \mathbb{N}$ set

$$\tilde{d}_\ell(n, m) = \min_{1 \leq i,j \leq \ell} |q_i(n) - q_j(m)|$$

and consider the graph $\tilde{G}(N, l) = (V_N, \tilde{E}_{N,l})$ on $V_N = \{1, 2, ..., N\}$ where the set of edges is given by

$$\tilde{E}_{N,l} = \{(n, m) \in V_N \times V_N : \tilde{d}_\ell(n, m) < l\}.$$

We begin with imposing restrictions on q_i's which guarantee that the cardinalities of the sets $N_n = \{1 \leq m \leq N : \tilde{d}_\ell(n, m) < l\}$ are of order l. First notice that

$$N_n = \bigcup_{1 \leq i,j \leq \ell} V_N \cap J_{i,j}(n)$$

where $J_{i,j}(n) = \left([q_j^{-1}(q_i(n) - l)], [q_j^{-1}(q_i(n) + l)] \right)$.

Assumption 1.7.1. There exists $Q \geq 1$ such that for any $a, b \geq 1$,

$$|q_j^{-1}(a) - q_j^{-1}(b)| \leq Q(1 + |a - b|) \qquad (1.7.2)$$

for any $1 \leq j \leq \ell$.

When (1.7.2) holds true then

$$|J_{i,j}(n)| \le Q(1 + 2l)$$

and we deduce that

$$|N_n| \le 2\ell^2 Q(1 + l).$$

From the above inequality we derive that the order of the cardinality of balls of radius 3 in the graph $\tilde{G}(N, l)$ is at most l^3. Condition (1.7.2) holds true, for instance, when all q_j's have the form $q_j(x) = [p_j(x)]$ where each p_j is a strictly increasing function whose inverse p_j^{-1} has bounded derivative on some ray $[K, \infty)$. For example we can take p_j's to be a polynomials with positive leading coefficient, exponential functions etc.

The following theorem is proved similarly to Theorem 1.3.7 *(i)*.

Theorem 1.7.2. *Suppose that Assumptions 1.3.1 and 1.3.2 are satisfied with $b = 5$. Furthermore, assume that (1.7.2) holds true and that the limit $\sigma^2 = \lim_{N\to\infty} \mathbb{E} Z_N^2$, $Z_N = \mathcal{Z}_N(1)$ exists and is positive. Then Z_N converges in distribution to a centered normal random variable with variance σ^2 and there exists a constant c which depends only on ℓ so that for any $N \ge 1$,*

$$\max\left(d_W(\mathcal{L}(\sigma^{-1}Z_N), \mathcal{N}), d_K(\mathcal{L}(\sigma^{-1}Z_N), \mathcal{N})\right)$$

$$\le cQ^4 C_1^{\frac{3}{2}}(1 + R) \max(R^{\frac{5}{6}}, R^2) N^{-\frac{1}{2}} \ln^4(N + 1) + 4\sigma^{-2}|\mathbb{E} Z_N^2 - \sigma^2|$$

where C_1 and R are defined in Theorem 1.3.7 (i). Furthermore, if the limits $b(t, s)$ appearing in Lemma 1.6.1 exist then the random function $\mathcal{Z}_N(\cdot)$ weakly converges in the Skorokhod space D toward a centered continuous Gaussian process η with covariance function $b(\cdot, \cdot)$.

A corresponding version of Theorem 1.3.7 *(ii)* follows, as well, but the purpose of this section is only to demonstrate that Stein's method in the nonconventional setup is not restricted to linear indexes. Some convergence rate in the functional CLT can be given, as well.

In [45] existence of the limits $b(t, s)$ was proved when for some $1 \le k \le \ell$ we have $q_i(n) = in$ for any $i \le k$ and $n \in \mathbb{N}$, while for any $k < i \le \ell$ and $\varepsilon > 0$,

$$\lim_{n\to\infty} (q_i(n + 1) - q_i(n)) = \infty \text{ and}$$

$$\liminf_{n\to\infty} (q_i(\varepsilon n) - q_{i-1}(n)) > 0$$

which imply that

$$\lim_{n\to\infty} (q_i(\varepsilon n) - q_{i-1}(n)) = \infty.$$

In [31] the authors showed that the limits $b(t, s)$ exist when all q_j's are polynomials. In the situation of [45], convergence rate can be obtained assuming that there exist constants $c, \alpha, \gamma_1 > 0$ and $0 < \gamma_2 < 1$ such that for any $i > k$ and $n \in \mathbb{N}$,

$$q_i(n+1) - q_i(n) \geq cn^\alpha \text{ and } q_i(n^{\gamma_2}) - q_{i-1}(n) \geq cn^{\gamma_1} \quad (1.7.3)$$

while in the setup of [31] convergence rates toward these limits can be easily recovered from the arguments in the proof of their existence.

1.7.2   *Another example with nonlinear indexes*

For general indexes $q_1(n), q_2(n), ..., q_\ell(n)$ existence of the limits $b(t, s)$ (and thus the CLT) depends on properties of solutions of equations of the form $q_i(n) = q_j(m) + u$, where $u \in \mathbb{N}$ and the pair $(n, m) \in \mathbb{N} \times \mathbb{N}$ satisfy some restriction of the form $m = cn + k$ for some c and k. When all q_i's are polynomials then the related number theory problem is solved in Section 4 of [31].

Instead of providing technical conditions for the existence of these limits, we will provide a demonstrative example. Recall that two positive numbers a, b are multiplicatively independent if $\frac{\ln a}{\ln b} \notin \mathbb{Q}$. Let $\lambda_1, \lambda_2 > 1$ and consider the functions $q_1(x) = \lambda_1^x$ and $q_2(x) = \lambda_1^x \lambda_2^x$. In order to avoid complications, we will not replace q_i's with their integer part and instead we consider a family of random variables $\{\xi_u : u \geq 0\}$ on the probability space $(\Omega, \mathcal{F}, P)$ which satisfies the conditions from Section 1.3.1. Namely, we assume that the distribution of ξ_u does not depend on u, that distribution of (ξ_u, ξ_v) depends only on $u - v$ and that there exists a nested family $\mathcal{F}_{u,v}$, $0 \leq u \leq v$ of sub-σ-algebras of $\mathcal{F}$ such that (1.3.10) holds true for any $\rho > 0$ in place of n where ϕ_ρ and $\beta_{q,\rho}$ are defined similarly to ϕ_n and $\beta_{q,n}$ but with nonnegative ρ's in place of natural n's. Let μ be the distribution of ξ_u and for any $a \geq 1$ set $\gamma_a^a = \|\xi_u\|_{L^a}^a = \int |x|^a d\mu$. Suppose that $\bar{F}$ in (1.3.7) vanishes, and consider the normalized random function

$$\mathcal{Z}_N(t) = N^{-\frac{1}{2}} \sum_{n=1}^{[Nt]} F(\xi_{\lambda_1^n}, \xi_{\lambda_1^n \lambda_2^n}).$$

Theorem 1.7.3. *In addition to the latter assumptions, suppose that Assumption 1.3.1 holds true with* $b = 5$.

(i) The random function $\mathcal{Z}_N(\cdot)$ *converges in the Skorokhod space* D *to a centered Gaussian process* $\eta(\cdot)$. *When* λ_1 *and* λ_2 *are multiplicatively independent then the covariances of* η *are given by*

$$Cov(\eta(t), \eta(s)) = \lim_{N \to \infty} \mathbb{E}\mathcal{Z}_N(t)\mathcal{Z}_N(s) = \min(t, s) \int F^2(x, y) d\mu(x) d\mu(y)$$

and in particular η has stationary and independent increments. When λ_1 and λ_2 are not multiplicatively independent then the covariances of η are given by

$$Cov(\eta(t), \eta(s)) = \lim_{N \to \infty} \mathbb{E}\mathcal{Z}_N(t)\mathcal{Z}_N(s) = \min(t, s) \int F^2(x, y)d\mu(x)d\mu(y)$$
$$+a_{q_1, p_1}(t, s) + a_{q_1, p_1}(s, t)$$

where $\frac{\ln \lambda_1}{\ln \lambda_2} = \frac{p_1}{q_1}$ for some coprime integers q_1 and p_1 and

$$a_{q_1, p_1}(t, s)$$
$$= \min(\frac{1}{q_1}, \frac{t}{p_1}, \frac{s}{p_1 + q_1}) \int F(x_1, x_2)F(x_2, x_3)d\mu(x_1)d\mu(x_2)d\mu(x_3).$$

(ii) Moreover, for any $0 \le t, s \le 1$ and $N \ge 1$,

$$|\mathbb{E}\mathcal{Z}_N(t)\mathcal{Z}_N(s) - b(t, s)| \le CN^{-\frac{1}{2}} \ln^{\frac{1}{2}}(N + 1) \tag{1.7.4}$$

and when $\sigma^2 = Var(\eta(1)) > 0$ then with $Z_N = \mathcal{Z}_N(1)$,

$$\max \left(d_W(\mathcal{L}(\sigma^{-1}Z_N), \mathcal{N}), d_K(\mathcal{L}(\sigma^{-1}Z_N), \mathcal{N}) \right)$$
$$\le C(1 + \sigma^{-1})\max(\sigma^{-\frac{5}{6}}, \sigma^{-2})N^{-\frac{1}{2}}\ln^4(N + 1)$$

where $\mathcal{N}$ is the standard normal law and C is a constant which depends only on $K, \theta, c, d, \ell, \gamma_m^\iota$ (and this dependence can be recovered from the proof).

Proof. Similarly to the situation of Theorem 1.7.2, in order to prove Theorem 1.7.3 we only have to show that the limits $b(t, s)$ exist and obtain convergence rates towards them. We first claim that there exists a constant $C > 0$ such that

$$\mathbb{E}Z_N^2 \le C \tag{1.7.5}$$

for any $N \in \mathbb{N}$. Indeed, for any natural n and m set

$$E_{n,m} = \mathbb{E}F(\xi_{\lambda_1^n}, \xi_{\lambda_1^n \lambda_2^n})F(\xi_{\lambda_1^m}, \xi_{\lambda_1^m \lambda_2^m}).$$

When $m = n + k$ for some $k > 0$ then the minimal difference between the numbers $\lambda_1^n, \lambda_1^n\lambda_2^n, \lambda_1^m$ and $\lambda_1^m\lambda_2^m$ equals

$$\lambda_1^n \min\left(\lambda_2^n - 1, |\lambda_1^k - \lambda_2^n|\right) := \rho(n, k).$$

When $\rho(n, k)$ vanishes then $\lambda_1^m = \lambda_1^n\lambda_2^n$ and otherwise these numbers are distinct. It follows from Corollary 1.3.14 that

$$|E_{n,m}| = |E_{n,m} - (\bar{F})^2| \le C_1 \tau_{\rho(n,k)} \tag{1.7.6}$$

for some constant $C_1 > 0$, where $\tau_\rho = \phi_\rho^{1-\frac{1}{b}} + \beta_{q,\rho}^\kappa$ and we recall that under our version of Assumption 1.3.2 we have

$$\tau_\rho \leq 2d\rho^{-\theta_b}, \quad \forall \rho > 0 \tag{1.7.7}$$

where $\theta_b = \theta(1 - \frac{1}{b})$. Next, write

$$\mathbb{E}Z_N^2 = N^{-1} \sum_{n=1}^{N} E_{n,n} + 2N^{-1} \sum_{1 \leq n < m \leq N} E_{n,m} := I_1 + I_2.$$

By Assumption 1.3.1, (1.3.6) and the Cauchy-Schwarz inequality,

$$|E_{n_1,n_2}| \leq M_1 := K^2(1 + 2\ell\gamma_\iota^\iota + \ell^2\gamma_{2\iota}^{2\iota}) \tag{1.7.8}$$

for any $n_1, n_2 \in \mathbb{N}$, and therefore

$$|I_1| \leq M_1. \tag{1.7.9}$$

In order to bound I_2, we first write

$$I_2 = 2N^{-1} \sum_{n=1}^{N} \sum_{k \in \mathcal{R}_{n,N}} E_{n,n+k} + 2N^{-1} \sum_{n=1}^{N} \sum_{k \in \mathcal{J}_{n,N}} E_{n,n+k} := J_1 + J_2$$

where $\mathcal{R}_{n,N} = \{1 \leq k \leq N - n : |\lambda_1^k - \lambda_2^n| \leq \frac{1}{2}\}$ and

$$\mathcal{J}_{n,N} = \{1 \leq k \leq N - n : |\lambda_1^k - \lambda_2^n| > \frac{1}{2}\}.$$

There exists $c \in \mathbb{N}$ which depends only on λ_1 and λ_2 such that for any $n \geq 1$ there exist at most c natural numbers k such that $|\lambda_1^k - \lambda_2^n| \leq \frac{1}{2}$, and therefore

$$|J_1| \leq 2cN^{-1} \sum_{n=1}^{N} M_1 \leq 2cM_1.$$

On the other hand, by (1.7.6) and (1.7.7) we have

$$|J_2| \leq 2\sum_{n=1}^{\infty} \tau_{c_0\lambda_1^n} < 4dc_0^{-\theta_b} \sum_{n=1}^{\infty} \lambda_1^{-n\theta_b} < R_1$$

for some constants $R_1 > 0$ and $c_0 > 0$, and the proof of (1.7.5) is complete.

Next, let $N \in \mathbb{N}$, $t, s \in [0, 1]$ and let M be of the form $M = A\ln(N+1)$ where $A > 0$ is a constant. For the proof of Theorem 1.7.3 *(i)* we can take any constant $A > 0$ but in order to prove Theorem 1.7.3 *(ii)* we will consider A as a parameter. Set

$$\tilde{Z}_N(u) = N^{-\frac{1}{2}} \sum_{M < n \leq [Nu]} F(\xi_{\lambda_1^n}, \xi_{\lambda_1^n\lambda_2^n}), \quad u \in [0, 1].$$

Then

$$|\mathbb{E}\mathcal{Z}_N(t)\mathcal{Z}_N(s) - \mathbb{E}\tilde{\mathcal{Z}}_N(t)\tilde{\mathcal{Z}}_N(t)|$$
$$\leq N^{-\frac{1}{2}}M^{\frac{1}{2}}\left(\mathbb{E}|\mathcal{Z}_N(t)Z_M| + \mathbb{E}|\tilde{\mathcal{Z}}_N(s)Z_M|\right).$$

We conclude from (1.7.5) and the Cauchy-Schwarz inequality that

$$|\mathbb{E}\mathcal{Z}_N(t)\mathcal{Z}_N(s) - \mathbb{E}\tilde{\mathcal{Z}}_N(t)\tilde{\mathcal{Z}}_N(t)| \leq 2CN^{-\frac{1}{2}}M^{\frac{1}{2}}. \qquad (1.7.10)$$

Thus, in order to show that the limits $b(t,s)$ exist it is sufficient to prove that the limits

$$\tilde{b}(t,s) = \lim_{N\to\infty} \mathbb{E}\tilde{\mathcal{Z}}_N(t)\tilde{\mathcal{Z}}_N(s)$$

exist. For this purpose, write

$$\mathbb{E}\tilde{\mathcal{Z}}_N(t)\tilde{\mathcal{Z}}_N(s) = N^{-1}\sum_{n=M}^{[Nt]}\sum_{m=M}^{[Ns]}E_{n,m}$$

$$= N^{-1}\sum_{n=M}^{[N\min(t,s)]}E_{n,n} + N^{-1}\sum_{k=1}^{N}\sum_{(n,m)\in\mathcal{I}_{N,k,t,s}}E_{n,m}$$

$$+N^{-1}\sum_{k=1}^{N}\sum_{(m,n)\in\mathcal{I}_{N,k,s,t}}E_{m,n} := R_{1,N}(t,s) + R_{2,N}(t,s) + R_{3,N}(t,s)$$

where

$$\mathcal{I}_{N,k,t,s} = \{(n,m) : M \leq n \leq [Nt], M \leq m \leq [Ns], m = n + k\}.$$

By Corollary 1.3.14 and (1.7.7),

$$\left|E_{n,n} - \int F^2(x_1,x_2)d\mu(x_1)d\mu(x_2)\right| \leq C_0\tau_{c_0}\lambda_1^n \leq C_1\lambda_1^{-n\theta_b} \qquad (1.7.11)$$

for some constants $C_0, C_1, c_0 > 0$, and therefore $R_{1,N}(t,s)$ converges to $\min(t,s)\int F^2(x_1,x_2)$ as $N \to \infty$. Next, for any $\varepsilon \in (0,1)$ we have

$$R_{2,N}(t,s) = N^{-1}\sum_{k=1}^{N}\sum_{(n,m)\in\mathcal{J}_{N,k,t,s,\varepsilon}}E_{n,m}$$

$$+N^{-1}\sum_{k=1}^{N}\sum_{(n,m)\in\mathcal{R}_{N,k,t,s,\varepsilon}}E_{n,m} := Q_{1,N}(\varepsilon,t,s) + Q_{2,N}(\varepsilon,t,s)$$

where

$$\mathcal{J}_{N,k,t,s,\varepsilon} = \mathcal{I}_{N,k,t,s} \cap \{|\lambda_1^k - \lambda_2^n| \geq \varepsilon\}$$

and

$$\mathcal{R}_{N,k,t,s,\varepsilon} = \mathcal{I}_{N,k,t,s} \cap \{|\lambda_1^k - \lambda_2^n| < \varepsilon\}.$$

Now, by (1.7.6) there exist constants $C_3, C_2, c_2 > 0$ such that

$$|Q_{1,N}(\varepsilon, t, s)| \qquad (1.7.12)$$

$$\leq N^{-1} \sum_{k=1}^{N} \sum_{(n,m) \in \mathcal{J}_{N,k,t,s,\varepsilon}} |E_{n,m}| \leq C_2 N^{-1} \sum_{k=1}^{N} \sum_{n=M}^{N} \tau_{c_2 \varepsilon \lambda_1^n}$$

$$= C_2 \sum_{n=M}^{N} \tau_{c_2 \varepsilon \lambda_1^n} \leq 2dC_2(c_2\varepsilon)^{-\theta_b} \sum_{n=M}^{\infty} \lambda_1^{-n\theta_b} = C_3(c_2\varepsilon)^{-\theta_b} \lambda_1^{-M\theta_b}.$$

Thus, when ε is fixed then $Q_{1,N}(\varepsilon, t, s)$ converges to 0 as $N \to \infty$.

Next, let $(n, m) \in \mathcal{R}_{N,k,t,s,\varepsilon}$, $k \in \mathbb{N}$, and assume that

$$\varepsilon < \min(1, \tfrac{1}{2}\lambda_1, \tfrac{1}{2}\lambda_1 \ln \lambda_2). \qquad (1.7.13)$$

The inequality

$$|\lambda_1^k - \lambda_2^n| < \varepsilon$$

together with the mean value theorem imply that

$$k \ln \lambda_1 - \frac{\varepsilon}{\lambda_1^k - \varepsilon} < n \ln \lambda_2 < k \ln \lambda_1 + \frac{\varepsilon}{\lambda_1^k}$$

and so with $\Lambda = \frac{\ln \lambda_1}{\ln \lambda_2}$,

$$|k\Lambda - n| \leq \frac{\varepsilon}{(\lambda_1^k - \varepsilon) \ln \lambda_2} \leq \frac{2\varepsilon}{\lambda_1 \ln \lambda_2} := \varepsilon' < 1. \qquad (1.7.14)$$

Suppose next that $\Lambda \notin \mathbb{Q}$. Then the sequence $\{k\Lambda : k \in \mathbb{N}\}$ is equidistributed modulo 1, and therefore by (1.7.8) and (1.7.14) the upper limit as $N \to \infty$ of

$$N^{-1} \sum_{k=1}^{N} \sum_{(n,m) \in \mathcal{R}_{N,k,t,s,\varepsilon}} |E_{n,m}| \qquad (1.7.15)$$

does not exceed $M_1 \varepsilon'$ and hence the upper limit of $R_{2,N}(t, s)$ as $N \to \infty$ does not exceed $M_1 \varepsilon'$. Letting $\varepsilon \to 0$ we deduce that $R_{2,N}(t, s)$ converges to 0 as $N \to \infty$. Exchanging (n, t) and (m, s) and repeating the above arguments we deduce similarly that $R_{3,N}(t, s)$ converges to 0 as $N \to \infty$. We conclude that the limit $b(t, s)$ exists and that

$$b(t, s) = \min(t, s) \int F^2(x_1, x_2) d\mu(x_1) d\mu(x_2).$$

Next, suppose that $\Lambda \in \mathbb{Q}$ and write $\Lambda = \frac{p_1}{q_1}$ for some coprime positive integers p_1 and q_1. Since $\frac{rp_1}{q_1}, r = 1, 2, ..., q_1 - 1$ is not an integer, there exists a constant $c_1 > 0$ such that either k divides q_1 or $|k\Lambda - n| > c_1$ for any $n \in \mathbb{N}$. When $k = jq_1$ for some natural j, then $|k\Lambda - n| < \varepsilon'$ only if $n = p_1 j$ and in this case $n = k\Lambda$ and $\lambda_1^{n+k} = \lambda_1^n \lambda_2^n$. We conclude that when $\varepsilon' < c_1$ and ε satisfies (1.7.13),

$$Q_{2,N}(\varepsilon, t, s) = N^{-1} \sum_{1 \leq k \leq N : q_1 | k} E_{\Lambda k, \Lambda k + k} \mathbb{I}\left(k \leq \min\left(\frac{Nt}{\Lambda}, \frac{Ns}{\Lambda + 1}\right)\right) \quad (1.7.16)$$

which converges as $N \to \infty$ to $a_{q_1, p_1}(t, s)$ defined in the statement of the theorem. Exchanging n and m and repeating the above arguments, it follows that $R_{3,N}(t, s)$ converges to

$$a_{q_1, p_1}(s, t)$$

as $N \to \infty$ and Theorem 1.7.3 *(i)* follows.

Next we explain how to obtain convergence rates in Theorem 1.7.3 *(ii)*. First, by (1.7.10),

$$|b_N(t, s) - \tilde{b}_N(t, s)| \leq A^{\frac{1}{2}} C N^{-\frac{1}{2}} \ln^{\frac{1}{2}}(N + 1).$$

Suppose that $\Lambda \in \mathbb{Q}$ and let $\varepsilon > 0$ be sufficiently small but fixed. Then we obtain from Corollary 1.3.14 and (1.7.7) that the convergence rates in all the limits computed earlier in this proof are of order $\lambda_1^{-c_0 M} = (N+1)^{-Ac_0 \ln \lambda_1}$ for an appropriate c_0 and when taking a sufficiently large A we obtain (1.7.4). When $\Lambda \notin \mathbb{Q}$, we first recall that the inequalities $|\lambda_1^k - \lambda_2^n| < \varepsilon$ and (1.7.13) imply that

$$|n - \Lambda k| < \frac{2\varepsilon}{\lambda_1 \ln \lambda_2} = \varepsilon'.$$

By the Erdős-Turán inequality (see Theorem 2.5 in [52]) for any $\varepsilon' > 0$ and $N \in \mathbb{N}$,

$$\frac{1}{N} |\mathcal{B}(\Lambda, \varepsilon') \cap [1, N]| \leq 2\varepsilon' + C_4 \frac{\ln N}{N}$$

where

$$\mathcal{B}(\Lambda, \varepsilon') = \{k \in \mathbb{N} : \Lambda k \bmod 1 \in [0, \varepsilon') \cup [1 - \varepsilon', 1)\}$$

and $C_4 > 0$ is an absolute constant. Taking $\varepsilon = \varepsilon_N = a N^{-\frac{1}{2}}$ for some sufficiently small $a > 0$, the rate of convergence of $|Q_{2,N}(\varepsilon_N)|$ to 0 in (1.7.12) is of order $N^{\frac{1}{2}\theta_b} \lambda_1^{-M\theta_b}$ which for a sufficiently large A is at most of order $N^{-\frac{1}{2}}$. We conclude that (1.7.4) holds true when Λ is irrational, as well. The convergence rates in the CLT stated in Theorem 1.7.3 follow now using the arguments in Section 1.3.6. $\square$

Chapter 2

Nonconventional local central limit theorem

2.1 Introduction

The classical De Moivre-Laplace theorem states that if $X_1, X_2, X_3, \ldots$ are independent identically distributed (i.i.d.) $0-1$ Bernoulli random variables taking on 1 with probability p and $S_N = \sum_{n=1}^{N} X_n$, then the probability $P\{S_N = k\}$ is equivalent as $N \to \infty$ to

$$(2\pi Npq)^{-\frac{1}{2}} \exp(-(k - Np)^2/2Npq), \quad q = 1 - p$$

uniformly in k such that $|k - Np| = o(Npq)^{2/3}$. The latter expression is the density of a normal distribution with mean $Np = \mathbb{E}S_N$ and variance $Npq = \text{Var}S_N$ evaluated at the point k, and so the De Moivre-Laplace theorem can be viewed as a local (central) limit theorem (LLT) for the sums S_N. Modern versions of the local limit theorem include the situation when the summands X_n's are not lattice valued, where in this situation the asymptotics of expectations of the form $\mathbb{E}g(S_N - u)$ is determined for continuous functions g with compact support.

Among the main situations when an LTT holds true is the case when $S_N = \sum_{n=1}^{N} g(\zeta_n)$ where $\zeta_1, \zeta_2, \zeta_3, \ldots$ is a stationary Markov chain whose transition operator has a spectral gap when acting on a Banach space $\mathcal{B}$ which contains the function g and satisfies certain conditions (see [28]) such as the Lasota-Yorke inequality. The situation when the transition probability of $\zeta_1, \zeta_2, \zeta_3, \ldots$ satisfies certain regularity condition (a version of the Doeblin condition) can be considered, as well (see [50] and [51]). As an application of the first situation we can consider sums of the form $S_N = \sum_{n=0}^{N-1} g(T^n Z_0)$, where T ranges over a special class of dynamical systems (e.g. one sided topologically mixing subshift of finite type), Z_0 is distributed according to a special T-invariant Gibbs measure μ and g is a Hölder continuous function. For instance, when T is a (locally) distance ex-

panding map on some compact space then the sums $S_N = \sum_{n=0}^{N-1} g(T^n Z_0)$
and $\tilde{S}_N = \sum_{n=1}^{N} g(\zeta_n)$ are identically distributed, where $\zeta_1, \zeta_2, \zeta_3, \dots$ is a
stationary Markov chain with initial distribution μ whose transition opera-
tor is the dual of the Koopman operator $f \to f \circ T$ with respect to the space
$L^1(\mu)$ (such operators are often referred to as "transfer operators"). Relying on the Ruelle-Perron-Frobenius theorem for such operators (see [10]
and [53]), these transfer operators indeed have a spectral gap when acting
on spaces of (locally) Hölder continuous functions. Extensions to other (invertible) dynamical system such as two sided topologically mixing subshifts
of finite type and C^2 Axiom A diffeomorphisms in a neighborhood of an
attractor follow, as well (see Lemma 1.6 and Sections 3-4 in [10]).

In [29] extensions of the LLT to nonconventional sums of the form

$$S_N = \sum_{n=1}^{N} F(\zeta_n, \zeta_{2n}, \dots, \zeta_{\ell n})$$

were obtained in the case of a Markov chain whose transition probability
satisfies a version of the Doeblin condition, where F is a square integrable
function. This condition guarantees that the Markov chain under consideration is sufficiently well ψ-mixing, and, in fact, under certain conditions the
proof from [29] can be modified to the case of a general ψ-mixing Markov
chain (see Remark 2.4.2). The main result of this chapter is an extension of
the above nonconventional LLT to the case when $\zeta = \{\zeta_n : n \geq 0\}$ is a stationary Markov chain generated by a (locally) distance expanding map T on
a compact space $\mathcal{X}$ considered with a special T-invariant initial (Gibbs) distribution μ and F is a (locally) Hölder continuous function which satisfies
some additional regularity conditions related to a prescribed periodic point
of T. Namely, we will prove an LLT for these (nonconventional) sums when
ζ is a stationary Markov chain whose transition probabilities are given by

$$P(\zeta_1 \in \Gamma | \zeta_0 = x) = \sum_{y \in \Gamma : Ty = x} e^{f(y)}$$

where $f : \mathcal{X} \to \mathbb{R}$ is an appropriately chosen function, assuming that the
process $\{T^n Z_0 : n \geq 0\}$ satisfies certain mixing and approximation conditions when Z_0 is distributed according to μ.

Extensions to dynamical system will be discussed, as well. More precisely, let T be either a two sided subshift of finite type or a C^2 Axiom A
diffeomorphism (in particular, Anosov) in a neighborhood of an attractor.
Then we derive an LLT for sums of the form

$$S_N(x) = \sum_{n=1}^{N} F(T^n x, T^{2n} x, \dots, T^{\ell n} x) \tag{2.1.1}$$

when x is drawn at random with respect to an appropriate Gibbs measure.

The structure of this chapter is as follows. In the next section we will describe the general Fourier analysis approach for proving an LLT, and in Section 2.3 we will show that, in general, a nonconventional LLT for Markov chains follows from appropriate estimates of norms of products of certain random Fourier operators together with some quantitative mixing properties. For readers' convenience, in Section 2.4 we will repeat the main arguments of the proof from [29] and then in Section 2.5 we will describe the setup of locally distance expanding maps, present the corresponding thermodynamic formalism theory and additional mixing assumptions and state and prove our main results. The proof relies on some (soft) version of the random complex Ruelle-Perron-Frobenius theorem for iterates of random transfer operators, which will follow from the more general results of Part 2. The proof will involve also a "periodic point approach" to random dynamics which will be discussed in Section 2.10 and will be elaborated in Chapter 7. In the last section of this chapter we will explain how to modify the proof from Section 2.5 in order to prove the nonconventional LLT in the dynamical systems case (i.e. for the sums (2.1.1)).

2.2 Local central limit theorem via Fourier analysis

Let $S_1, S_2, S_3, ..$ be a sequence of random variables. We will say that the sequence satisfies the local (central) limit theorem (LTT) with respect to a measure ν on $\mathbb{R}$, a centralizing constant $m \in \mathbb{R}$ and a normalizing constant $\sigma > 0$ if for any real continuous function g on $\mathbb{R}$ with compact support,

$$\lim_{N \to \infty} \sup_{u \in supp\,\nu} \left| \sigma\sqrt{2\pi N}\mathbb{E}g(S_N - u) - e^{-\frac{(u - mN)^2}{2N\sigma^2}} \int g d\nu \right| = 0. \qquad (2.2.1)$$

Let $\varphi_N : \mathbb{R} \to \mathbb{C}$ be the characteristic function of S_N given by $\varphi_N(t) = \mathbb{E}e^{itS_N}$. As in many expositions of the LLT, we will distinguish between lattice and non-lattice cases. We call the case a lattice one when there exists $h > 0$ such that with probability one $S_N, N \in \mathbb{N}$ take values on the lattice $h\mathbb{Z} := \{hk : k \in \mathbb{Z}\}$, or equivalently when $\varphi_N(t) = \varphi_N(t + \frac{2\pi}{h})$ for any $t \in \mathbb{R}$ and $N \in \mathbb{N}$. When there exists no h with this property then we call the case non-lattice one.

The Fourier-analysis proof of the LLT is based on the following growth properties of the characteristic functions φ_N.

Assumption 2.2.1. For any $\delta > 0$,

$$\lim_{N \to \infty} N^{\frac{1}{2}} \sup_{t \in J_\delta} |\varphi_N(t)| = 0$$

where in the lattice case $J_\delta = [-\frac{\pi}{h}, \frac{\pi}{h}] \setminus (-\delta, \delta)$ while in the non-lattice case $J_\delta = [-\delta^{-1}, -\delta] \cup [\delta, \delta^{-1}]$.

Assumption 2.2.2. There exist $\delta_0 \in (0, 1)$, positive constants c_0 and d_0 and a sequence $(b_n)_{n=1}^\infty$ of real numbers such that $\lim_{N \to \infty} n^{\frac{1}{2}} b_n = 0$ and

$$|\varphi_N(t)| \le c_0 e^{-d_0 N t^2} + b_N$$

for any $N \in \mathbb{N}$ and $t \in [-\delta_0, \delta_0]$.

Theorem 2.2.3. *Suppose that $N^{-\frac{1}{2}}(S_N - mN)$ converges in distribution as $N \to \infty$ to a centered normal random variable with variance $\sigma^2 > 0$ and that Assumptions 2.2.1 and 2.2.2 hold true. Then in the lattice case the LLT holds true with the measure ν_h which assigns mass h to each point of the lattice $h\mathbb{Z} = \{hk : k \in \mathbb{Z}\}$, while in the non-lattice the LLT holds with the Lebesgue measure ν_0.*

Proof. First, by Theorem 10.7 in [12] (see also Section 10.4 there and Lemma IV.5 together with arguments of Section VI.4 in [28]) it suffices to prove (2.2.1) for all continuous complex-valued functions g on $\mathbb{R}$ such that

$$\int_{-\infty}^\infty |g(x)| dx < \infty \qquad (2.2.2)$$

and having Fourier transform

$$\hat{g}(\lambda) = \int_{-\infty}^\infty e^{-i\lambda x} g(x) dx, \ \lambda \in \mathbb{R} \qquad (2.2.3)$$

vanishing outside a finite interval $[-L, L]$. Then, in particular the inversion formula

$$g(x) = \frac{1}{2\pi} \int_{-\infty}^\infty e^{i\lambda x} \hat{g}(\lambda) d\lambda \qquad (2.2.4)$$

holds true.

Let g be a function with the above properties and let $u \in \mathbb{R}$. Then by (2.2.4) we have

$$\mathbb{E}g(S_N - u) = \frac{1}{2\pi} \int_{-\infty}^\infty \varphi_N(\lambda) e^{-i\lambda u} \hat{g}(\lambda) d\lambda \qquad (2.2.5)$$

where $\varphi_N(\lambda) = \mathbb{E}e^{i\lambda S_N}$ is the characteristic function of S_N. Changing variables $s = \lambda \sigma \sqrt{N}$ we obtain

$$\sigma \sqrt{2\pi N} \mathbb{E}g(S_N - u) = \frac{1}{\sqrt{2\pi}} \int_{-\infty}^\infty \varphi_N\left(\frac{s}{\sigma\sqrt{N}}\right) e^{-\frac{isu}{\sigma\sqrt{N}}} \hat{g}\left(\frac{s}{\sigma\sqrt{N}}\right) ds. \quad (2.2.6)$$

On the other hand, from the formula for the characteristic function of the Gaussian distribution and the Fourier inversion formula it follows that

$$e^{-\frac{(u-mN)^2}{2N\sigma^2}} \int g d\nu_h = \frac{\int g d\nu_h}{\sqrt{2\pi}} \int_{-\infty}^{\infty} \exp\left(-\frac{i\lambda(u-mN)}{\sigma\sqrt{N}} - \frac{\lambda^2}{2}\right) d\lambda \quad (2.2.7)$$

where in the lattice case ν_h is the measure which assign mass h to each point of the lattice $h\mathbb{Z} = \{hk : k \in \mathbb{Z}\}$, while in the non-lattice case we set $h = 0$ and denote the Lebesgue measure by ν_0.

Now, in the non-lattice case by (2.2.6) and (2.2.7) for any $\delta, T > 0$ we can write

$$\left| \sigma\sqrt{2\pi N} \mathbb{E}g(S_N - u) - e^{-\frac{(u-mN)^2}{2N\sigma^2}} \int g d\nu_0 \right| \quad (2.2.8)$$

$$\leq I_1(N,T) + I_2(T) + I_3(N,\delta) + I_4(N,\delta,T),$$

where

$$I_1(N,T) = \frac{1}{\sqrt{2\pi}} \int_{-T}^{T} \left| e^{-\frac{i\lambda mN}{\sigma\sqrt{N}}} \varphi_N\left(\frac{\lambda}{\sigma\sqrt{N}}\right) \hat{g}\left(\frac{\lambda}{\sigma\sqrt{N}}\right) - e^{-\frac{\lambda^2}{2}} \int g d\nu_0 \right| d\lambda,$$

$$I_2(T) = \frac{|\int g d\nu_0|}{\sqrt{2\pi}} \int_{|\lambda|>T} e^{-\frac{\lambda^2}{2}} d\lambda,$$

$$I_3(N,\delta) = \frac{\|\hat{g}\|_\infty}{\sqrt{2\pi}} \int_{\delta\sigma\sqrt{N} \leq |\lambda| \leq L\sigma\sqrt{N}} \left| \varphi_N\left(\frac{\lambda}{\sigma\sqrt{N}}\right) \right| d\lambda \quad \text{and}$$

$$I_4(N,\delta,T) = \frac{\|\hat{g}\|_\infty}{\sqrt{2\pi}} \int_{T<|\lambda|<\delta\sigma\sqrt{N}} \left| \varphi_N\left(\frac{\lambda}{\sigma\sqrt{N}}\right) \right| d\lambda$$

where $\|\hat{g}\|_\infty = \sup_\lambda |\hat{g}(\lambda)|$, and in writing $I_3(N,\delta)$ we used that $\hat{g}(s) = 0$ when $s \notin [-L, L]$.

In the lattice case we proceed in a slightly different way. First, observe that in this case for any $\lambda \in \mathbb{R}$ and $N \in \mathbb{N}$,

$$\varphi_N\left(\frac{\lambda}{\sigma\sqrt{N}} + \frac{2\pi k}{h}\right) = \varphi_N\left(\frac{\lambda}{\sigma\sqrt{N}}\right) \quad \text{for all } k \in \mathbb{Z}. \quad (2.2.9)$$

Set

$$r(v) = \sum_{k=-\infty}^{\infty} \hat{g}\left(v + \frac{2\pi k}{h}\right).$$

Taking into account that here $u \in \{kh : k \in \mathbb{Z}\}$, we can rewrite (2.2.6) in the following way

$$\sigma\sqrt{2\pi N}\mathbb{E}g(S_N - u) = \frac{1}{\sqrt{2\pi}} \int_{-\frac{\pi\sigma\sqrt{N}}{h}}^{\frac{\pi\sigma\sqrt{N}}{h}} \varphi_N\left(\frac{\lambda}{\sigma\sqrt{N}}\right) e^{-\frac{i\lambda u}{\sigma\sqrt{N}}} r\left(\frac{\lambda}{\sigma\sqrt{N}}\right) d\lambda. \quad (2.2.10)$$

This together with (2.2.7) yields

$$\left|\sigma\sqrt{2\pi N}\,\mathbb{E}g(S_N - u) - e^{-\frac{(u-mN)^2}{2N\sigma^2}}\int g\,d\nu_h\right| \qquad (2.2.11)$$

$$\leq J_1(N,T) + J_2(T) + J_3(N,\delta) + J_4(N,\delta,T)$$

where

$$J_1(N,T) = \frac{1}{\sqrt{2\pi}}\int_{-T}^{T}\left|e^{-\frac{i\lambda mN}{\sigma\sqrt{N}}}\varphi_N\left(\frac{\lambda}{\sigma\sqrt{N}}\right)r\left(\frac{\lambda}{\sigma\sqrt{N}}\right) - e^{-\frac{\lambda^2}{2}}\int g\,d\nu_h\right|d\lambda,$$

$$J_2(T) = \frac{|\int g\,d\nu_h|}{\sqrt{2\pi}}\int_{|\lambda|>T}e^{-\frac{\lambda^2}{2}}d\lambda,$$

$$J_3(N,\delta) = \frac{1}{\sqrt{2\pi}}\int_{\delta\sigma\sqrt{N}\leq|\lambda|\leq\frac{\pi\sigma\sqrt{N}}{h}}\left|\varphi_N\left(\frac{\lambda}{\sigma\sqrt{N}}\right)\right|\left\|r\left(\frac{\lambda}{\sigma\sqrt{N}}\right)\right|d\lambda \quad \text{and}$$

$$J_4(N,\delta,T) = \frac{1}{\sqrt{2\pi}}\int_{T<|\lambda|<\delta\sigma\sqrt{N}}\left|\varphi_N\left(\frac{\lambda}{\sigma\sqrt{N}}\right)\right|\left\|r\left(\frac{\lambda}{\sigma\sqrt{N}}\right)\right|d\lambda.$$

Next, since $\sigma^{-1}N^{-\frac{1}{2}}(S_N - mN)$ converges in distribution to the standard normal distribution,

$$\lim_{N\to\infty}e^{-\frac{i\lambda mN}{\sigma\sqrt{N}}}\varphi_N\left(\frac{\lambda}{\sigma\sqrt{N}}\right) = e^{-\frac{\lambda^2}{2}} \quad \text{for any } \lambda \in \mathbb{R}. \qquad (2.2.12)$$

Furthermore, since g is continuous and integrable on $\mathbb{R}$,

$$\lim_{N\to\infty}\hat{g}\left(\frac{\lambda}{\sigma\sqrt{N}}\right) = \int g\,d\nu_0 \quad \text{for any } \lambda \in \mathbb{R}. \qquad (2.2.13)$$

In the lattice case it follows from the so-called Poisson summation formula (see Ch. 10 in [12]) that

$$\lim_{N\to\infty}r\left(\frac{\lambda}{\sigma\sqrt{N}}\right) = \int g\,d\nu_h \quad \text{for any } \lambda \in \mathbb{R}. \qquad (2.2.14)$$

Now by (2.2.12)-(2.2.14) and the dominated convergence theorem we obtain that for any $T > 0$,

$$\lim_{N\to\infty}I_1(N,T) = \lim_{N\to\infty}J_1(N,T) = 0. \qquad (2.2.15)$$

Next, clearly,

$$\lim_{T\to\infty}I_2(T) = \lim_{T\to\infty}J_2(T) = 0. \qquad (2.2.16)$$

Finally, in the non-lattice case we deduce from Assumption 2.2.1 that

$$I_3(N,\delta) \leq \frac{\|\hat{g}\|_\infty}{\sqrt{2\pi}}L\sigma\left(\sqrt{N}\sup_{\delta\leq|t|\leq\delta^{-1}}|\varphi_N(t)|\right) \to 0 \text{ as } N\to\infty \qquad (2.2.17)$$

assuming that $\max(L,1) < \delta^{-1}$, and from Assumption 2.2.2 that

$$I_4(N,\delta,T) \le \frac{\|\hat{g}\|_\infty}{\sqrt{2\pi}}\Big(\int_{|\lambda|>T} c_0 e^{-\frac{d_0\lambda^2}{\sigma^2}}d\lambda + \delta\sigma\sqrt{N}b_N\Big) \to 0 \text{ as } N,T \to \infty$$

(2.2.18)

assuming that $\delta < \delta_0$. In the lattice case we deduce similarly that

$$J_3(N,\delta) \le \frac{R}{\sqrt{2\pi}}\cdot\frac{\pi\sigma}{h}\Big(\sqrt{N}\sup_{t\in[-\frac{\pi}{h},\frac{\pi}{h}]\backslash(-\delta,\delta)}|\varphi_N(t)|\Big) \to 0 \text{ as } N \to \infty \quad (2.2.19)$$

and

$$J_4(N,\delta,T) \le \frac{R}{\sqrt{2\pi}}\Big(\int_{|\lambda|>T} c_0 e^{-\frac{d_0\lambda^2}{\sigma^2}}d\lambda + \delta\sigma\sqrt{N}b_N\Big) \to 0 \text{ as } N,T \to \infty$$

(2.2.20)

assuming that $\delta < \min(\frac{\pi}{h},\delta_0)$, where

$$R = \sup_{|t|\le\frac{\pi}{h}}|r(t)| \le (2+hL)\|\hat{g}\|_\infty < \infty.$$

The proof of the theorem is complete in both cases by first taking a sufficiently small δ, then letting $N \to \infty$ and then $T \to \infty$. $\qquad\square$

2.3 Nonconventional LLT for Markov chains by reduction to random dynamics

Our setup consists of a probability space $(\Omega,\mathcal{F},P)$ together with a stationary Markov chain $\zeta = \{\zeta_n : n \ge 0\}$ evolving on a compact space $\mathcal{X}$ equipped with the Borel σ-algebra $\mathcal{B}_\mathcal{X}$. Let

$$P(x,\Gamma) = P(\zeta_1 \in \Gamma|\zeta_0 = x), \ x \in \mathcal{X}, \Gamma \in \mathcal{B}_\mathcal{X}$$

be the transition probabilities of the Markov chain ζ and let μ be its stationary distribution which satisfies $\int P(x,\Gamma)d\mu(x) = \mu(\Gamma)$ for any $\Gamma \in \mathcal{B}_\mathcal{X}$. Let $F = F(x_1,...,x_\ell)$, $\ell \ge 1$ be a Borel function on $\mathcal{X}^\ell = \mathcal{X} \times \cdots \times \mathcal{X}$ such that

$$b^2 = \int F^2(x_1,...,x_\ell)d\mu(x_1)...d\mu(x_\ell) < \infty. \qquad (2.3.1)$$

Consider the sums

$$S_N = \sum_{n=1}^{N} F(\zeta_n,\zeta_{2n},...,\zeta_{\ell n}), \ N \in \mathbb{N}$$

and set

$$\bar{F} := \int F(x_1,x_2,...,x_\ell)d\mu(x_1)d\mu(x_2)...d\mu(x_\ell). \qquad (2.3.2)$$

In what follows the centralizing constant m will always be $m = \bar{F}$.

As we have seen, the LLT follows from the CLT together with some control on the decay rates of φ_N as $N \to \infty$. In this section we will reduce the problem of approximation of the characteristic functions φ_N of S_N to approximation of norms of certain products of random Fourier operators. Let $P(\ell, \cdot, \cdot)$ be the ℓ step transition probability which is given by $P(\ell, x, \Gamma) = P(\zeta_\ell \in \Gamma | \zeta_0 = x)$. It will be more convenient to write

$$P(\ell, \cdot, \cdot) = P_\ell(\cdot, \cdot).$$

The first step of the reduction goes as follows. Let $1 \leq M \leq N$ and $t \in \mathbb{R}$. Then

$$|\varphi_N(t)| = |\mathbb{E}e^{itS_N}| = |\mathbb{E}\mathbb{E}[e^{itS_N}|\zeta_1, \zeta_2, ..., \zeta_{\ell M}]| \qquad (2.3.3)$$
$$\leq \mathbb{E}|\mathbb{E}[e^{it(S_N - S_M)}|\zeta_1, \zeta_2, ..., \zeta_{\ell M}]|.$$

Set $M = M_\ell(N) = N - 2[\frac{N - N_\ell}{2}]$, where $N_\ell = [N(1 - \frac{1}{2\ell})] + 1$. Then

$$\ell M - (\ell - 1)N \geq \frac{N}{2}$$

and therefore by the Markov property,

$$\mathbb{E}[e^{it(S_N - S_M)}|\zeta_1, \zeta_2, ..., \zeta_{\ell M}] = \prod_{k=M+1}^{N} Q_{t, \zeta_k, \zeta_{2k}, ..., \zeta_{(\ell-1)k}} \mathbf{1}(\zeta_{\ell M}) \qquad (2.3.4)$$

where $\mathbf{1}$ is the function which takes the constant value 1 and for any $\bar{x} = (x^{(1)}, ..., x^{(\ell-1)}) \in \mathcal{X}^{\ell-1} = \mathcal{X} \times \cdots \times \mathcal{X}$ and $t \in \mathbb{R}$ the Fourier operator $Q_{t, \bar{x}}$ is given by

$$Q_{t, \bar{x}} g(y) = \int e^{itF(\bar{x}, z)} g(z) P_\ell(y, dz) = \mathbb{E}[e^{itF(\bar{x}, \zeta_\ell)} g(\zeta_\ell)|\zeta_0 = y]$$

for any bounded Borel measurable function $g : \mathcal{X} \to \mathbb{C}$. The right-hand side of (2.3.4) consists of a product of random Fourier operators, and Assumptions 2.2.1 and 2.2.2 will follow from appropriate estimates of the norms of these random products. Still, the process $\{(\zeta_n, \zeta_{2n}, ..., \zeta_{(\ell-1)n}) : n \geq 1\}$ is not stationary (unless $\ell = 2$) and thus, in contrary to the classical situation of products of random operators, these random products are not taken along paths of a measure preserving system, namely they do not have the form $A^{\vartheta^{M+1} \omega_1} \circ A^{\vartheta^{M+2} \omega_1} \circ \cdots \circ A^{\vartheta^N \omega_1}$ for some family $\{A_{\omega_1}\}$ of random operators and a measure preserving system $(\Omega_1, \mathcal{F}_1, P_1, \vartheta)$.

In order to overcome this difficulty, let $\zeta^{(i)} = \{\zeta_n^{(i)} : n \geq 0\}$, $i = 1, 2, ..., \ell - 1$ be $\ell - 1$ independent copies of the process $\zeta = \{\zeta_n : n \geq 0\}$ and set

$$\Theta_n = (\zeta_n^{(1)}, \zeta_{2n}^{(2)}, ..., \zeta_{(\ell-1)n}^{(\ell-1)}), \ n \geq 0. \qquad (2.3.5)$$

Then the process $\Theta = \{\Theta_n : n \geq 0\}$ is stationary and, in fact, forms a Markov chain. Consider the sets of indexes

$$\mathcal{M}_i = \mathcal{M}_{i,N} = \{in : M < n \leq N\}, \ i = 1, 2, ..., \ell - 1.$$

Then for each i, the random variables $\{\zeta_m : m \in \mathcal{M}_i\}$ are the only ones appearing in (2.3.4) in the i-th coordinate inside F. For the sake of convenience set $\mathcal{M}_\ell = \mathcal{M}_{\ell,N} = \{\ell M\}$. Observe that

$$jM - (j-1)N \geq \frac{N}{2}, \ j = 1, 2, ..., \ell \qquad (2.3.6)$$

which means that the sets $\mathcal{M}_1, ..., \mathcal{M}_{\ell-1}$ are ordered so that $m_i + \frac{N}{2} \leq m_{i+1}$ for any $i = 1, 2, ..., \ell - 1$ and $m_i \in \mathcal{M}_i, i = 1, ..., \ell - 1$. Under appropriate mixing conditions and regularity assumptions on F (which will be introduced later depending on the case) this large distance between the $\mathcal{M}_i$'s implies that the random variables $\zeta(\mathcal{M}_i) = \{\zeta_m : m \in \mathcal{M}_i\}, i = 1, 2, ..., \ell$ are weakly dependent and as a consequence we will derive that

$$\left| \mathbb{E} \left| \prod_{k=M+1}^{N} Q_{t,\zeta_k,\zeta_{2k},...,\zeta_{(\ell-1)k}} \mathbf{1}(\zeta_{\ell M}) \right| - \mathbb{E} \left| \prod_{k=M+1}^{N} Q_{t,\Theta_k} \mathbf{1}(\zeta_{\ell M}^{(\ell)}) \right| \right| \leq w(t)c_N$$
$$(2.3.7)$$

where $\zeta^{(\ell)}$ is another copy of ζ which is independent of the rest of the copies, $w : \mathbb{R} \to \mathbb{R}$ is a continuous function and $(c_N)_{N=1}^{\infty}$ is a sequence satisfying $\lim_{N\to\infty} N^{\frac{1}{2}} c_N = 0$. Assuming the validity of (2.3.7) and relying on the previous estimates, the goal now is to estimate

$$\mathbb{E} \left| \prod_{k=M+1}^{N} Q_{t,\Theta_n} \mathbf{1}(\zeta_{\ell M}^{(\ell)}) \right|.$$

It will be convenient in Section 2.8 to represent this expression in the following way. Let $(\Omega_\Theta, \mathcal{F}_\Theta, P_\Theta, \vartheta)$ be an invertible measure preserving system corresponding to the process Θ (where ϑ is the path shift) and let $p_0 : \Omega_\Theta \to \mathcal{X}^{\ell-1}$ be a measurable function so that the processes Θ and $\{p_0 \circ \vartheta^n : n \geq 0\}$ have the same distribution (see Section 2.7.1). Then

$$\mathbb{E} \left| \prod_{k=M+1}^{N} Q_{t,\Theta_n} \mathbf{1}(\zeta_{\ell M}^{(\ell)}) \right| = \int \int |\mathbf{Q}_{it}^{\vartheta^{M+1}\omega, N-M} \mathbf{1}(x)| d\mu(x) dP_\Theta(\omega) \quad (2.3.8)$$

$$= \int \int |\mathbf{Q}_{it}^{\omega, N-M} \mathbf{1}(x)| d\mu(x) dP_\Theta(\omega)$$

where $\mathbf{Q}_{it}^{\omega} = Q_{it,p_0(\omega)}$ and for any $n \in \mathbb{N}$,

$$\mathbf{Q}_{it}^{\omega,n} = \mathbf{Q}_{it}^{\omega} \circ \mathbf{Q}_{it}^{\vartheta\omega} \circ \cdots \circ \mathbf{Q}_{it}^{\vartheta^{n-1}\omega}$$

and we arrive at the classical situation of products of random operators, which in our situation are evaluated at the function $\mathbf{1}$.

2.4 Markov chains with densities

In this section we will describe the proof of the LLT from [29] in the case when the ℓ-step transition probabilities $P_\ell(\cdot, \cdot) = P(\ell, \cdot, \cdot)$ have densities with respect to some probability measure, which are bounded and bounded away from 0.

2.4.1 *Basic assumptions and CLT*

We assume here that there exists a Borel measure η on $\mathcal{X}$ and a constant $\gamma \in (0, 1)$ such that for any Borel measurable set $\Gamma \subset \mathcal{X}$ and $x \in \mathcal{X}$,

$$\gamma \eta(\Gamma) \leq P(\ell, x, \Gamma) \leq \gamma^{-1} \eta(\Gamma). \qquad (2.4.1)$$

This assumption is a specific form of the two sided Doeblin condition. The right-hand side already implies that there exists a stationary distribution μ and that the geometric ergodicity condition

$$\|P(n, x, \cdot) - \mu\| \leq \beta^{-1} e^{-\beta n}, \ \beta > 0$$

holds true, where $\|\cdot\|$ is the total variation norm. Inequality (2.4.1) implies that the measures $P(\ell, x, \cdot)$ are absolutely continuous with respect to η, and that the corresponding densities

$$p^{(\ell)}(x, \cdot) := \frac{dP(\ell, x, \cdot)}{d\eta} \qquad (2.4.2)$$

satisfy

$$\gamma \leq p^{(\ell)}(x, y) \leq \gamma^{-1} \qquad (2.4.3)$$

for any x and η-a.a. y, where a.a. stands for almost all.

Next, in [29] the case when F is a function of the first $\ell - 1$ variables $\mu^\ell = \mu \times \cdots \times \mu$-almost surely is excluded according to the following reasoning. Consider the function $F_\ell : \mathcal{X}^\ell \to \mathbb{R}$ given by

$$F_\ell(x) = F(x) - \int F(x_1, ..., x_{\ell-1}, y) d\mu(y), \ x = (x_1, ..., x_\ell). \qquad (2.4.4)$$

Then $F(x)$ depends only on $x_1, ..., x_{\ell-1}$ for μ^ℓ-a.a. x if and only if F_ℓ vanishes μ^ℓ-a.s., where a.s. stands for almost surely. Since μ is the stationary measure of $P(\cdot, \cdot)$ the distribution $\mu_{n,\ell}$ of $(\zeta_n, \zeta_{2n}, ..., \zeta_{\ell n})$ is absolutely continuous with respect to μ^ℓ, for any $n \in \mathbb{N}$. We refer the readers to the end of the proof of Theorem 2.4 in [29] for the details. As a consequence, when F_ℓ vanishes μ^ℓ-a.s. then with $G_\ell = F - F_\ell$ for any $N \in \mathbb{N}$ we have

$$S_N = \sum_{n=1}^{N} G_\ell(\zeta_n, \zeta_{2n}, ..., \zeta_{(\ell-1)n}), \ P\text{-a.s.}$$

This leads to the reduction from ℓ to $\ell - 1$ and we can assume from the beginning that $\ell = j$ is the maximal number of variables for which F is a function of j variables, μ^ℓ-a.s.

Consider the family $\mathcal{F}_{m,n}, m \leq n$ of σ-algebras given by $\mathcal{F}_{m,n} = \sigma\{\zeta_m, ..., \zeta_n\}$ and set $\mathcal{F}_{m,\infty} = \sigma\{\zeta_s : s \geq m\}$. The ψ-mixing coefficients associated with the family $\mathcal{F}_{m,n}$ is given by

$$\psi_n \qquad (2.4.5)$$
$$= \sup\left\{ \left| \frac{P(A \cap B)}{P(A)P(B)} - 1 \right| : A \in \mathcal{F}_{0,k}, B \in \mathcal{F}_{k+n,\infty}, P(A)P(B) > 0, k \geq 0 \right\}.$$

By [4] inequality (2.4.1) implies the exponentially fast mixing rate

$$\psi_n \leq ab^n, \, a > 0, \, b \in (0,1). \qquad (2.4.6)$$

In fact (see [4]) this mixing condition implies that (2.4.1) holds true with some n_0 in place of ℓ and for μ-a.a. x. By Theorems 2.2 and 2.3 in [29] (which rely on (2.4.6)) the limit

$$\sigma^2 = \lim_{N \to \infty} \frac{1}{N} \mathbb{E}(S_N - \bar{F})^2$$

exists and $N^{-\frac{1}{2}}(S_N - N\bar{F})$ converges in distribution to the centered normal distribution with variance σ^2. For a different proof of the CLT, see Chapter 1. Henceforth, we assume that $\sigma^2 > 0$, and we refer the readers to Theorems 2.3 and 2.4 in [29] for equivalent conditions for positivity of σ^2.

Next, we will consider here the following particular lattice and non-lattice cases. For any $v = (v_1, ..., v_{\ell-1}) \in \mathcal{X}^{\ell-1}$ consider the set

$$B_v = \{h \geq 0 : F(v,x) - F(v,y) \in h\mathbb{Z} \text{ for } \mu^2\text{-a.a. } (x,y) \in \mathcal{X}^2\} \qquad (2.4.7)$$

where $h\mathbb{Z} = \{hk : k \in \mathbb{Z}\}$. We call the case a lattice one if there exists $h > 0$ such that $F(x) \in h\mathbb{Z}$ for μ^ℓ-a.a. x, and

$$h = \sup\{u \geq 0 : u \in B_v\} \text{ for } \mu^{\ell-1}\text{-a.a. } v.$$

Since the distribution of $(\zeta_n, \zeta_{2n}, ..., \zeta_{\ell n})$ is absolutely continuous with respect to μ^ℓ for any $n \in \mathbb{N}$, it follows in this lattice case that S_N takes values on $h\mathbb{Z}$ for any N. We call the case a non-lattice case if

$$\mu^{\ell-1}\{v : B_v = \emptyset\} > 0.$$

Note that there are other cases beyond what we designated as a lattice and a non-lattice case.

2.4.2 *Characteristic functions estimates*

Relying on (2.4.6), it is shown in [29] that (2.3.7) holds true with $w \equiv 1$ and $c_N = (\ell - 1)\psi_{[\frac{N}{2}]-2}$. In fact, this approximation follows also from Corollary 1.3.11, taking into account (2.3.6) and relying on the relation $\phi_n \leq \psi_n$, where ϕ_n is defined by (1.3.1). Consider the Fourier operators $Q_{\bar{x},t}$ from (2.3.5) where $\bar{x} = (x_1, ..., x_{\ell-1})$. For each $\bar{x} \in \mathcal{X}^{\ell-1}$ and $t \in \mathbb{R}$ consider the function $\rho : \mathcal{X}^{\ell-1} \to [0, 1]$ given by

$$\rho_t(\bar{x}) = \sup_{y \in \mathcal{X}} \int d\eta(v) \Big| \int p^{(\ell)}(y, z)p^{(\ell)}(z, v)\exp\big(itF(\bar{x}, z)\big)d\eta(z)\Big|$$

where $p^{(\ell)}(\cdot, \cdot)$ is the ℓ-th step transition density defined by (2.4.2). Then for any $\bar{x}_1, \bar{x}_2 \in \mathcal{X}^{\ell-1}$ and $t \in \mathbb{R}$,

$$\|Q_{\bar{x}_1,t}Q_{\bar{x}_2,t}\|_\infty = \sup_{f:\|f\|_\infty=1} \sup_{y \in \mathcal{X}} \Big| \int d\eta(v)\exp\big(itF(\bar{x}_2, v)\big)f(v) \quad (2.4.8)$$

$$\times \int p^{(\ell)}(y, z)p^{(\ell)}(z, v)\exp\big(itF(\bar{x}_1, z)\big)d\eta(z)\Big| \leq \rho_t(\bar{x}_1)$$

and therefore by the submultiplicativity of norms of operators,

$$\mathbb{E}\Big| \prod_{k=M+1}^N Q_{t,\Theta_n} \mathbf{1}(\zeta_{\ell M}^{(\ell)})\Big| \leq \mathbb{E}\Big\| \prod_{k=M+1}^N Q_{t,\Theta_n}\Big\|_\infty \quad (2.4.9)$$

$$\leq \mathbb{E} \prod_{j=1}^{[\frac{N-N_\ell}{2}]} \Big\| \prod_{k=M+2(j-1)+1}^{M+2j} Q_{t,\Theta_n}\Big\|_\infty \leq \mathbb{E} \prod_{j=1}^{[\frac{N-N_\ell}{2}]} \rho_t(\Theta_{M+2j-1}).$$

Next, we check that Assumptions 2.2.1 and 2.2.2 hold true we will use the following. Let κ be a probability measure on some measure space and let f be a measurable function. Then by (5.2) from [29],

$$1 - \Big| \int e^{if(x)}d\kappa(x)\Big| \geq \frac{1}{4} \int \int |e^{if(x)} - e^{if(y)}|^2 d\kappa(x)d\kappa(y). \quad (2.4.10)$$

In order to show that Assumption 2.2.1 holds true, in the above notations, let Γ_1 and Γ_2 be two subsets of the unit circle such that

$$\inf_{\gamma_i \in \Gamma_i, i=1,2} |\gamma_1 - \gamma_2| = \delta > 0 \text{ and } \min_{j=1,2} \kappa\{x : e^{if(x)} \in \Gamma_i\} = \varepsilon > 0. \quad (2.4.11)$$

Then with $G_j = \{x : e^{if(x)} \in \Gamma_j\}$, $j = 1, 2$,

$$\int \int |e^{if(x)} - e^{if(y)}|^2 d\kappa(x)d\kappa(y) \quad (2.4.12)$$

$$\geq \int_{G_1} \int_{G_2} |e^{if(x)} - e^{if(y)}|^2 d\kappa(x)d\kappa(y) \geq \delta^2\varepsilon^2.$$

Next, fix $\bar{x} \in \mathcal{X}^{\ell-1}$ and for any $t \in \mathbb{R}$ consider the function $g_{\bar{x},t}$ given by $g_{\bar{x},t}(y) = \exp(itF(\bar{x}, y))$. Then either $g_{\bar{x},t}(y)$ does not depend on y η-a.s. or there exist Borel subsets $\Gamma_1 = \Gamma_{1,t,\bar{x}}$ and $\Gamma_2 = \Gamma_{2,t,\bar{x}}$ of the unit circle such that (2.4.11) holds true with $\kappa = \eta$ and some $\delta = \delta_{\bar{x},t}$ and $\varepsilon = \varepsilon_{\bar{x},t}$. In the latter case we consider the Borel probability measures $\kappa_{y,v}$, $y, v \in \mathcal{X}$ on $\mathcal{X}$ given by

$$\kappa_{y,v}(G) = \frac{1}{p^{(2\ell)}(y,v)} \int_G p^{(\ell)}(y,z) p^{(\ell)}(z,v) d\eta(z)$$

where $p^{(2\ell)}(\cdot, \cdot)$ are the densities of transition probabilities of $P(2\ell, \cdot, \cdot)$ with respect to the measure η. Then by (2.4.1) and the definition of $\kappa_{y,v}$ for $j = 1, 2$ we have

$$\kappa_{y,v}\{z : g_{\bar{x},t}(z) \in \Gamma_j\} \geq \frac{\gamma^2}{p^{(2\ell)}(y,v)} \eta\{z : g_{\bar{x},t}(z) \in \Gamma_j\} \geq \frac{\gamma^2 \varepsilon_{\bar{x},t}}{p^{(2\ell)}(y,v)}.$$
$$(2.4.13)$$

This together with (2.4.10)-(2.4.12) applied now with the measure $\kappa_{y,v}$ yields

$$c_t(\bar{x}, y, v) := p^{(2\ell)}(y,v) - \left| \int p^{(\ell)}(y,z) p^{(\ell)}(z,v) g_{\bar{x},t}(z) d\eta(z) \right| \quad (2.4.14)$$

$$= p^{(2\ell)}(y,v) \left(1 - \left| \int g_{\bar{x},t}(z) d\kappa_{y,v}(z) \right| \right) \geq \frac{\gamma^4 \delta_{\bar{x},t}^2 \varepsilon_{\bar{x},t}^2}{4 p^{(2\ell)}(y,v)}.$$

Set $U_y = \{v : p^{(2\ell)}(y,v) > 2\}$. Then $\eta(U_y) \leq \frac{1}{2} \int_{U_y} p^{(2\ell)}(y,v) d\eta(v) \leq \frac{1}{2}$, and so

$$\int_{\mathcal{X}} \frac{d\eta(v)}{p^{(2\ell)}(y,v)} \geq \int_{\mathcal{X} \setminus U_y} \frac{d\eta(v)}{p^{(2\ell)}(y,v)} \geq \frac{1}{4}. \quad (2.4.15)$$

Combining this with (2.4.14) we deduce that

$$\rho_t(\bar{x}) \leq 1 - \inf_y \int c_t(\bar{x}, y, v) d\eta(v) \leq 1 - \frac{1}{16} \gamma^4 \delta_{\bar{x},t}^2 \varepsilon_{\bar{x},t}^2. \quad (2.4.16)$$

In the non-lattice case $g_{\bar{x},t}(y)$ cannot be η-a.s. constant in y if $t \neq 0$ and $B_{\bar{x}} = \emptyset$, where $B_{\bar{x}}$ is defined in (2.4.7), while in the lattice case it cannot be η-a.s. constant in y if $0 < |t| < \frac{2\pi}{h}$. Let $\delta > 0$. In the non-lattice case set $J_\delta = [-\delta^{-1}, -\delta] \cup [\delta, \delta^{-1}]$ while in the lattice case set $J_\delta = [-\frac{\pi}{h}, \frac{\pi}{h}] \setminus (-\delta, \delta)$. Since $\rho_t(\bar{x})$ is continuous in t, it follows from (2.4.16) that there exists $c_\delta(\bar{x}) > 0$ such that

$$\sup_{t \in J_\delta} \rho_t(\bar{x}) \leq 1 - c_\delta(\bar{x}) \quad (2.4.17)$$

where in the non-lattice case we assume that $B_{\bar{x}} = \emptyset$. We conclude that for any $\delta > 0$ there exists $c_\delta > 0$ and a Borel set $G \subset \mathcal{X}^{\ell-1}$ such that for any $t \in J_\delta$,

$$\rho_t(\bar{x}) \leq 1 - c_\delta \text{ for all } \bar{x} \in G \text{ and } \mu^{\ell-1}(G) \geq \varepsilon > 0. \tag{2.4.18}$$

We return now to (2.4.9). The transition probabilities $P_\Theta(\cdot, \cdot)$ of the stationary Markov chain $\Theta = \{\Theta_n : n \geq 0\}$ on $\mathcal{X}^{\ell-1}$ are determined by the formula

$$P_\Theta(\bar{x}, \bar{\Gamma}) = \prod_{j=1}^{\ell-1} P(j, x^{(j)}, \Gamma_j)$$

where $\bar{x} = (x^{(1)}, ..., x^{(\ell-1)}) \in \mathcal{X}^{\ell-1}$ and $\bar{\Gamma} = \Gamma_1 \times \cdots \times \Gamma_{\ell-1}$ is a Borel set of $\mathcal{X}^{\ell-1}$. The measure $\mu^{\ell-1} = \mu \times \cdots \times \mu$ is the stationary distribution of Θ and its ℓ-th step transition probabilities $P_\Theta(\ell, \bar{x}, \cdot)$ have transition densities $p_\Theta^{(\ell)}(\cdot)$ with respect to $\eta^{\ell-1}$ satisfying

$$\gamma^{\ell-1} \leq p_\Theta^{(\ell)}(\bar{x}, \cdot) \leq \gamma^{-(\ell-1)}$$

where γ is the same as in (2.4.1). Next, observe that by (2.4.18),

$$P\{\Theta_n \in G\} = P\{\Theta_1 \in G\} = \mu^{\ell-1}(G) \geq \varepsilon \tag{2.4.19}$$

for any $n \in \mathbb{N}$. Consider the counting function

$$V(N) = \sum_{n=1}^{[\frac{1}{2}(N-N_\ell)]} \mathbb{I}_G(\Theta_{2n-1})$$

and the events

$$\Gamma(N) = \{V(N) < \frac{\varepsilon N}{9\ell^2}\}.$$

Since $[\frac{1}{2}(N - N_\ell)] \geq \frac{N}{4\ell} - 2$, it follows from (2.4.19) and the large deviation results from [8] together with [20] applied to the Markov chain $\{\Theta(2\ell n - 1), n \geq 1\}$ that

$$P(\Gamma(N)) \leq c^{-1} e^{-cN} \tag{2.4.20}$$

for some $c > 0$ independent of N. We refer the readers to the paragraph preceding (5.18) from [29] for the exact details. We conclude by (2.4.18) that

$$\prod_{j=1}^{[\frac{N-N_\ell}{2}]} \rho_t(\Theta_{M+2j-1}) \leq \mathbb{I}_{\Gamma(N)} + (1 - c_\delta)^{\frac{N\varepsilon}{9\ell^2}}$$

which together with (2.4.9), (2.4.20) and the estimates from Section 2.3 complete the proof that Assumption 2.2.1 holds true.

We show now that Assumption 2.2.2 holds true, as well. Similarly to the above arguments, we start with the estimate (2.4.8) and the first inequality from (2.4.16), but now employing the Taylor reminder formula we can represent $c_t(\bar{x}, y, v)$ from there for $|t|$ small enough in the following way (cf. Ch. 8 in [12]),

$$c_t(\bar{x}, y, v) = \frac{1}{2}t^2 \int_{\mathcal{X}} p^{(\ell)}(y, z)p^{(\ell)}(z, v)D(\bar{x}, y, z, v)d\eta(z) + t^2\hat{\varphi}_{\bar{x}, y, v}(t)$$

(2.4.21)

where for some constant $C > 0$,

$$\hat{\varphi}_{\bar{x}, y, v}(t) \leq C\hat{\varphi}_{\bar{x}}(t) \to 0 \text{ as } t \to 0 \text{ and}$$

$$D(\bar{x}, y, z, v) = \left(F(\bar{x}, z) - \frac{1}{p^{(2\ell)}(y, v)}\int p^{(\ell)}(y, z)p^{(\ell)}(z, v)F(\bar{x}, z)d\eta(z)\right)^2.$$

Now, either $F(\bar{x}, z)$ does not depend on z η-a.s., i.e. $F_\ell(\bar{x}) = 0$ η-a.s. which is excluded, or there exist Borel subsets $U_1 = U_{1,\bar{x}}$ and $U_2 = U_{2,\bar{x}}$ of the real line $\mathbb{R}$ such that

$$\inf_{z \in U_1, w \in U_2} |z - w| = \delta_{\bar{x}} > 0 \text{ and } \min_{j=1,2} \eta(G_j(\bar{x})) = \varepsilon_{\bar{x}}$$

where $G_j(\bar{x}) = \{z : F(\bar{x}, z) \in U_j\}$, $j = 1, 2$. In this case

$$\int_{\mathcal{X}} p^{(\ell)}(y, z)p^{(\ell)}(z, v)D(\bar{x}, y, z, v)d\eta(z) \qquad (2.4.22)$$

$$\geq \gamma^2 \int_{\mathcal{X}} D(\bar{x}, y, z, v)d\eta(z)$$

$$\geq \gamma^2 \inf_c \int_{G_1(\bar{x}) \cup G_2(\bar{x})} \left(F(\bar{x}, z) - c\right)^2 d\eta(z)$$

$$\geq \gamma^2 \varepsilon_{\bar{x}} \inf_c \inf_{a \in U_1, b \in U_2} \left((a - c)^2 + (b - c)^2\right) = \frac{1}{2}\gamma^2 \varepsilon_{\bar{x}} \delta_{\bar{x}}^2.$$

Now by (2.4.16), (2.4.21) and (2.4.22) for $|t|$ small enough,

$$\rho_t(\bar{x}) \leq 1 - \frac{t^2}{4}\gamma^2 \varepsilon_{\bar{x}} \delta_{\bar{x}}^2 + C\varphi_{\bar{x}}(t). \qquad (2.4.23)$$

Observe that by (2.3.1),

$$\hat{D}(\bar{x}) := \sup_y \int D(\bar{x}, y, z, v)d\eta(z)d\eta(v)$$

$$\leq 2(1 + \gamma^{-3})\int F^2(\bar{x}, z)d\eta(z) < \infty, \quad \mu^{\ell-1}\text{-a.s.}$$

Hence

$$\lim_{L\to\infty} \mu^{\ell-1}\{\bar{x} : \hat{D}(\bar{x}) \le L\} = 1.$$

This together with (2.4.23) yield that there exist $c > 0$ and a Borel set $G \subset \mathcal{X}^{\ell-1}$ such that $\mu^{\ell-1}(G) = \varepsilon > 0$ and for all $|t|$ small enough,

$$\rho_t(\bar{x}) \le e^{-ct^2} \text{ whenever } \bar{x} \in G. \tag{2.4.24}$$

Considering, again, the counting function

$$W(N) = \sum_{n=1}^{[\frac{1}{2}(N-N_\ell)]} \mathbb{I}_G(\Theta_{2n-1})$$

and the events

$$\Gamma(N) = \{W(N) < \varepsilon(\frac{N}{8\ell^2} - 2)\}.$$

we complete the proof that Assumption 2.2.2 holds true relying again on the large deviation results from [8] together with [20].

Remark 2.4.1. In the proofs that Assumptions 2.2.1 and 2.2.2 hold true we could have avoided using large deviation estimates. Indeed, inequalities of the form

$$\mathrm{Var}V(N) \le CN \text{ and } \mathrm{Var}W(N) \le CN$$

together with the Markov inequality are sufficient for obtaining estimates of the form

$$P(V(N)) \le C_1 N^{-1} \text{ and } P(W(N)) \le C_2 N^{-1}$$

which is sufficient for Assumptions 2.2.1 and 2.2.2 to hold. Such upper bounds of the variances of $V(N)$ and $W(N)$ follow from the mixing condition (2.4.6), see Remark 5.2 from [29].

Remark 2.4.2. The situation that (2.4.1) holds true with some n_0 in place of ℓ can also be considered. In this case (2.4.1) will hold true with ℓn_0, as well, and imposing appropriate restrictions on functions of the form $\sum_{j=1}^{n_0} F(\bar{x}_j, y_j)$ our proof proceeds similarly. In fact, the assumption that (2.4.1) holds true for any x can be relaxed to μ-a.a. x, and so our proof can be modified to the general situation of a ψ-mixing stationary Markov chain, taking into account Theorem 5 in [4].

2.5 Markov chains related to dynamical systems

2.5.1 *Locally distance expanding maps and transfer opera-. tors*

Let $(\mathcal{X}, \rho)$ be a compact metric space normalized in size so that $diam_\rho(\mathcal{X}) \leq 1$. We consider $\mathcal{X}$ with its Borel σ-algebra $\mathcal{B}_\mathcal{X}$ and denote by $B(x, r)$ an open ball of radius $r > 0$ around $x \in \mathcal{X}$. Let $T : \mathcal{X} \to \mathcal{X}$ be a continuous and surjective map. We further assume that there exist $\xi, \eta > 0$ and $\gamma > 1$ such that the following conditions hold true.

Assumption 2.5.1 (Uniform Openness). For any $z \in \mathcal{X}$,

$$B(Tz, \xi) \subset TB(z, \eta).$$

Assumption 2.5.2 (Topological exactness). There exists a constant $n_\xi \in \mathbb{N}$ such that

$$T^{n_\xi} B(z, \xi) = \mathcal{X} \text{ for any } z \in \mathcal{X}.$$

Assumption 2.5.3 (Locally distance expanding). For any $z_1, z_2 \in \mathcal{X}$,

$$\rho(Tz_1, Tz_2) \geq \gamma\rho(z_1, z_2), \quad \text{if} \quad \rho(z_1, z_2) < \eta.$$

We remark that the locally distance expanding condition implies that $T|_{B(z,\eta)}$ is injective for any $z \in \mathcal{X}$. Together with the compactness of the space $\mathcal{X}$ this yields that

$$\deg T := \sup_{x \in \mathcal{X}} |T^{-1}\{x\}| < \infty$$

where $|T^{-1}\{x\}|$ is the number of preimages of a point $x \in \mathcal{X}$. Given $f : \mathcal{X} \to \mathbb{C}$, we can define the transfer operator $\mathcal{L}_f$ acting on functions $g : \mathcal{X} \to \mathbb{C}$ by the formula

$$\mathcal{L}_f g(x) = \sum_{y \in T^{-1}\{x\}} e^{f(y)} g(y). \tag{2.5.1}$$

When f takes real values and $\mathcal{L}_f 1 = 1$ then $\mathcal{L}_f$ is a Markov operator which defines transition probabilities via the formula

$$P_f(x, \Gamma) = \mathcal{L}_f \mathbb{I}_\Gamma(x) = \sum_{y \in \Gamma : Ty = x} e^{f(y)}, \quad x \in \mathcal{X}, \Gamma \in \mathcal{B}_\mathcal{X}.$$

In what follows we will prove a nonconventional LLT for Markov chains generated by transition probabilities of the above form together with a natural stationary measure $\mu = \mu^{(f)}$ (see Section 2.5.3).

Next, let $0 < \alpha \leq 1$. For any $g : \mathcal{X} \to \mathbb{C}$ set

$$v_{\alpha,\xi}(g) = \inf\{R : |g(x) - g(x')| \leq R\rho^\alpha(x, x') \text{ if } \rho(x, x') < \xi\}$$
$$\text{and} \quad \|g\|_{\alpha,\xi} = \|g\|_\infty + v_{\alpha,\xi}(g)$$

where $\|g\|_\infty$ stands for the supremum of the absolute value of a function $g : \mathcal{X} \to \mathbb{C}$. Denote by $\mathcal{H}^{\alpha,\xi} = \mathcal{H}^{\alpha,\xi}(\mathcal{X}, \rho)$ the space of all functions $g : \mathcal{X} \to \mathbb{C}$ such that $\|g\|_{\alpha,\xi} < \infty$. In what follows we will always assume that the function f in the definition of the transfer operator is a member of $\mathcal{H}^{\alpha,\xi}$.

2.5.2 *Inverse branches, the pairing property and periodic points*

The following two properties of the iterates $T^n, n \geq 1$ of the map T are discussed in [53] in the more general setup of iterates of random expanding maps, and are given here for readers' convenience.

First, the locally distance expanding property guarantees that the map T is injective when it is restricted to open balls with radius η. This together with the uniform openness property implies that for any $y \in \mathcal{X}$ there exists a unique continuous inverse branch

$$T_y^{-1} : B(Ty, \xi) \to B(y, \eta)$$

of T sending $T(y)$ to y. By the locally distance expanding property we have

$$\rho(T_y^{-1}z_1, T_y^{-1}z_2) \leq \gamma^{-1}\rho(z_1, z_2) \quad \text{for any } z_1, z_2 \in B(Ty, \xi) \qquad (2.5.2)$$

and thus, in fact,

$$T_y^{-1}B(Ty, \xi) \subset B(y, \min(\gamma^{-1}\xi, \eta)) \subset B(y, \xi).$$

As a consequence, for any $n \in \mathbb{N}$ the map

$$T_y^{-n} := T_y^{-1} \circ \cdots \circ T_{T^{n-1}y}^{-1} : B(T^n y, \xi) \to B_\omega(y, \gamma^{-n}\xi)$$

is well defined and is a continuous inverse branch of T^n sending $T^n y$ to y such that

$$T^n \circ T_y^{-n} = Id\big|_{B(T^n y, \xi)}, \quad T_y^{-n}(T^n z) = z$$

and

$$\rho(T_y^{-n}z_1, T_y^{-n}z_2) \leq \gamma^{-n}\rho(z_1, z_2) \qquad (2.5.3)$$

for any $z_1, z_2 \in B(T^n y, \xi)$.

Secondly, let $x, x' \in \mathcal{X}$ be such that $\rho(x, x') < \xi$ and let $n \in \mathbb{N}$. For any $y \in T^{-n}\{x\}$ the point $y' = T_y^{-n}x'$ satisfies $T^n y' = x'$ and so by (2.5.3),

$$\rho(y, y') = \rho(T_y^{-n}x, T_y^{-n}x') \leq \gamma^{-n}\rho(x, x').$$

Since $T_{y'}^{-n}x = y$, exchanging the roles x and x' we can write

$$T^{-n}\{x'\} = \{y' = T_y^{-n}x' : y \in T^{-n}\{x\}\}.$$

Henceforth we will refer the property described above as the *paring property*.

Next, the following lemma describes the structure of periodic points of the map T.

Lemma 2.5.4. *Let $z_0 \in \mathcal{X}$ and $0 < r \leq \xi$. Let k_r be the smallest nonnegative integer k such that $\gamma^{-k}\xi \leq r$. Then for any $M \geq n_\xi + k_r$ there exists $x_0 \in \mathcal{X}$ such that*

$$T^M x_0 = x_0 \quad and \quad \rho(z_0, x_0) \leq \frac{r\gamma}{\gamma - 1}.$$

In particular periodic points are dense in $\mathcal{X}$.

Proof. Let $z_0 \in \mathcal{X}$. We first consider the case when $r = \xi$. Let $M \geq n_\xi$. Then by the topological exactness property,

$$T^M B(z_0, \xi) = \mathcal{X}$$

and we have used that T is surjective, as well. Therefore, there exists $z_1 \in B(z_0, \xi)$ such that $T^M z_1 = z_0$. Set $z_2 = T_{z_1}^{-M}z_1$, which is well defined since $z_1 \in B(z_0, \xi) = B(T^M z_1, \xi)$. Then by (2.5.3),

$$\rho(z_2, z_1) \leq \gamma^{-M}\rho(z_1, z_0) < \gamma^{-M}\xi < \xi$$

since $z_1 = T_{z_1}^{-M}T^M z_1 = T_{z_1}^{-M}z_0$. Thus, we can define recursively a sequence $\{z_k\}_{k=0}^\infty$ such that $\rho(z_k, z_{k-1}) < \xi$ and $z_{k+1} = T_{z_k}^{-M}z_k$, $k \in \mathbb{N}$. Then $T^M z_{k+1} = z_k$ and we claim that $\{z_k\}_{k=0}^\infty$ is a Cauchy sequence. Indeed, for any $k \geq 1$ we have

$$\rho(z_{k+1}, z_k) = \rho(T_{z_k}^{-M}z_k, T_{z_k}^{-M}T^M z_k) \qquad (2.5.4)$$
$$= \rho(T_{z_k}^{-M}z_k, T_{z_k}^{-M}z_{k-1}) \leq \gamma^{-M}\xi$$

where in the last inequality we used (2.5.3) and that $\rho(z_k, z_{k-1}) < \xi$. This completes the proof that $\{z_k\}_{k=0}^\infty$ is a Cauchy sequence, since $\gamma > 1$. We conclude that $\lim_{k\to\infty} z_k = x_0$ exists, and since $T^M z_{k+1} = z_k$, it follows by the continuity of T^M that $T^M x_0 = x_0$. The inequality

$$\rho(z_0, x_0) \leq \frac{\gamma\xi}{\gamma - 1}$$

is a consequence of (2.5.4). Next, let $r < \xi$ and set $k = k_r$. The above proof will proceed similarly with r in place of ξ if we show that

$$T^{n_\xi + k_r} B(z_0, r) = \mathcal{X} \qquad (2.5.5)$$

for any $z_0 \in \mathcal{X}$. Indeed, since

$$T_{z_0}^{-k}(B(T^k z_0, \xi)) \subset B(z_0, \gamma^{-k}\xi) \subset B(z_0, \xi)$$

we derive that

$$B(T^k z_0, \xi) \subset T^k B(z_0, \gamma^{-k}\xi) = T^k B(z_0, \gamma^{-k_r}\xi)$$

and (2.5.5) follows by substituting both sides in T^{n_ξ} and using the topological exactness assumption and that $r \geq \gamma^{-k_r}\xi$. $\square$

Example 2.5.5. Let $(\mathcal{X}, T)$ be a one sided topologically mixing subshift of finite type (see [10] and Section 2.11). Namely, let $A = (A_{i,j})$ be a matrix of size $n \times n$ with $0 - 1$ entries such that A^M has only positive entries for some $M \geq 1$. Consider the space

$$\mathcal{X} = \mathcal{X}_A = \{x = (x_i)_{i=1}^\infty \in \mathcal{A}^{\mathbb{N}} : A_{x_i, x_{i+1}} = 1 \ \forall i \in \mathbb{N}\}, \ \mathcal{A} = \{1, 2, ..., n\}$$
$$(2.5.6)$$

and let $T : \mathcal{X} \to \mathcal{X}$ be the one sided shift operator. The distance between two distinct points $x, y \in \mathcal{X}$ is given by

$$\rho(x, y) = 2^{-\min\{i \in \mathbb{N} : x_i \neq y_i\}}.$$

Let $x \in \mathcal{X}$. For any $K \geq M$ and $x_1 \in \mathcal{A}$ we have $A_{x_1, x_1}^K > 0$ and so there exist $x_2, ..., x_K \in \mathcal{A}$ such that $A_{x_i, x_{i+1}} = 1$ for any $i = 1, 2, ..., K$, where $x_{K+1} = x_1$. Then the periodic point $\bar{x} \in \mathcal{X}$ given by $\bar{x} = (x_1, x_2, ..., x_K, x_1, x_2, ..., x_K, ...)$ satisfies $T^K \bar{x} = \bar{x}$ and $\rho(x, \bar{x}) \leq 2^{-K}$.

2.5.3 Thermodynamic formalism constructions and the associated Markov chains

We will state here the main results from [53] in the case of deterministic distance expanding maps. Note that, in fact, those results are a particular case of the results from Part 2.

Let $f \in \mathcal{H}^{\alpha, \xi}$ be a real valued function and consider the transfer operator $\mathcal{L}_f$ given by (2.5.1). Since f is bounded and $\deg T < \infty$ the operator $\mathcal{L}_f$ acts continuously on the space $B(\mathcal{X})$ of all bounded Borel measurable functions on $\mathcal{X}$. As a consequence, the dual operator $\mathcal{L}_f^*$ which acts on the space of measures on $\mathcal{X}$ by the formula

$$\mathcal{L}_f^* \kappa = \kappa \circ \mathcal{L}_f$$

is well defined. We begin with the following Ruelle-Perron-Frobenius type theorem (see, for instance, [53]).

Theorem 2.5.6. *There exists a unique triplet* (λ, h, ν) *consisting of a positive number* $\lambda = \lambda^{(f)}$, *a strictly positive function* $h = h^{(f)} \in \mathcal{H}^{\alpha, \xi}$ *and a probability measure* $\nu = \nu^{(f)}$ *on* $\mathcal{X}$ *such that*

$$\mathcal{L}_f h = \lambda h, \ \mathcal{L}_f^* \nu = \lambda \nu \ and \ \nu(h) = 1.$$

An exponential convergence result is established, as well (see [53]).

Theorem 2.5.7. *There exist constants* $A > 0$ *and* $c \in (0, 1)$ *such that for any* $g \in \mathcal{H}^{\alpha, \xi}$ *and* $n \in \mathbb{N}$,

$$\left\| \lambda^{-n} \mathcal{L}_f^n(g) - \nu(g) h \right\|_\infty \le A\big(\nu|g| + v_{\alpha, \xi}(g)\big) c^n. \tag{2.5.7}$$

We remark that a stronger version of Theorem 2.5.7 follows from Theorem 4.2.2 in Chapter 4 of Part 2.

Next, consider the (Gibbs) measure $\mu = \mu^{(f)} = h\nu$. Then μ is T invariant since for any bounded Borel function $g : \mathcal{X} \to \mathbb{R}$,

$$\mu(g \circ T) = \nu(h \cdot (g \circ T)) = \lambda^{-1} \mathcal{L}_f^* \nu(h \cdot (g \circ T))$$
$$= \lambda^{-1} \nu(\mathcal{L}_f(h \cdot (g \circ T))) = \lambda^{-1} \nu(g \mathcal{L}_f(h)) = \nu(hg) = \mu(g)$$

where we used that $\mathcal{L}_f(g_1 \cdot (g_2 \circ T)) = g_2 \cdot \mathcal{L}_f g_1$ for any $g_1, g_2 : \mathcal{X} \to \mathbb{C}$. The following decay of correlations and Gibbs property are established in [53], as well.

Proposition 2.5.8. *There exist constants* $A > 0$ *and* $c \in (0, 1)$ *such that for any* $g_1 \in \mathcal{H}^{\alpha, \xi}$, $g_2 \in L^1(\mathcal{X}, \mu)$ *and* $n \in \mathbb{N}$,

$$|\mu(g_1 \cdot (g_2 \circ T^n)) - \mu(g_1)\mu(g_2)| \le A\|g_1\|_{\alpha, \xi} \mu(|g_2|) c^n.$$

As a consequence, for any $g_1, g_2 \in L^1(\mathcal{X}, \mu)$,

$$\lim_{n \to \infty} \mu(g_1 \cdot (g_2 \circ T^n)) = \mu(g_1)\mu(g_2)$$

and therefore the measure preserving system $(\mathcal{X}, \mathcal{B}_\mathcal{X}, \mu, T)$ *is mixing.*

Proposition 2.5.9. *There exist constants* $C_1, C_2 > 0$ *such that for any* $y \in \mathcal{X}$ *and* $n \in \mathbb{N}$,

$$C_1 \le \frac{\mu\big(T_y^{-n} B(T^n y, \xi)\big)}{\lambda^n e^{S_n f(y)}} \le C_2. \tag{2.5.8}$$

The following important property of ν and μ follows from here. Let $y \in \mathcal{X}$ and $n \in \mathbb{N}$. Since

$$T_y^{-n} B(T^n y, \xi) \subset B(y, \gamma^{-n} \xi)$$

we deduce that for any $0 < r < \xi$,

$$\mu(B(y, r)) \geq C_1 e^{-k_r \|f\|_\infty} \lambda^{k_r}$$

where k_r is the smallest nonnegative integer k satisfying $r \geq \xi \gamma^{-k}$, and in particular μ and ν assign positive mass to open sets.

Next, given a real valued $f \in \mathcal{H}^{\alpha, \xi}$ we consider the function

$$\tilde{f} = f + \ln h - \ln h \circ T - \ln \lambda \qquad (2.5.9)$$

where $h = h^{(f)}$ and $\lambda = \lambda^{(f)}$. Then

$$\mathcal{L}_{\tilde{f}} 1 = 1, \ \lambda^{(\tilde{f})} = 1, \ h^{(\tilde{f})} \equiv 1 \text{ and } \nu^{(\tilde{f})} = \mu^{(\tilde{f})} = \mu^{(f)}.$$

Henceforth, we will always replace f by $\tilde{f}$, namely, it will be assumed that $\mathcal{L}_f 1 = 1$, $\lambda^{(f)} = 1$ and $h^{(f)} \equiv 1$. In these circumstances the transfer operator $\mathcal{L}_f$ defines transition probabilities via the formula

$$P_f(x, \Gamma) = \mathcal{L}_f \mathbb{I}_\Gamma(x) = \sum_{y \in \Gamma : T(y) = x} e^{f(y)}, \ x \in \mathcal{X}, \Gamma \in \mathcal{B}_\mathcal{X}. \qquad (2.5.10)$$

Let $\{\zeta_n : n \geq 0\}$ be the Markov chain with initial distribution $\mu = \mu^{(f)}$ and transition probabilities $P_f(\cdot, \cdot)$ and let ℓ be a positive integer. In this section we will prove a nonconventional local (central) limit theorem for sums of the form

$$S_N = \sum_{n=1}^{N} F(\zeta_n, \zeta_{2n}, ..., \zeta_{\ell n})$$

where $F : \mathcal{X}^\ell \to \mathbb{R}$ is a function with some regularity conditions yet to be specified.

2.5.4   Relations between dynamical systems and Markov chains

In the "conventional" case $\ell = 1$ the topic of limit theorems for sums of the form $\sum_{k=0}^{n-1} G \circ T^k$ taken with the measure $\mu = \mu^{(f)}$ is well studied (see, for instance [23] and [28]). One of the main strategies for proving such limit theorems is as follows. Let Z_0 be a random element of $\mathcal{X}$ which is drawn according to μ, and consider the stationary $\mathcal{X}$-valued process $Z = \{Z_k, k \geq 0\}$, where $Z_k = T^k Z_0$. Then by Lemma XI.3 in [28] for any $n \in \mathbb{N}$,

$$(\zeta_{n-1}, \zeta_{n-2}, ..., \zeta_0) \overset{d}{=} (Z_0, Z_1, ..., Z_{n-1}) \qquad (2.5.11)$$

where $X \overset{d}{=} Y$ means that X and Y have the same distribution. As a consequence,

$$\sum_{k=0}^{n-1} G(Z_k) \overset{d}{=} \sum_{k=1}^{n} G(\zeta_k) \qquad (2.5.12)$$

and therefore any distributional limit theorem for the sums on the left-hand side will follow from the corresponding limit theorem for the sums on the right-hand side. Now the general theory of limit theorems for quasi-compact Markov operators (see Chap. XII in [28]) can be applied, taking into account Theorems 2.5.6 and 2.5.7.

Consider now the nonconventional situation when $\ell > 1$. First, it follows from (2.5.11) that

$$S_N = \sum_{n=1}^{N} F(\zeta_n, \zeta_{2n}, ..., \zeta_{\ell n}) \overset{d}{=} \sum_{n=1}^{N} F(Z_{\ell N-n}, Z_{\ell N-2n}, ..., Z_{\ell N-n\ell}) \qquad (2.5.13)$$

and so an LLT for the sums from the above right-hand side follows from an LLT for the sums on the left-hand side. In fact, since Z is stationary there exists a two sided stationary process $\mathbf{Z} = \{\mathbf{Z}_n : n \in \mathbb{Z}\}$ on some probability space $(\Omega, \mathcal{F}, P)$ and a measurable function $\pi : \Omega \to \mathcal{X}$ such that

$$\{\pi \circ \mathbf{Z}_n : n \geq 0\} \overset{d}{=} Z. \qquad (2.5.14)$$

Therefore, for any $N \in \mathbb{N}$,

$$S_N = \sum_{n=1}^{N} F(\zeta_n, \zeta_{2n}, ..., \zeta_{\ell n}) \overset{d}{=} \sum_{n=1}^{N} F \circ \pi(\mathbf{Z}_{-n}, \mathbf{Z}_{-2n}..., \mathbf{Z}_{-n\ell}) \qquad (2.5.15)$$

and again an LLT for the sums from the above right-hand side follows from the corresponding LLT for S_N. Still, in Section 2.11 we will extend this type of nonconventional LLT to sums of the form

$$\sum_{n=1}^{N} F(T^n x, T^{2n} x, ..., T^{\ell n} x)$$

when T is either a topologically mixing two sided subshift of finite type or an Axiom A diffeomorphism (in a neighborhood of an attractor) and x is drawn according to a Gibbs measure corresponding to some Hölder continuous (potential) function.

2.5.5 *Mixing and approximation assumptions*

Let $\mathcal{F}_{m,n}$, $-\infty \leq m \leq n \leq \infty$ be a family of sub-σ-algebras of $\mathcal{B}_{\mathcal{X}}$ such that $\mathcal{F}_{k,l} \subset \mathcal{F}_{k',l'}$ if $k' \leq k$ and $l' \geq l$. Consider the $\mathcal{X}$-valued stationary process $\{Z_k : k \geq 0\}$, where Z_0 is distributed according to μ and $Z_k = T^k Z_0$. For any $n, r \geq 0$ let $Z_{n,r}$ be a $\mathcal{F}_{n-r,n+r}$-measurable random element of $\mathcal{X}$. Let $d \in \mathbb{N}$ and consider the metric on $\mathcal{X}^d = \mathcal{X} \times \mathcal{X} \times \cdots \times \mathcal{X}$ given by

$$\rho_{d,\infty}(x, y) = \max_{1 \leq i \leq d} \rho(x_i, y_i)$$

for any $x = (x_1, ..., x_d)$ and $y = (y_1, ..., y_d)$ in $\mathcal{X}^d$. For any $H : \mathcal{X}^d \to \mathbb{C}$ set

$$v_{\alpha,\xi}(H) := \inf\{R \geq 0 : |H(x) - H(y)| \leq R(\rho_{d,\infty}(x, y))^\alpha$$
$$\text{for all } x, y \text{ with } \rho_{d,\infty}(x, y) < \xi\}$$

and let $\mathcal{H}_d = \mathcal{H}^{\alpha,\xi}(\mathcal{X}^d, \rho_{d,\infty})$ be the space of all functions $H : \mathcal{X}^d \to \mathbb{C}$ so that $v_{\alpha,\xi}(H) < \infty$. Note that H is continuous when $v_{\alpha,\xi}(H) < \infty$ and therefore $\|\cdot\|_{\alpha,\xi} = \|\cdot\|_\infty + v_{\alpha,\xi}(\cdot)$ defines a norm on $\mathcal{H}_d$.

For each $q \geq 1$ consider the approximation coefficients $\beta_{q,\alpha,\xi}(r)$, $r \geq 0$ given by

$$\beta_{q,\alpha,\xi}(r) \tag{2.5.16}$$
$$= \sup \left\{ \frac{\|H(Z_1, Z_2, ..., Z_d) - H(Z_{1,r}, Z_{2,r}, ..., Z_{d,r})\|_{L^q}}{v_{\alpha,\xi}(H)} : H \in \mathcal{H}_d, d \in \mathbb{N} \right\}.$$

In most of this section we will assume that $q = \infty$ and this case

$$\beta_{\infty,\alpha,\xi}(r) \leq \left(\beta_\infty(r)\right)^\alpha$$

if $\beta_{\infty,\alpha,\xi}(r) < \xi$, where

$$\beta_\infty(r) = \sup_{n \geq 0} \|\rho(Z_n, Z_{n,r})\|_{L^\infty}. \tag{2.5.17}$$

Our results will rely on the following

Assumption 2.5.10.

$$\sum_{n=1}^{\infty} n(\phi_n + \beta_{\infty,\alpha,\xi}(n)) < \infty$$

where ϕ_n was defined in (1.3.1).

Assumption 2.5.10 holds true in the situation when T is a one sided topologically mixing subshift of finite type, as well as in the case when T is an expanding C^2 endomorphism of a Riemannian manifold. In the first case $\mathcal{F}_{m,n}$ is the σ-algebra generated by cylinder sets with fixed coordinates.

Similarly, in the second case $\mathcal{F}_{m,n}$ is generated by partitions of the form $\bigcap_{j=m}^{n} T^{-j}(\mathcal{M})$ where $\mathcal{M}$ is a Markov partition with a sufficiently small diameter.

We remark that the results stated in this section hold true under some conditions involving the L^q-approximation coefficients $\beta_{q,\alpha,\xi}$ for some $1 \leq q < \infty$ rather than $\beta_{\infty,\alpha,\xi}$ (see Remark 2.8.7 for the precise details).

Next, let $F : \mathcal{X}^\ell \to \mathbb{R}$ satisfy that

$$|F(x) - F(y)| \leq C_F\big(\rho_{\ell,\infty}(x,y)\big)^\alpha \qquad (2.5.18)$$

for some constant $C_F > 0$ and all $x = (x_1, ..., x_\ell)$ and $y = (y_1, ..., y_\ell)$ in $\mathcal{X}^\ell$ so that $\rho_{\ell,\infty}(x,y) < \xi$. For any $N \in \mathbb{N}$ consider the sum

$$S_N = \sum_{n=1}^{N} F(\zeta_n, \zeta_{2n}, ..., \zeta_{\ell n}).$$

Our main result is a local limit theorem for these sums (and we refer the readers to Section 2.11 for extensions to certain dynamical systems).

2.5.6 *Asymptotic variance and the CLT*

Let $\zeta^{(i)} = \{\zeta_n^{(i)} : n \geq 0\}$, $i = 1, ..., \ell$ be independent copies of $\zeta = \{\zeta_n : n \geq 0\}$ and set

$$U_N = \sum_{n=1}^{N} F(\zeta_n^{(1)}, \zeta_{2n}^{(2)}, ..., \zeta_{\ell n}^{(\ell)}).$$

Theorem 2.5.11. *Suppose that Assumption 2.5.10 holds true. Then the limits*

$$\sigma^2 = \lim_{N \to \infty} \frac{1}{N} \mathbb{E}\big(S_N - N\bar{F}\big)^2 \quad and \quad s^2 = \lim_{N \to \infty} \frac{1}{N} \mathbb{E}\big(U_N - N\bar{F}\big)^2$$

exist, where $\bar{F}$ is defined in (2.3.2). Moreover, $\sigma^2 > 0$ if and only if $s^2 > 0$ and the latter two are positive if and only if there exists no $g \in L^2(\mathcal{X}^\ell, \mu^\ell)$ such that

$$F \circ (T \times T^2 \times \cdots \times T^\ell) = \bar{F} + g \circ (T \times T^2 \times \cdots \times T^\ell) - g, \quad \mu^\ell\text{-a.s.} \quad (2.5.19)$$

where a.s. stands for almost surely. Furthermore, $N^{-\frac{1}{2}}(S_N - N\bar{F})$ converges in distribution towards a centered normal random variable with the variance σ^2.

Proof. Existence of σ^2 is proved similarly to [45] relying on (2.5.11) and on Corollary 1.3.14 instead of Lemma 4.3 from there. The characterization of positivity of σ^2 follows from the arguments in the proof of Theorem 2.3 in [30], relying on (2.5.11) and on Corollary 1.3.14 instead of Lemma 3.4 from there. The proof of the CLT proceeds similarly to Chapter 1 first by using (2.5.11) and then using Assumption 2.5.10 in order to obtain an appropriate version of Corollary 1.3.16. $\qquad\square$

We remark that when (2.5.19) holds true then we can replace g there by a member of $\mathcal{H}^{\alpha,\xi}$ (and in this case the equality will hold true for any $x \in \mathcal{X}^\ell$). Indeed, this is a consequence of the arguments in the proof of Lemma III.1 in [23] and the appropriate versions of Theorems 2.5.6 and 2.5.7 for the map $T \times T^2 \times \cdots \times T^\ell$.

Next, consider the function $F_\ell : \mathcal{X}^\ell \to \mathbb{R}$ given by

$$F_\ell(x_1, ..., x_\ell) = F(x_1, ..., x_\ell) - \int F(x_1, ..., x_{\ell-1}, y)d\mu(y). \qquad (2.5.20)$$

Then F_ℓ satisfies (2.5.18) with some constant C_{F_ℓ} and

$$\int F_\ell(y_1, ..., y_{\ell-1}, x)d\mu(x) = 0 \quad \forall y_1, ..., y_{\ell-1} \in \mathcal{X}$$

and in particular $\bar{F}_\ell = 0$ where $\bar{F}_\ell$ is defined similarly to $\bar{F}$ in (2.3.2) but with F_ℓ in place of F. Theorem 2.5.11 holds true also with F_ℓ in place of F and in particular the limit

$$\sigma_\ell^2 = \lim_{N\to\infty} \frac{1}{N}\mathbb{E}\Big(\sum_{n=1}^{N} F_\ell(\zeta_n, \zeta_{2n}, ..., \zeta_{\ell n})\Big)^2 \qquad (2.5.21)$$

exists. If g satisfies (2.5.19) then the function g_ℓ which is defined similarly to F_ℓ but with g in place of F satisfies (2.5.19) with F_ℓ in place of F. Therefore, $\sigma^2 > 0$ when $\sigma_\ell^2 > 0$. Henceforth, we will assume that $\sigma_\ell^2 > 0$ which is equivalent to (2.5.19) not being satisfied with any $g \in L^2(\mathcal{X}^\ell, \mu^\ell)$ and F replaced with F_ℓ.

2.6 Statement of the local limit theorem

Let $x_0 \in \mathcal{X}$ be a periodic point of T and let $m_0 \in \mathbb{N}$ be so that $T^{m_0}x_0 = x_0$. Consider the point $\bar{x}_0 = (x_0, ..., x_0) \in \mathcal{X}^{\ell-1}$ and the points

$$\hat{T}^k\bar{x}_0 = (T^kx_0, T^{2k}x_0, ..., T^{k(\ell-1)}x_0) \in \mathcal{X}^{\ell-1}, \quad 0 \le k < m_0.$$

For any $u \in \mathcal{X}^{\ell-1}$ let the function $F_u : \mathcal{X} \to \mathbb{R}$ be given by $F_u(x) = F(u, x)$. Our results will rely on the following regularity assumption on F_u around the above points.

Assumption 2.6.1. The function $u \to F_u$ is continuous at the points $u = \hat{T}^k \bar{x}_0$, $k = 0, 1, ..., m_0 - 1$ when considered as a function from $\mathcal{X}^{\ell-1}$ to $\left(H^{\alpha,\xi}(\mathcal{X}), \| \cdot \|_{\alpha,\xi} \right)$.

Since F is a Hölder continuous function this assumption is equivalent to the assumption that the function $u \to v_{\alpha,\xi}(F_u - F_v)$ is continuous at these points. This continuity assumption holds true when the following assumption is satisfied.

Assumption 2.6.2. There exist constants $b, K_F > 0$ such that with any $v = \hat{T}^k \bar{x}_0, k = 0, 1, ..., m_0 - 1$ we have

$$\left| \left(F_u(r) - F_v(r) \right) - \left(F_u(r') - F_v(r') \right) \right| \leq K_F \left(\rho_{\ell-1,\infty}(u,v) \right)^b \left(\rho(r,r') \right)^\alpha$$

$$\tag{2.6.1}$$

for any $r, r' \in \mathcal{X}$ and $u \in \mathcal{X}^{\ell-1}$ such that $\rho(r,r') < \xi$ and $\rho_{\ell-1,\infty}(u,v) < \xi$.

Next, consider the map $T_\ell = \hat{T} \times T^\ell = T \times T^2 \times \cdots \times T^\ell$ and let the function $F_{x_0, m_0} : \mathcal{X} \to \mathbb{R}$ be given by

$$F_{x_0, m_0}(x) = \sum_{k=0}^{m_0-1} F(T^k x_0, T^{2k} x_0, ..., T^{(\ell-1)k} x_0, T^{\ell k} x) \tag{2.6.2}$$

$$= \sum_{k=0}^{m_0-1} F \circ T_\ell^k(\bar{x}_0, x).$$

We call the case a non-lattice one if for any $t \in \mathbb{R} \setminus \{0\}$ there exist no nonzero $g \in \mathcal{H}^{\alpha,\xi}$ and $\lambda \in \mathbb{C}$, $|\lambda| = 1$ such that

$$e^{itF_{x_0,m_0}} g = \lambda g \circ T^{\ell m_0}, \quad \mu\text{-a.s.} \tag{2.6.3}$$

where we recall that $\mu = \mu^{(f)}$. The non-lattice condition means that the function F_{x_0, m_0} is non-arithmetic (or aperiodic) with respect to the map $\tau_0 = T^{m_0 \ell}$ in the classical sense of [23] and [28], which according to Lemma XII.7 in [28] means that the spectral radius of the corresponding Fourier operator defined in (2.9.5) is strictly less than 1 for any real $t \neq 0$.

Theorem 2.6.3. *Suppose that Assumption 2.5.10 holds true, and let F be a non-arithmetic function satisfying (2.5.18) and Assumption 2.6.1. Assume in addition that $\sigma_\ell^2 > 0$. Then for any continuous function $g : \mathbb{R} \to \mathbb{R}$ with a compact support,*

$$\lim_{N \to \infty} \sup_{u \in \mathbb{R}} \left| \sigma \sqrt{2\pi N} \mathbb{E}g(S_N - u) - e^{-\frac{(u - N\bar{F})^2}{2N\sigma^2}} \int g(x) dx \right| = 0.$$

Next, we call the case a lattice one if the function F takes values on some lattice $h\mathbb{Z} = \{hk : k \in \mathbb{Z}\}$, $h > 0$ and the following assumption holds true.

Assumption 2.6.4. The function F_{x_0,m_0} cannot be written in the form

$$F_{x_0,m_0} = a + \beta - \beta \circ T^{\ell m_0} + h'\mathbf{k}, \quad \mu\text{-a.s.} \tag{2.6.4}$$

for some $h' > h$, $a \in \mathbb{R}$, $\beta : \mathcal{X} \to \mathbb{R}$ such that $e^{i\beta} \in \mathcal{H}^{\alpha,\xi}$ and an integer valued function $\mathbf{k} : \mathcal{X} \to \mathbb{Z}$.

Assumption 2.6.4 is equivalent to the statement that for any $t \in (-\frac{2\pi}{h}, \frac{2\pi}{j}) \setminus \{0\}$ there exists no $g \in \mathcal{H}^{\alpha,\xi} \setminus \{0\}$ and $a \in \mathbb{R}$ such that

$$ge^{itF_{x_0,m_0}} = e^{ia}g \circ T^{\ell m_0}, \quad \mu\text{-a.s.} \tag{2.6.5}$$

Indeed, when such g and a exist then by ergodicity of $(\mathcal{X}, \mathcal{B}_\mathcal{X}, \mu, T^{\ell m_0})$ the function $|g|$ is μ-a.s. constant, and dividing by $|g|$ we arrive at (2.6.4) with $h' = \frac{2\pi}{|t|} > h$, while the other direction is trivial.

Theorem 2.6.5. *Suppose that Assumption 2.5.10 holds true, and let F be a lattice valued function as described above so that (2.5.18) and Assumption 2.6.1 hold true. Assume in addition that $\sigma_\ell^2 > 0$. Then for any continuous function $g : \mathbb{R} \to \mathbb{R}$ with a compact support,*

$$\lim_{N\to\infty} \sup_{u\in h\mathbb{Z}} \left| \sigma\sqrt{2\pi N}\mathbb{E}g(S_N - u) - he^{-\frac{(u-N\bar{F})^2}{2N\sigma^2}} \sum_{hk\in h\mathbb{Z}} g(hk) \right| = 0.$$

Remark that, in general, the natural choice of m_0 is $m_0 = n_\xi$, where n_ξ is specified in Assumption 2.5.2. Still, in some situations there exist periodic points with smaller order. For instance, when $(\mathcal{X}, T)$ is a (topologically mixing) one sided subshift of finite type and $A_{i,i} = 1$ for some i then $T\bar{i} = \bar{i}$, where $\bar{i} = (iii...)$ is the word which consists only of the letter i, and the above conditions concern the function $F(\bar{i}, \bar{i}, ..., \bar{i}, \cdot)$ and the ℓ-th step shift T^ℓ.

In the following sections we will show that Assumptions 2.2.1 and 2.2.2 hold true in both lattice and non-lattice cases, which in view of Theorems 2.2.3 and 2.5.11 will imply that the statements of both Theorems 2.6.3 and 2.6.5 hold true.

2.7   The associated random transfer operators

In this section we will make the first step towards the proof that Assumption 2.2.2 holds true.

2.7.1 *Random complex RPF theorem*

Let $F : \mathcal{X}^\ell \to \mathbb{R}$ be a function satisfying (2.5.18). In this section we provide precise details of the construction of the random Fourier operators $\mathbf{Q}_{it}^\omega$ defined after (2.3.8) which in our circumstances are random transfer operators denoted by $\mathbf{L}_{it}^\omega$. Let $\zeta^{(i)}$, $i = 1, ..., \ell - 1$ be independent copies of ζ. Consider the process $\Theta = \{\Theta_n : n \geq 0\}$ given by

$$\Theta_n = (\zeta_n^{(1)}, \zeta_{2n}^{(2)}, ..., \zeta_{(\ell-1)n}^{(\ell-1)}).$$

Then Θ is the stationary Markov chain with initial distribution $\mu^{\ell-1} = \mu \times \cdots \times \mu$ and transition operator $\hat{\mathcal{L}}_{\hat{f}}$ which is the transfer operator defined by the map $\hat{T} = T \times T^2 \cdots \times T^{\ell-1}$ and the function $\hat{f}(x_1, ..., x_{\ell-1}) = \sum_{j=1}^{\ell-1} f(x_j)$. Let $M(\Theta) = (\Omega_\Theta, \mathcal{B}_\Theta, P_\Theta, \vartheta)$ be the natural invertible measure preserving system (MPS) associated with the process Θ. This means that

$$\Omega_\Theta = (\mathcal{X}^{\ell-1})^{\mathbb{Z}}$$

ϑ the left shift map and the processes Θ and $\{p_0 \circ \vartheta^n, n \geq 0\}$ have the same distribution, where p_0 is the 0-th coordinate projection given by

$$p_0(s) = s_0, \quad s = \{s_n : n \in \mathbb{Z}\} \in (\mathcal{X}^{\ell-1})^{\mathbb{Z}}.$$

Note that by Theorem 2.5.8 the MPS $(\mathcal{X}, \mathcal{B}_\mathcal{X}, \mu, T)$ is mixing. Therefore, the MPS's $(\mathcal{X}^{\ell-1}, \mathcal{B}_{\mathcal{X}^{\ell-1}}, \mu^{\ell-1}, \hat{T})$ and $M(\Theta)$ are mixing, as well. Consider now the random function $\mathbf{F}_\omega(\cdot)$ on $\mathcal{X}$ given by

$$\mathbf{F}_\omega(\cdot) = F(p_0(\omega), \cdot), \quad \omega \in \Omega_\Theta. \tag{2.7.1}$$

The operators $\mathbf{Q}_{it}^\omega = \mathbf{L}_{it}^\omega$ introduced below (2.3.8) are the random transfer operators $\mathbf{L}_z^\omega$ defined by the map T^ℓ and the (potential) function $S_\ell f + z\mathbf{F}_\omega$ which act on functions $g : \mathcal{X} \to \mathbb{C}$ by the formula

$$\mathbf{L}_z^\omega g(x) = \sum_{y \in T^{-\ell}\{x\}} e^{S_\ell f(y) + z\mathbf{F}_\omega(y)} g(y).$$

The following theorem plays a crucial role in Section 2.8 and it concerns only the above random transfer operators where additional assumptions needed for more general results of Chapters 4 and 5 are automatically satisfied (see Section 5.4).

Theorem 2.7.1. *(i) There exists a constant $r > 0$ such that for P_Θ-almost every $\omega \in \Omega_\Theta$ and any $z \in \mathbb{C}$ with $|z| < r$ there exists a triplet $\lambda_\omega(z)$, $h_\omega(z)$ and $\nu_\omega(z)$ consisting of a nonzero complex number $\lambda_\omega(z)$, a* *complex function $h_\omega(z) \in \mathcal{H}^{\alpha,\xi}$ and a complex continuous linear functional $\nu_\omega(z) : \mathcal{H}^{\alpha,\xi} \to \mathbb{C}$ such that*

$$\boldsymbol{L}_z^\omega h_{\vartheta\omega}(z) = \lambda_\omega(z) h_\omega(z), \quad \left(\boldsymbol{L}_z^\omega \nu_\omega(z)\right)^* = \lambda_\omega(z)\nu_{\vartheta\omega}(z) \text{ and } \quad (2.7.2)$$

$$\nu_\omega(z)h_\omega(z) = \nu_\omega(z)\boldsymbol{1} = 1.$$

When $z = t \in \mathbb{R}$ then $\lambda_\omega(t) > 0$, the function $h_\omega(t)$ is strictly positive, $\nu_\omega(t)$ is a probability measure and the equality $\nu_\omega(t)\left(\boldsymbol{L}_t^\omega g\right) = \lambda_\omega(t)\nu_{\vartheta\omega}(t)(g)$ holds true for any bounded Borel function $g : \mathcal{X} \to \mathbb{C}$.

(ii) Moreover, the above triplet is measurable. Namely, for each $z \in B(0,r) = \{\alpha \in \mathbb{C} : |\alpha| < r\}$ the maps

$$\omega \to \lambda_\omega(z), \quad \omega \to h_\omega(z) \quad \text{and } \omega \to \nu_\omega(z)$$

are measurable.

(iii) Furthermore, this triplet is (strongly) analytic and uniformly bounded. Namely, the maps

$$\lambda_\omega(\cdot) : B(0,r) \to \mathbb{C}, h_\omega(\cdot) : B(0,r) \to \mathcal{H}^{\alpha,\xi} \text{ and } \nu_\omega(\cdot) : B(0,r) \to \left(\mathcal{H}^{\alpha,\xi}\right)^*$$

are analytic, where $(\mathcal{H}^{\alpha,\xi})^*$ is the dual space of $\mathcal{H}^{\alpha,\xi}$, and for any $k \geq 0$ there exists $C_k > 0$ such that P_Θ-a.s. for any $z \in B(0,r)$,

$$\max\left(|\lambda_\omega^{(k)}(z)|, \|h_\omega^{(k)}(z)\|_{\alpha,\xi}, \|\nu_\omega^{(k)}(z)\|_{\alpha,\xi}\right) \leq C_k$$

where $g^{(k)}$ stands for the k-th derivative of a function on the complex plane which takes values in some Banach space.

When $z = 0$ then $\boldsymbol{L}^\omega = \mathcal{L}_f^\ell$ and therefore by the uniqueness part of Theorem 2.5.6 P-a.s.,

$$h_\omega(0) \equiv \boldsymbol{1}, \ \lambda_\omega(0) = 1 \ \text{ and } \ \bar{\nu}_\omega(0) = \mu_\omega(0) = \mu^{(f)} = \mu \qquad (2.7.3)$$

where we recall our assumption that $\mathcal{L}_f\boldsymbol{1} = 1$ which implies that $\lambda^{(f)} = 1$, $h^{(f)} \equiv 1$ and $\nu^{(f)} = \mu^{(f)}$.

The following proposition is established in Part 2 in a more general situation, as well.

Proposition 2.7.2. *There exist constants $r, A > 0$ and $c \in (0,1)$ such that P_Θ-a.s. for any $n \in \mathbb{N}$, $g \in \mathcal{H}^{\alpha,\xi}$ and $z \in \mathbb{C}$ with $|z| < r$,*

$$\left\| \frac{\boldsymbol{L}_z^{\omega,n}g}{\lambda_{\omega,n}(z)} - \left(\nu_{\vartheta^n\omega}(z)g\right)h_\omega(z) \right\|_{\alpha,\xi} \leq Ac^n$$

where

$$\boldsymbol{L}_z^{\omega,n} = \boldsymbol{L}_z^\omega \circ \boldsymbol{L}_z^{\vartheta\omega} \circ \cdots \circ \boldsymbol{L}_z^{\vartheta^{n-1}\omega} \ \text{ and } \ \lambda_{\omega,n}(z) = \prod_{k=0}^{n-1} \lambda_{\vartheta^k\omega}(z).$$

In Part 2 we will obtain several additional results concerning the random transfer operators $\boldsymbol{L}_z^\omega$ such as exponential decay of correlations and results concerning Lyapunov exponents.

2.7.2 Distortion properties

We will need the following.

Lemma 2.7.3. Let $\{g_\omega\}_{\omega\in\Omega_\Theta}$ be a family of functions from $\mathcal{H}^{\alpha,\xi}$. Let $\omega \in \Omega$, $n \in \mathbb{N}$ and consider the function

$$S^\omega_{n,\ell}g = \sum_{k=0}^{n-1} g_{\vartheta^k\omega} \circ T^{k\ell}$$

on $\mathcal{X}$. Let $x, x' \in \mathcal{X}$ be such that $\rho(x,x') < \xi$. Then for any $y \in T^{-n\ell}\{x\}$,

$$|S^\omega_{n,\ell}g(y) - S^\omega_{n,\ell}g(y')| \le \rho^\alpha(x,x') \sum_{j=0}^{n-1} v_{\alpha,\xi}(g_{\vartheta^j\omega})\gamma^{-\alpha\ell(n-j)} \qquad (2.7.4)$$

where $y' = T_y^{-n\ell}\{x'\}$ and $T_y^{-n\ell}$ is the inverse branch of $T^{n\ell}$ around $T^{n\ell}y = x$ introduced in Section 2.5.2.

Proof. Let $x, x' \in \mathcal{X}$ be such that $\rho(x,x') < \xi$ and let y and y' be as in the statement of the lemma. Observe that for any $j < n$,

$$T^{\ell j}T_y^{-\ell n}x = T_{T^{\ell j}y}^{-\ell(n-j)}x \quad \text{and} \quad T^{\ell j}T_y^{-\ell n}x' = T_{T^{\ell j}y}^{-\ell(n-j)}x'.$$

Since $y = T_y^{-\ell n}x$, it follows that

$$|S^\omega_{n,\ell}g(y) - S^\omega_{n,\ell}g(y')| \le \sum_{j=0}^{n-1} |g_{\vartheta^j\omega}(T^{\ell j}y) - g_{\vartheta^j\omega}(T^{\ell j}y')|$$

$$= \sum_{j=0}^{n-1} |g_{\vartheta^j\omega}(T_{T^{\ell j}y}^{-\ell(n-j)}x) - g_{\vartheta^j\omega}(T_{T^{\ell j}y}^{-\ell(n-j)}x')|$$

$$\le \sum_{j=0}^{n-1} v_{\alpha,\xi}(g_{\vartheta^j\omega})\rho^\alpha(T_{T^{\ell j}y}^{-\ell(n-j)}x, T_{T^{\ell j}y}^{-\ell(n-j)}x')$$

and (2.7.4) follows from (2.5.3). $\qquad\qquad\square$

The following random Lasota-Yorke type inequality follows.

Lemma 2.7.4. Suppose that $\mathcal{L}_f\mathbf{1} = \mathbf{1}$. Then there exist constants $B \ge 1$ and $c \in (0,1)$ such that for any $\omega \in \Omega_\Theta$, $n \in \mathbb{N}$, $g \in \mathcal{H}^{\alpha,\xi}$ and $z \in \mathbb{C}$,

$$v_{\alpha,\xi}(\mathbf{L}_z^{\omega,n}g) \le Be^{n|\Re(z)|\|F\|_\infty}(v_{\alpha,\xi}(g)c^n + \|g\|_\infty)$$

and

$$\|\mathbf{L}_z^{\omega,n}g\|_{\alpha,\xi} \le 2Be^{n|\Re(z)|\|F\|_\infty}(c^n\|g\|_{\alpha,\xi} + (1+|z|)\|g\|_\infty). \qquad (2.7.5)$$

In particular, there exists a constant $C > 0$ such that for any $n \in \mathbb{N}$ and $\omega \in \Omega_\Theta$,

$$\|\mathbf{L}_z^{\omega,n}\|_{\alpha,\xi} \le C(1+|z|)e^{n|\Re(z)|\|F\|_\infty}.$$

Proof. First, for any $z \in \mathbb{C}$ consider the random function $\mathbf{f}_z$ given by

$$\mathbf{f}_{z,\omega} = S_\ell f + z\mathbf{F}_\omega = \sum_{j=0}^{\ell-1} f \circ T^j + zF(p_0(\omega), \cdot).$$

Let $n \in \mathbb{N}$ and $z \in \mathbb{C}$. In order to estimate $v_{\alpha,\xi}(\mathbf{L}_z^{\omega,n})$, let $x, x' \in \mathcal{X}$ be such that $\rho(x, x') < \xi$. Then by the pairing property discussed in Section 2.5.2 we can write

$$T^{-n\ell}\{x'\} = \{y' : y \in T^{-n\ell}\{x\}\}$$

where $y' = T_y^{-n\ell}\{x'\}$, which satisfies

$$\rho(y, y') \le \gamma^{-n\ell}\rho(x, x').$$

Therefore,

$$\left| \mathbf{L}_z^{\omega,n}g(x) - \mathbf{L}_z^{\omega,n}g(x') \right| = \Big| \sum_{y \in T^{-\ell n}\{x\}} \left(e^{S_{n,\ell}^\omega \mathbf{f}_z(y)}g(y) - e^{S_{n,\ell}^\omega \mathbf{f}_z(y')}g(y') \right) \Big|$$

$$\le \sum_{y \in T^{-\ell n}\{x\}} e^{S_{n,\ell}^\omega \mathbf{f}_{\Re(z)}(y)} |e^{i\Im(z)S_{n,\ell}^\omega \mathbf{F}(y)}g(y) - e^{i\Im(z)S_{n,\ell}^\omega \mathbf{F}(y')}g(y')|$$

$$+ \sum_{y \in T^{-\ell n}\{x\}} |e^{i\Im(z)S_{n,\ell}^\omega \mathbf{F}(y')}g(y')||e^{S_{n,\ell}^\omega \mathbf{f}_{\Re(z)}(y)} - e^{S_{n,\ell}^\omega \mathbf{f}_{\Re(z)}(y')}| := I_1 + I_2.$$

In order to estimate I_1, let $y \in T^{-\ell n}\{x\}$. Then

$$|e^{i\Im(z)S_{n,\ell}^\omega \mathbf{F}(y)}g(y) - e^{i\Im(z)S_{n,\ell}^\omega \mathbf{F}(y')}g(y')|$$

$$\le |g(y) - g(y')| + |g(y)| \cdot |e^{i\Im(z)S_{n,\ell}^\omega \mathbf{F}(y)} - e^{i\Im(z)S_{n,\ell}^\omega \mathbf{F}(y')}| := J_1 + J_2.$$

Since $\rho(y, y') \le \gamma^{-\alpha n}\rho(x, x') < \xi$ we have

$$J_1 \le v_{\alpha,\xi}(g)\rho^\alpha(y, y') \le v_{\alpha,\xi}(g)\gamma^{-\alpha\ell n}\rho^\alpha(x, x').$$

By (2.5.18) there exists a constant $Q_1 > 0$ such that $\|\mathbf{F}_{\vartheta^k\omega}\|_{\alpha,\xi} \le Q_1$ for any $\omega \in \Omega_\Theta$ and $k \in \mathbb{N}$. It follows from Lemma 2.7.3 together with the mean value theorem that there exists a constant $Q \ge 1$ such that

$$J_2 \le Q\|g\|_\infty |\Im(z)|\rho^\alpha(x, x'),$$

and so

$$I_1 \le \mathbf{L}_{\Re(z)}^{\omega,n}\mathbf{1}(x)\big(v_{\alpha,\xi}(g)\gamma^{-\alpha\ell n} + Q\|g\|_\infty|\Im(z)|\big)\rho^\alpha(x, x').$$

Finally, observe that

$$\mathbf{L}_{\Re(z)}^{\omega,n}\mathbf{1}(x) \le \mathcal{L}_f^{\ell n}\mathbf{1}(x)e^{|\Re(z)|\|S_{n,\ell}^\omega \mathbf{F}\|_\infty} = e^{|\Re(z)|\|S_{n,\ell}^\omega \mathbf{F}\|_\infty} \le e^{n|\Re(z)|\|F\|_\infty}$$

$$\tag{2.7.6}$$

and we conclude that

$$I_1 \leq e^{n|\Re(z)|\|F\|_\infty} \left(v_{\alpha,\xi}(g)\gamma^{-\alpha \ell n} + Q\|g\|_\infty |\Im(z)| \right) \rho^\alpha(x, x').$$

Next, we approximate I_2. First, similarly to arguments in the proof of Lemma 2.7.3 we see that

$$|S_{n\ell}f(y) - S_{n\ell}f(y')| \leq v_{\alpha,\xi}(f)\rho^\alpha(x, x') \sum_{j=0}^{\ell n - 1} \gamma^{-\alpha j}. \tag{2.7.7}$$

Therefore by the mean value theorem and Lemma 2.7.3 there exists a constant Q_2 such that

$$|e^{S^\omega_{n,\ell}\mathbf{f}_{\Re(z)}(y)} - e^{S^\omega_{n,\ell}\mathbf{f}_{\Re(z)}(y')}|$$
$$\leq Q_2(1 + |\Re(z)|) \max\{e^{S^\omega_{n,\ell}\mathbf{f}_{\Re(z)}(y)}, e^{S^\omega_{n,\ell}\mathbf{f}_{\Re(z)}(y')}\} \rho^\alpha(x, x'),$$

and so

$$I_2 \leq Q_2(1 + |\Re(z)|)\|g\|_\infty (\mathbf{L}^{\omega,n}_{\Re(z)}\mathbf{1}(x) + \mathcal{L}^{\omega,n}_{\Re(z)}\mathbf{1}(x')) \rho^\alpha(x, x')$$
$$\leq 2Q_2(1 + |\Re(z)|)\|g\|_\infty e^{n|\Re(z)|\|F\|_\infty} \rho^\alpha(x, x')$$

where in the last inequality we used (2.7.6). We conclude that there exists a constant $Q_3 > 0$ such that

$$v_{\alpha,\xi}(\mathbf{L}^{\omega,n}_z g) \leq e^{n|\Re(z)|\|F\|_\infty} \left(v_{\alpha,\xi}(g)\gamma^{-\alpha \ell n} + (1 + |z|)Q_3\|g\|_\infty \right).$$

Finally, in order to approximate $\|\mathbf{L}^{\omega,n}_z g\|_\infty$, we observe that

$$\|\mathbf{L}^{\omega,n}_z g\|_\infty \leq \|g\|_\infty \|\mathcal{L}^{\ell n}_f \mathbf{1}\|_\infty \cdot e^{|\Re(z)|\|S^\omega_{n,\ell}\mathbf{F}\|_\infty} = \|g\|_\infty e^{|\Re(z)|\|S^\omega_{n,\ell}\mathbf{F}\|_\infty}$$

which together with the previous estimates implies (2.7.5), and the proof of the lemma is complete. $\qquad\square$

2.7.3 *Reduction to random dynamics*

In this section we will show that (2.3.7) holds true with an appropriate function w and a sequence $(c_n)_{n=1}^\infty$. More precisely, we claim that there exists a constant $C_2 > 0$ such that for any $t \in \mathbb{R}$ and $N \in \mathbb{N}$,

$$\Big| \mathbb{E}| \prod_{k=M+1}^N Q_{it,\zeta_k,\zeta_{2k},\dots,\zeta_{(\ell-1)k}} \mathbf{1}(\zeta_{\ell M})| \tag{2.7.8}$$

$$- \mathbb{E}| \prod_{k=M+1}^N Q_{it,\Theta_k} \mathbf{1}(\zeta^{(\ell)}_{\ell M})| \Big| \leq NC_2(1 + |t|)\left(\phi_{\frac{N}{6}} + \beta_{\infty,\alpha,\xi}(\frac{N}{6})\right)$$

where $\phi_s = \phi_{[s]}$ and $\beta_{\infty,\alpha,\xi}(s) = \beta_{\infty,\alpha,\xi}([s])$ for any $s \geq 0$, and we recall that $M = M(N) = N - 2[\frac{N-N_\ell}{2}]$ and $N_\ell = [N(1 - \frac{1}{2\ell})] + 1$. Before proving (2.7.8) we assume its validity and bound from above the characteristic function of S_N. We first notice that in our situation the operator $Q_{it,\bar{x}}$ is the transfer operator generated by the map T^ℓ and the (potential) function $S_\ell f + itF(\bar{x}, \cdot)$. Therefore,

$$\mathbb{E}|\prod_{k=M+1}^{N} Q_{it,\Theta_k} \mathbf{1}(\zeta_{\ell M}^{(\ell)})| \tag{2.7.9}$$

$$= \int\int |\mathbf{L}_{it}^{\vartheta^{M+1}\omega,N-M} \mathbf{1}(x)|d\mu(x)dP_\Theta(\omega)$$

$$= \int\int |\mathbf{L}_{it}^{\omega,N-M} \mathbf{1}(x)|d\mu(x)dP_\Theta(\omega).$$

We conclude from (2.7.8), (2.7.9), (2.3.3) and (2.3.4) that for any $t \in \mathbb{R}$,

$$|\varphi_N(t)| = |\mathbb{E}e^{itS_N}| \leq C_2(1+|t|)c_N + \int\int |\mathbf{L}_{it}^{\omega,N-M} \mathbf{1}(x)|d\mu(x)dP_\Theta(\omega) \tag{2.7.10}$$

where $c_N = N(\phi(\frac{N}{6}) + \beta_{\infty,\alpha,\xi}(\frac{N}{6}))$ which under Assumption 2.5.10 satisfies $\lim_{n\to\infty}\sqrt{n}c_n = 0$.

In order to prove (2.7.8), consider the function $H : (\mathcal{X}^{\ell-1})^{N-M} \times \mathcal{X} \to \mathbb{R}$ given by

$$H(x) = H_{N,t}(x) = \prod_{k=M+1}^{N} Q_{it,x_k,x_{2k},...,x_{(\ell-1)k}} \mathbf{1}(x_0)$$

where $x = (\bar{x}, x_0)$ and $\bar{x} = \{(x_k, x_{2k}, ..., x_{(\ell-1)k}) : M+1 \leq k \leq N\}$. Set $\tilde{X} = (\mathcal{X}^{\ell-1})^{N-M} \times \mathcal{X}$ and let the metric ρ_∞ on $\tilde{X}$ be defined by

$$\rho_\infty(x,z) = \max_i \rho(x_i, z_i).$$

We assert that $H \in \mathcal{H}^{\alpha,\xi}(\tilde{X}, \rho_\infty)$ and that there exists a constant $C_2 > 0$ such that for any $N \in \mathbb{N}$ and $t \in \mathbb{R}$,

$$\|H\|_{\alpha,\xi} \leq NC_2(1+|t|). \tag{2.7.11}$$

Observe that the proof of Lemma 1.3.10 proceeds similarly in the case when V_i's appearing there take values in some compact metric spaces. Therefore, Corollary 1.3.11 also holds true for random variable taking values in compact metric spaces and the estimate (2.7.8) now follows from (2.7.11), the definition of $\beta_{\infty,\alpha,\xi}$, (2.3.6) and the above version of Corollary 1.3.11.

In order to prove (2.7.11), first, it is clear that $|H(x)| \leq \mathcal{L}_f^{\ell(M-N)}\mathbf{1}(x_0) = 1$, for any $x \in \tilde{X}$. In order to estimate $v_{\alpha,\xi}(H)$, let $x, z \in \tilde{X}$, set $\varepsilon = \rho_\infty(x,z)$ and suppose that $\varepsilon < \xi$. By the pairing property discussed in Section 2.5.2 we can write

$$T^{-\ell(N-M)}\{z_0\} = \{y_0' : y_0 \in T^{-\ell(N-M)}\{x_0\}\}$$

where $y_0' = T_{y_0}^{-\ell(N-M)} z_0$ which satisfies

$$\rho(T^m y_0, T^m y_0') \leq \gamma^{-(\ell(N-M)-m)} \rho(x_0, z_0)$$

for any $0 \leq m \leq \ell(N-M)$. Therefore, we can write

$$H(x) - H(z) = \sum_{y_0 \in T^{-\ell(N-M)}\{x_0\}} \left(D(\bar{x}, y_0) - D(\bar{z}, y_0')\right) \qquad (2.7.12)$$

where for any $(\bar{b}, a) \in \tilde{X}$,

$$D(\bar{b}, a) = e^{S_{\ell(N-M)} f(a) + it \sum_{k=M+1}^{N} F(b_k, b_{2k}, \ldots, b_{(\ell-1)k}, T^{\ell(k-M-1)} a)}.$$

Next, it follows from (2.7.7) that there exists a constant $C > 0$ so that

$$|S_{\ell(N-M)} f(y_0) - S_{\ell(N-M)} f(y_0')| \leq C v_{\alpha,\xi}(f) \rho^{\alpha}(x_0, z_0)$$

and by (2.5.18) that for any k,

$$|F(x_k, \ldots, x_{(\ell-1)k}, T^{\ell(k-M-1)} y_0) - F(z_k, \ldots, z_{(\ell-1)k}, T^{\ell(k-M-1)} y_0')|$$
$$\leq C_F v_{\alpha,\xi}(F) \varepsilon^{\alpha}.$$

It follows from the triangle inequality and the mean value theorem that

$$|D(\bar{x}, y_0) - D(\bar{z}, y_0')| \leq |D(\bar{x}, y_0) - D(\bar{z}, y_0)| + |D(\bar{z}, y_0) - D(\bar{z}, y_0')|$$
$$\leq A(1 + |t|)(e^{S_{\ell(N-M)} f(y_0)} + e^{S_{\ell(N-M)} f(y_0')})(N\varepsilon^{\alpha} + \rho^{\alpha}(x_0, z_0))$$

for some constant $A > 0$. Since $\mathcal{L}_f 1 = 1$ we conclude that

$$|H(x) - H(z)| \leq 2A(1 + |t|)(N\varepsilon^{\alpha} + \rho^{\alpha}(x_0, z_0)) \qquad (2.7.13)$$
$$\leq 4A(1 + |t|)N\varepsilon^{\alpha} = 4A(1 + |t|)N\rho_{\infty}^{\alpha}(x, z)$$

and the proof of (2.7.11) is complete. $\qquad \square$

2.8 Decay of characteristic functions for small t's

In this section we will prove that Assumption 2.2.2 holds true under Assumption 2.5.10 when the function F satisfies (2.5.18) and σ_{ℓ}^2 defined in (2.5.21) is positive.

2.8.1 *The random pressure function*

We introduce the k-th step random pressure functions and show that they are analytic functions of the parameter z around 0.

Proposition 2.8.1. *There exists a constant $s > 0$ such that P_{Θ}-a.s. for any $k \in \mathbb{N}$ there exists an analytic function $\Pi_{\omega,k} : B(0, s) \to \mathbb{C}$ so that $\Pi_{\omega,k}(0) = 0$ and for any $z \in B(0, s)$,*

$$e^{\Pi_{\omega,k}(z)} = \lambda_{\omega,k}(z) \quad and \quad |\Pi_{\omega,k}(z)| \leq k(\ln 2 + \pi).$$

Before proving the proposition we need the following Taylor reminder estimates.

Lemma 2.8.2. *Let $(X, \|\cdot\|)$ be a complex Banach space and let $Q : U \to X$ be an analytic function defined on an open set $U \subset \mathbb{C}$ which contains a closed ball $\bar{B}(z_0, \delta)$ of radius $\delta > 0$ around some $z_0 \in \mathbb{C}$. Suppose that there exists $r > 0$ such that $\|Q(z)\| \le r$ for any $z \in \bar{B}(z_0, \delta)$. Let $k \ge 0$ be an integer and denote by Q_{k,z_0} the Taylor polynomial of Q of order k around z_0. Then for any $z \in \bar{B}(z_0, \delta)$,*

$$\|Q(z) - Q_{k,z_0}(z)\| \le \frac{(k+2)r|z - z_0|^{k+1}}{\delta^{k+1}}.$$

Proof. We first recall that as a consequence of the complex version of the Hahn-Banach theorem, for any $x \in X$ we have

$$\|x\| = \sup_{f \in X^*: \|f\| = 1} |f(x)|$$

where X^* is the space of all continuous complex linear functionals equipped with the operator norm. Plugging in $x = Q(z) - Q_{k,z_0}(z)$ and then considering complex valued analytic functions φ of the form $\varphi = f \circ Q$ we see that it is sufficient to prove the lemma in the case when $Q : U \to \mathbb{C}$.

Let $k \ge 0$ and consider the analytic function $G : U \to \mathbb{C}$ given by

$$G(z) = \frac{Q(z) - Q_{k,z_0}(z)}{(z - z_0)^{k+1}}$$

for any $z \ne z_0$ and $(k+1)! G(z_0) = Q^{(k+1)}(z_0) = \frac{\partial^{k+1} Q(z)}{dz^{k+1}}\big|_{z=z_0}$. By the maximum modulus principle,

$$\sup_{z \in \bar{B}(z_0, \delta)} |G(z)| \le \sup_{z: |z - z_0| = \delta} |G(z)| = \frac{\sup_{z: |z-z_0| = \delta} |Q(z) - Q_{k,z_0}(z)|}{\delta^{k+1}}.$$

In order to bound the numerator from above, by the Cauchy integral formula for any $s \in \mathbb{N}$ we have

$$Q^{(s)}(z_0) = \frac{s!}{2\pi i} \int_{\{z: |z - z_0| = \delta\}} \frac{Q(z)}{(z - z_0)^{s+1}} dz.$$

Therefore, since $|Q(z)| \le r$ for any $z \in \bar{B}(z_0, \delta)$,

$$|Q_{k,z_0}(z)| \le \sum_{s=0}^{k} \frac{|Q^{(s)}(z_0)|}{s!} |z - z_0|^s \le \sum_{s=0}^{k} \frac{2\pi\delta}{2\pi\delta^{s+1}} \cdot r \cdot \delta^s = (k+1)r$$

assuming that $|z - z_0| = \delta$, and the lemma follows. $\qquad \square$

We remark that another consequence of the Hahn-Banach theorem is that the maximum modulus principle holds true for analytic functions taking values in a Banach space, and so, in fact, the proof of the lemma for the case when Q takes complex values proceeds similarly.

Proof of Proposition 2.8.1. It is sufficient to prove the proposition when $k = 1$ since then we can define

$$\Pi_{\omega,k}(z) = \sum_{j=0}^{k-1} \Pi_{\vartheta^j \omega, 1}(z).$$

Observe now that P_Θ-a.s.,

$$\lambda_\omega(z) = \lambda_\omega(z) \cdot \nu_{\vartheta\omega}(z)\mathbf{1} = \nu_\omega(z)\big(\mathbf{L}_z^\omega \mathbf{1}\big)$$

for any $z \in B(0,r) = \{\alpha \in \mathbb{C} : |\alpha| < r\}$, where r is specified in Theorem 2.7.1. By Theorem 2.7.1 *(iii)* there exists $M \geq 1$ such that $\|\nu_\omega(z)\|_{\alpha,\xi} \leq M$ for P_Θ-a.a. ω and any $z \in B(0,r)$. Thus, by Lemma 2.7.4 there exists a constant $C_0 \geq 1$ such that

$$|\lambda_\omega(z)| = |\nu_\omega(z)\big(\mathbf{L}_z^\omega \mathbf{1}\big)| \leq M\|\mathbf{L}_z^\omega \mathbf{1}\|_{\alpha,\xi} \leq 2MC_0 e^{\|F\|_\infty} \tag{2.8.1}$$

for any $z \in B(0,r)$ so that $|\Re(z)| \leq 1$. Next, recall that $\lambda_\omega(0) = 1$ (see (2.7.3)). Applying Lemma 2.8.2 with $z_0 = 0$ and $k = 0$ we deduce that for any $z \in B(0, \frac{1}{2}\min(1,r))$,

$$|\lambda_\omega(z) - \lambda_\omega(0)| = |\lambda_\omega(z) - 1| \leq \delta^{-1}|z| \tag{2.8.2}$$

where $\delta^{-1} = \frac{4MC_0}{\min(1,r)}$. Set $s = \frac{1}{2}\delta$. We conclude that there exists an analytic function $\Pi_{\omega,1}(\cdot) : B(0,s) \to \mathbb{C}$ such that $e^{\Pi_{\omega,1}(z)} = \lambda_\omega(z)$ (i.e. $\Pi_{\omega,1}$ is a branch of $\ln \lambda_\omega(\cdot)$), $\Pi_{\omega,1}(0) = \ln \lambda_\omega(0) = 0$, $|\Im(\Pi_{\omega,1}(z))| \leq \pi$ and

$$|\Pi_{\omega,1}(z)| \leq \ln(2\lambda_\omega(0)) + \pi = \ln 2 + \pi \tag{2.8.3}$$

for any $z \in B(0,s)$, and the proof of the proposition is complete. $\square$

2.8.2 The derivatives of the pressure

Lemma 2.8.3. P_Θ-a.s. for any $k \in \mathbb{N}$,

$$\Pi'_{\omega,k}(0) = \int S_{k,\ell}^\omega \boldsymbol{F} d\mu \quad and \tag{2.8.4}$$

$$\Pi''_{\omega,k}(0) = Var_\mu(S_{k,\ell}^\omega \boldsymbol{F}) + 2\int S_{k,\ell}^\omega \boldsymbol{F} h'_{\vartheta^k \omega}(0)d\mu + R_{\omega,k}$$

where $R_{k,\omega} = \int h''_{\vartheta^k \omega}(0)d\mu - \int h''_\omega(0)d\mu$ and $S_{k,\ell}^\omega = \sum_{j=0}^{k-1} \boldsymbol{F}_{\vartheta^j \omega} \circ T^{j\ell}$.

Proof. First, since $\lambda_{\omega,k}(z) = e^{\Pi_{\omega,k}(z)}$ and $\lambda_{\omega,k}(0) = 1$ it is sufficient to show that

$$\lambda'_{\omega,k}(0) = \int S^{\omega}_{k,\ell}\mathbf{F}d\mu \quad \text{and} \quad (2.8.5)$$

$$\lambda''_{\omega,k}(0) = \int (S^{\omega}_{k,\ell}\mathbf{F})^2 d\mu + 2\int S^{\omega}_{k,\ell}\mathbf{F}\, h_{\vartheta^k\omega}(0)d\mu + R_{\omega,n}$$

for any $k \in \mathbb{N}$, P_Θ-a.s. In order to prove (2.8.5), using the equality $\mathbf{L}^{\omega_1}_t h_{\vartheta\omega_1}(z) = \lambda_{\omega_1}(t)h_{\omega_1}(z)$ with $\omega_1 = \vartheta^j\omega, j = k-1, k-2, ..., 0$ we obtain that

$$\lambda_{\omega,k}(z)h_\omega(z) = \mathbf{L}^{\omega,k}_z h_{\vartheta^k\omega}(z). \quad (2.8.6)$$

Differentiating both sides with respect to z we infer that

$$\lambda'_{\omega,k}(z)h_\omega(z) + \lambda_{\omega,k}(z)h'_\omega(z) = \frac{d\mathbf{L}^{\omega,k}_z(h_{\vartheta^k\omega}(z))}{dz} \quad (2.8.7)$$

$$= \mathcal{L}^{\omega,k}_z\big(S^{\omega}_{k,\ell}\mathbf{F}\cdot h_{\vartheta^k\omega}(z) + h'_{\vartheta^k\omega}(z)\big).$$

Next, similarly to (2.8.6),

$$\big(\mathbf{L}^{\omega,k}_z\big)^* \nu_\omega(z) = \lambda_{\omega,k}(z)\nu_{\vartheta^k\omega}(z).$$

Note that when $z = t \in \mathbb{R}$ then the above equality is an equality in the space of measures on $\mathcal{X}$. Taking $z = 0$ and integrating both sides of (2.8.7) with respect to $\nu_\omega(0)$ we arrive at

$$\lambda'_{\omega,k}(0) = \mu(S^{\omega}_{k,\ell}\mathbf{F}) + G_{\vartheta^k\omega}(0) - G_\omega(0) \quad (2.8.8)$$

where $G_\omega(z) = \nu_\omega(z)h'_\omega(z)$ and we used that $\lambda_{\omega,k}(0) = 1$, $h_\omega(0) \equiv 1$ and $\nu_\omega(0) = \mu$ (see (2.7.3)). We claim now that $G_\omega(0) = 0$, P_Θ-a.s. Indeed, differentiating the equality $\nu_\omega(z)h_\omega(z) = 1$ with respect to z yields

$$G_\omega(z) + \nu'_\omega(z)h_\omega(z) = 0.$$

Since $h_\omega(0) \equiv 1$ and $\nu_\omega(z)\mathbf{1} = 1$,

$$\nu'_\omega(0)h_\omega(0) = \nu'_\omega(0)\mathbf{1} = \frac{d\nu_\omega(z)\mathbf{1}}{dz}(0) = 0$$

and it follows that $G_\omega(0) = 0$, yielding the first equality in (2.8.5).

In order to obtain the second equality in (2.8.5), we first differentiate both sides of (2.8.7) with respect to z and deduce that

$$\lambda''_{\omega,k}(z)h_\omega(z) + 2\lambda'_{\omega,k}(z)h'_\omega(z) + \lambda_{\omega,k}(z)h''_\omega(z)$$

$$= \mathbf{L}^{\omega,k}_z\big((S^{\omega}_{k,\ell}\mathbf{F})^2\cdot h_{\vartheta^k\omega}(z) + 2S^{\omega}_{k,\ell}\mathbf{F}\cdot h'_{\vartheta^k\omega}(z) + h''_{\vartheta^k\omega}(z)\big).$$

Taking $z = 0$ and integrating both sides of (2.8.7) with respect to $\nu_\omega(0)$, taking into account (2.7.3) and that $G_\omega(0) = 0$, we arrive at

$$\lambda''_{\omega,k}(0) = \int (S^{\omega}_{k,\ell}\mathbf{F})^2 d\mu + 2\int S^{\omega}_{k,\ell}\mathbf{F}\, h'_{\vartheta^k\omega}(0)d\mu + R_{n,\omega}$$

where $R_{n,\omega} = F_{\vartheta^k\omega}(0) - F_\omega(0)$ and $F_\omega(0) = \nu_\omega(0)h''_\omega(0)$, which completes the proof of the lemma. $\qquad\square$

2.8.3 *Pressure near* 0

We begin with bounding the terms $R_{k,\omega}$ and $\int S^\omega_{k,\ell} \mathbf{F} h'_{\vartheta^k \omega}(0) d\mu$ defined in Lemma 2.8.3. First by Theorem 2.7.1 there exists $C_2 > 0$ such that $\|h''_\omega(0)\|_\infty \le C_2$ for P_Θ-a.a. ω. Therefore,

$$\left| \int h''_\omega(0) d\mu \right| \le \|h''_\omega(0)\|_\infty \le C_2$$

P_Θ-a.s., and we conclude that for any $k \ge 0$,

$$|R_{k,\omega}| \le 2C_2, \quad P_\Theta\text{-a.s.} \tag{2.8.9}$$

Next, we will estimate $\int S^\omega_{k,\ell} \mathbf{F} h'_{\vartheta^k \omega}(0) d\mu$. First, by Theorem 2.5.8 there exist constants $A > 0$ and $c \in (0,1)$ such that for any $m, k \ge 0$

$$\left| \mu\left((\mathbf{F}_{\vartheta^m \omega} \circ T^{\ell m}) \cdot h'_{\vartheta^k \omega}(0) \right) \right| \le Ac^{\ell m} \|F\|_\infty \|h'_{\vartheta^k \omega}(0)\|_{\alpha, \xi}$$

where we used that $G_{\vartheta^k \omega}(0) = \mu(h'_{\vartheta^k \omega}(0)) = 0$. By Theorem 2.7.1 *(iii)* there exists a constant $B > 0$ such that $\|h'_{\vartheta^k \omega}(0)\|_{\alpha, \xi} \le B$ P_Θ-a.s. for any $k \ge 0$. We conclude that there exists $\Gamma > 0$ such that P_Θ-a.s. for any $k \ge 0$,

$$\left| \int S^\omega_{k,\ell} \mathbf{F} h'_{\vartheta^k \omega}(0) d\mu \right| \le \Gamma. \tag{2.8.10}$$

It follows that there exists a constant $R > 0$ so that for any k and for P_Θ-a.a. ω,

$$|\Pi''_{\omega,k}(0) - \mathrm{Var}_\mu(S^\omega_{k,\ell}\mathbf{F})| \le R.$$

Applying Lemma 2.8.2 with the neighborhood $U = B(0, s)$ specified in Proposition 2.8.1, taking into account that $|\Pi_{\omega,k}| \le k(\pi + \ln 2)$, we deduce that for any $z \in B(0, \frac{1}{2}s)$,

$$\left| \Pi_{\omega,k}(z) - z\mu(S^\omega_{k,\ell}\mathbf{F}) - \frac{1}{2}z^2 \mathrm{Var}_\mu(S^\omega_{k,\ell}\mathbf{F}) \right| \le c_0(k|z|^3 + R|z|^2)$$

where $c_0 > 0$ is a constant which does not depend on ω, z and k. It follows that for any $c_1 > 0$ there exists $d > 0$ and (a sufficiently small) $t_0 > 0$ so that for any $t \in [-t_0, t_0]$,

$$\Re(\Pi_{\omega,k}(it)) \le -kdt^2 \tag{2.8.11}$$

for any ω and k such that $\mathrm{Var}_\mu(S^\omega_{k,\ell}\mathbf{F}) \ge c_1 k > 4c_0 R$.

2.8.4   *Norms estimates: employing the pressure*

We will complete here the proof that Assumption 2.2.2 holds true. First, we claim that there exists a constant $A_1 > 0$ such that P_Θ-a.s.,

$$\|\mathbf{L}_{it}^{\omega,m}\|_{\alpha,\xi} \leq A_1 |\lambda_{\omega,m}(it)| \tag{2.8.12}$$

for any $m \in \mathbb{N}$ and $t \in (-r,r)$, where r is specified in Theorem 2.7.1. Indeed, by Theorem 2.7.1 *(iii)* there exists a constant $A_2 > 0$ such that for P_Θ-a.a. ω,

$$\|\nu_{\vartheta^m\omega}(z)(g) \cdot h_\omega(z)\|_{\alpha,\xi} = |\nu_{\vartheta^m\omega}(z)(g)| \cdot \|h_\omega(z)\|_{\alpha,\xi} \leq A_2 \|g\|_{\alpha,\xi}$$

for any $m \in \mathbb{N}$, $g \in \mathcal{H}^{\alpha,\xi}$ and $z \in \mathbb{C}$ such that $|z| < r$, and (2.8.12) follows now by Proposition 2.7.2.

Next, by (2.8.12) and the definition of $\Pi_{\omega,m}(\cdot)$,

$$\|\mathbf{L}_{it}^{\omega,m}\|_{\alpha,\zeta} \leq A_1 |\lambda_{\omega,m}(it)| = A_1 e^{\Re(\Pi_{\omega,m}(it))}, \quad P_\Theta\text{-a.s.} \tag{2.8.13}$$

for any $t \in (-s,s)$, where s is specified in Proposition 2.8.1. For any $c_1 > 0$ and $n \in \mathbb{N}$ set

$$\Gamma_{n,c_1} = \{\omega \in \Omega_\Theta : \mathrm{Var}_\mu(S_{\omega,\ell}^n \mathbf{F}) \geq c_1 n\} \tag{2.8.14}$$

and $\Gamma_{n,c_1}^{\mathbf{c}} = \Omega_\Theta \setminus \Gamma_{n,c_1}$. By (2.8.13) and (2.8.11) for any $c_1 > 0$ there exist $t_0 > 0$ and $d > 0$ such that for any $\omega \in \Gamma_{N-M,c_1}$ and $t \in [-t_0, t_0]$,

$$\int |\mathbf{L}_{it}^{\omega,N-M}\mathbf{1}(x)|d\mu(x) \leq A_1\|\mathbf{L}_{it}^{\omega,N-M}\|_{\alpha,\xi} \leq A_1 e^{-dt^2(N-M)}$$

assuming that $N - M \geq \frac{4c_0 R}{c_1}$. Observe that $N - M \geq a_0 N$ for any $N > 3$, where $a_0 > 0$ is a constant which depends only on ℓ. We conclude that for any $t \in [-t_0, t_0]$ and $N \geq J_0 := \max(4, (4c_0 R)(c_1 a_0)^{-1})$,

$$\int\int |\mathbf{L}_{it}^{\omega,N-M}\mathbf{1}(x)|d\mu(x)dP_\Theta(\omega) \leq A_1 e^{-da_0 t^2 N} + P_\Theta(\Gamma_{N-M,c_1}^{\mathbf{c}}). \tag{2.8.15}$$

Next, we claim that there exists $c_1 > 0$ such that

$$\lim_{n\to\infty} \sqrt{n} P_\Theta(\Gamma_{n,c_1}^{\mathbf{c}}) = 0. \tag{2.8.16}$$

Before proving this claim we complete the proof that Assumption 2.2.2 holds true relying on (2.8.16). Indeed, by (2.7.10) and (2.8.15) for any $t \in [-t_0, t_0]$ and $N \geq J_0$,

$$|\varphi_N(t)| \leq \sqrt{N}c_N + \int\int |\mathbf{L}_{it}^{\omega,N-M}\mathbf{1}(x)|d\mu(x)dP_\Theta(\omega)$$

$$\leq \sqrt{N}c_N + A_1 e^{-da_0 t^2 N} + \sqrt{N}P_\Theta(\Gamma_{N-M,c_1}^{\mathbf{c}})$$

where $(c_n)_{n=1}^\infty$ is the sequence on the right-hand side of (2.7.8), which satisfies $\lim_{n\to\infty} \sqrt{n}c_n = 0$. Since $N - M \geq a_0 N$ for any $N > 3$, the third term in the above right-hand side converges to 0 as $N \to \infty$ and the proof that Assumption 2.2.2 holds true is complete.

In order to prove (2.8.16), let $k \in \mathbb{N}$ and consider the random variable $\hat{V}_k$ given by

$$\hat{V}_k(\omega) = \mathrm{Var}_\mu(S_{k,\ell}^\omega \mathbf{F}) = \mathrm{Var}_\mu\Big(\sum_{j=0}^{k-1} \mathbf{F}_{\vartheta^k \omega} \circ T^{j\ell}\Big).$$

Then $\hat{V}_k(\omega) = \mu(S_{k,\ell}^\omega \mathbf{F}_\ell)^2$ where the random function $\mathbf{F}_\ell$ is given by $\mathbf{F}_{\ell,\omega}(\cdot) = F_\ell(p_0(\omega), \cdot)$. For any $n \geq 1$, $\varepsilon_0 > 0$ and $c > 0$ set

$$\Gamma_{n,\varepsilon_0,c,k} = \Big\{\omega \in \Omega_\Theta : \frac{1}{n}\sum_{j=1}^n \mathbb{I}_{\{\hat{V}_k(\vartheta^{jk}\omega)\geq ck\}} < \varepsilon_0\Big\}$$

and for the sake of convenience we set $\Gamma_{0,\varepsilon_0,c,k} = \Omega_\Theta$. Equality (2.8.16) will follow from the following two claims.

Claim 2.8.4. For any $c_2 > 0$ and $\varepsilon_0 > 0$ there exists k_0 such that for any natural n and $k > k_0$,

$$\Omega_\Theta \setminus \Gamma_{[\frac{n}{k}],\varepsilon_0,c_2,k} \subset \Gamma_{n,c_1} \qquad (2.8.17)$$

where $c_1 = \frac{1}{4}\varepsilon_0 c_2$.

Claim 2.8.5. There exist $c_2, \varepsilon_0 > 0$ such that for any sufficiently large k,

$$\lim_{n\to\infty} \sqrt{n}P_\Theta(\Gamma_{n,\varepsilon_0,c_2,k}) = 0. \qquad (2.8.18)$$

Indeed, we obtain (2.8.16) by applying Claim 2.8.4 with sufficiently large k and using (2.8.18) with $[\frac{n}{k}]$ in place of n on the left-hand side.

Proof of Claim 2.8.4. First, by the definitions of $\mathbf{F}$ and $\hat{V}_k$,

$$\hat{V}_m(\omega) = \mathrm{Var}\sum_{j=1}^m F(p_0(\vartheta^j\omega), \zeta_{\ell j}) \qquad (2.8.19)$$

for any $m \geq 1$. Next, the process $\{Z_m : m \geq 0\}$ (defined in Section 2.5.4) is stationary since T preserves μ, and for any $k, s \geq 0$ the pairs (ζ_s, ζ_{s+k}) and (Z_k, Z_0) have the same distribution (see (2.5.11)). Thus, for any $j, m \geq 0$ and $\omega \in \Omega_\Theta$,

$$\mathrm{Cov}\big(F(p_0(\vartheta^j\omega), \zeta_{\ell j}), F(p_0(\vartheta^{j+m}\omega), \zeta_{\ell(j+m)})\big)$$
$$= \mathrm{Cov}\big(F(p_0(\vartheta^j\omega), Z_{m\ell}), F(p_0(\vartheta^{j+m}\omega), Z_0)\big)$$
$$= \mu\big(F(p_0(\vartheta^j\omega), T^{m\ell}(\cdot))F(p_0(\vartheta^{j+m}\omega), \cdot)\big)$$
$$- \mu\big(F(p_0(\vartheta^j\omega), \cdot)\big) \cdot \mu\big(F(p_0(\vartheta^{j+m}\omega), \cdot)\big).$$

For any $\omega \in \Omega_\Theta$ we have $\|F(p_0(\omega), \cdot)\|_{\alpha,\xi} \leq \|F\|_{\alpha,\xi}$, and therefore by Proposition 2.5.8 there exist constants $C > 0$ and $a \in (0,1)$ such that for any $j, m \geq 0$ and $\omega \in \Omega_\Theta$,

$$\left|\mathrm{Cov}(F(p_0(\vartheta^j\omega), \zeta_{\ell j}), F(p_0(\vartheta^{j+m}\omega), \zeta_{\ell(j+m)})\right| \leq Ca^{\ell m}. \qquad (2.8.20)$$

Next, let $\varepsilon_0, c_2 > 0$ and $\omega \in \Omega_\Theta$. In order to prove (2.8.17) let $n, k \in \mathbb{N}$ be such that $n \geq k > 1$ and set

$$B_r = \sum_{j=(r-1)k+1}^{rk} F(p_0(\vartheta^j\omega)), \zeta_{\ell j})$$

$r = 1, ..., [\frac{n}{k}]$ and

$$B_{[\frac{n}{k}]+1} = \sum_{j=1}^{n} F(p(\vartheta^j\omega), \zeta_{\ell j}) - \sum_{r=1}^{[\frac{n}{k}]} B_r.$$

We assert that there exists a constant $A_1 > 0$ which does not depend on n, k and ω such that for any $r = 1, ..., [\frac{n}{k}] + 1$,

$$\left|\mathrm{Cov}(B_r, \sum_{s>r} B_s)\right| \leq A_1. \qquad (2.8.21)$$

Indeed, set

$$h(m) = Ca^{\ell m}, \ m \in \mathbb{N}$$

where C and a come from (2.8.20). Let $s > r$. Then by (2.8.20),

$$\left|\mathrm{Cov}(B_r, B_s)\right| \leq \sum_{j_1,j_2=1}^{k} h\big((s-1)k + j_1 - (r-1)k - j_2\big)$$

$$= \sum_{m=0}^{k-1}(k-m)h\big((s-r)k+m\big) + \sum_{m=1}^{k-1}(k-m)h\big((s-r)k-m\big)$$

$$= \sum_{u=1}^{k} uh\big((s-r)k+k-u\big) + \sum_{u=1}^{k-1} uh\big((s-r)k+u-k\big)$$

$$\leq 2\sum_{u=1}^{k} uh\big((s-r-1)k+u\big)$$

where in the last inequality we used that h is non-increasing. We conclude that

$$\left|\mathrm{Cov}(B_r, \sum_{s>r} B_s)\right| \leq 2\sum_{j=1}^{[\frac{n}{k}]}\sum_{u=1}^{k} uh((j-1)k+u)$$

$$\leq 2\sum_{j=1}^{\infty}\sum_{u=1}^{k}((j-1)k+u)h((j-1)k+u) = 2\sum_{m=1}^{\infty} mh(m) := A_1 < \infty$$

and (2.8.21) follows.

Finally, since $S_{n,\ell}^{\omega}\mathbf{F} = \sum_{r=1}^{[n/k]+1} B_r$ we deduce from (2.8.21) that

$$\hat{V}_n(\omega) = \mathrm{Var}_\mu(S_{n,\ell}^{\omega}\mathbf{F}) \geq \sum_{j=1}^{[\frac{n}{k}]} \hat{V}_k(\vartheta^{jk}\omega) - 2A_1\left(\left[\frac{n}{k}\right]+1\right).$$

By the definition of the set $\Gamma_{[n/k],\varepsilon_0,c_2,k}$, if $\omega \notin \Gamma_{[n/k],\varepsilon_0,c_2,k}$ then the first sum in the above right-hand side is not less than $c_2\varepsilon k[n/k]$, and hence

$$\mathrm{Var}_\mu(S_{n,\ell}^{\omega}\mathbf{F}) \geq c_2\varepsilon_0 k[n/k] - 2A_1([n/k]+1)$$
$$\geq c_2\varepsilon_0 k[n/k] - 4A_1[n/k] = [n/k](c_2\varepsilon_0 k - 4A_1)$$

where we used that $[n/k] \geq 1$. If in addition $k > k_0 := \frac{8A_1}{c_2\varepsilon_0}$ then

$$\mathrm{Var}_\mu(S_{n,\ell}^{\omega}\mathbf{F}) \geq k\left[\frac{n}{k}\right] \cdot \frac{1}{2}c_2\varepsilon_0.$$

Claim 2.8.4 follows now with $c_1 = \frac{1}{4}c_2\varepsilon_0$ since $2[\frac{n}{k}] \geq \frac{n}{k}$ for any $n \geq k \geq 1$. $\qquad\square$

In order to prove Claim 2.8.5 we will need the following general result.

Lemma 2.8.6. *Let $V_i = (A_i, B_i) : (\mathcal{X}, \mathcal{B}_{\mathcal{X}}) \to X_i \times Y_i$, $i = 1, 2, ..., m$ be m pairs of measurable functions taking values in measurable spaces X_i and Y_i, respectively, so that each A_i is $\mathcal{F}_{-\infty,a_i}$-measurable and each B_i is $\mathcal{F}_{a_i+b_i,\infty}$-measurable for some $a_i \in \mathbb{Z}$ and $b_i \in \mathbb{N}$. Let $V_i^{(i)} = (A_i^{(i)}, B_i^{(i)})$, $i = 1, 2, ..., m$ be independent copies of the V_i's with respect to μ. Then for any measurable functions $G_1 : \prod_{j=1}^m X_i \to [-1,1]$ and $G_2 : \prod_{j=1}^m Y_i \to [-1,1]$,*

$$|Cov_\mu(G_1(A_1^{(1)}, ..., A_m^{(m)}), G_2(B_1^{(1)}, ..., B_m^{(m)}))| \leq 4\sum_{i=1}^m \alpha(b_i) \qquad (2.8.22)$$

where Cov_μ stands for the covariance with respect to μ and $\alpha(n)$ is defined similarly to ϕ_n (see (1.3.1)) but with the mixing coefficient α defined in (1.2.14) in place of ϕ.

Proof. The lemma goes by induction on m. First, the case when $m = 1$ follows from Theorem A.5 in [27]. Next, suppose that (2.8.22) holds true for $m = j$, all measurable spaces X_i, Y_i, $i = 1, 2, ..., j$, all measurable functions G_1 and G_2 as in the statement of the lemma and all couples of measurable functions $V_i = (A_i, B_i) : (\mathcal{X}, \mathcal{B}_{\mathcal{X}}) \to X_i \times Y_i$ which satisfy the conditions in the statement of the lemma with some $a_i \in \mathbb{Z}$, $b_i \in \mathbb{N}$, $i = 1, 2, ..., j$. Let $\mathcal{X}_i, \mathcal{Y}_i, i = 1, 2, ..., j+1$ be measurable spaces and for each i let $U_i =$

$(\mathcal{A}_i, \mathcal{B}_i) : (\mathcal{X}, \mathcal{B}_{\mathcal{X}}) \to \mathcal{X}_i \times \mathcal{Y}_i$ be a measurable function so that each $\mathcal{A}_i$ is $\mathcal{F}_{-\infty, \alpha_i}$-measurable and each $\mathcal{B}_i$ is $\mathcal{F}_{\alpha_i + \beta_i, \infty}$-measurable for some $\alpha_i \in \mathbb{Z}$ and $\beta_i \in \mathbb{N}$. Let $H_1 : \prod_{i=1}^{j+1} \mathcal{X}_i \to [-1, 1]$ and $H_2 : \prod_{i=1}^{j+1} \mathcal{Y}_i \to [-1, 1]$ be measurable functions and let $U_i^{(i)} = (\mathcal{A}_i^{(i)}, \mathcal{B}_i^{(i)}), i = 1, 2, ..., j + 1$ be independent copies of the U_i's with respect to μ. Then by the Fubini theorem we can write

$$\mathbb{E}_\mu H_1(\mathcal{A}_1^{(1)}, ..., \mathcal{A}_{j+1}^{(j+1)}) H_2(\mathcal{B}_1^{(1)}, ..., \mathcal{B}_{j+1}^{(j+1)})$$

$$= \int \mathbb{E}_\mu H_1(\mathcal{A}_1^{(1)}, ..., \mathcal{A}_j^{(j)}, x) H_2(\mathcal{B}_1^{(1)}, ..., \mathcal{B}_j^{(j)}, y) d\mathbf{m}(x, y)$$

where $\mathbf{m}$ is the distribution of U_{j+1}. Applying the induction hypothesis we deduce that

$$\left| \mathrm{Cov}_\mu \big(H_1(\mathcal{A}_1^{(1)}, ..., \mathcal{A}_j^{(j)}, x), H_2(\mathcal{B}_1^{(1)}, ..., \mathcal{B}_j^{(j)}, y) \big) \right| \leq 4 \sum_{i=1}^{j} \alpha(\beta_i)$$

for any x and y. On the other hand, by Theorem A.5 in [27] we have

$$\left| \int \mathbb{E}_\mu H_1(\mathcal{A}_1^{(1)}, ..., \mathcal{A}_j^{(j)}, x) \mathbb{E}_\mu H_2(\mathcal{B}_1^{(1)}, ..., \mathcal{B}_j^{(j)}, y) d\mathbf{m}(x, y) \right.$$

$$\left. - \int \mathbb{E}_\mu H_1(\mathcal{A}_1^{(1)}, ..., \mathcal{A}_j^{(j)}, x) \mathbb{E}_\mu H_2(\mathcal{B}_1^{(1)}, ..., \mathcal{B}_j^{(j)}, y) d\mathbf{k}(x, y) \right| \leq 4\alpha(\beta_{j+1})$$

where $\mathbf{k}$ is the product of the distributions of $\mathcal{A}_{j+1}$ and $\mathcal{B}_{j+1}$. The second integral in the above left-hand side is just the product of the expectations of $H_1(\mathcal{A}_1^{(1)}, ..., \mathcal{A}_{j+1}^{(j+1)})$ and $H_2(\mathcal{B}_1^{(1)}, ..., \mathcal{B}_{j+1}^{(j+1)})$ with respect to μ, and the proof of the induction step is complete. $\qquad \square$

Proof of Claim 2.8.5. First, by the definitions of $\hat{V}_k(\cdot)$ and the process $\{\Theta_n : n \geq 0\}$ and since P_Θ is ϑ-invariant,

$$\mathbb{E}_{P_\Theta} \hat{V}_k \circ \vartheta^{jk} = \mathbb{E}_{P_\Theta} \hat{V}_k = \mathbb{E}\big(\sum_{i=1}^{k} F_\ell(\zeta_i^{(1)}, \zeta_{2i}^{(2)}, ..., \zeta_{\ell i}^{(\ell)}) \big)^2$$

for any $j, k \in \mathbb{N}$, where $\{\zeta_n^{(i)} : n \geq 0\}, i = 1, 2, ..., \ell$ are independent copies of the Markov chain $\{\zeta_n : n \geq 0\}$. Moreover, it follows from (2.8.20) that there exists a constant $C_1 > 0$ such that $\hat{V}_k \leq C_1 k$, P_Θ-a.s. and therefore

$$\mathbb{E}_{P_\Theta} \hat{V}_k^2 \leq C_1^2 k^2. \tag{2.8.23}$$

Applying Theorem 2.5.11 with the function F_ℓ we obtain that

$$\lim_{k \to \infty} \frac{1}{k} \mathbb{E}\big(\sum_{i=1}^{k} F_\ell(\zeta_i^{(1)}, \zeta_{2i}^{(2)}, ..., \zeta_{\ell i}^{(\ell)}) \big)^2 > 0$$

since we assumed that $\sigma_\ell^2 > 0$. It follows that there exists a constant $b > 0$ such that

$$\mathbb{E}_{P_\Theta} \hat{V}_k \geq bk \qquad (2.8.24)$$

for any sufficiently large k. Before we proceed with the proof we need the following, so-called, Paley-Zygmund inequality. Let X be a nonnegative random variable and assume that $A_1^2 \leq (\mathbb{E}X)^2 \leq \mathbb{E}X^2 \leq A_2^2$ for some $A_1, A_2 > 0$. Then for any $\rho \in (0,1)$,

$$P(X \geq \rho A_1) \geq A_1^2 A_2^{-2} (1 - \rho)^2. \qquad (2.8.25)$$

Indeed, by the Cauchy-Schwarz inequality,

$$A_1 \leq \mathbb{E}X = \mathbb{E}X\mathbb{I}_{\{X \geq \rho A_1\}} + \mathbb{E}X\mathbb{I}_{\{X < \rho A_1\}}$$
$$\leq \sqrt{\mathbb{E}X^2}\sqrt{P(X \geq \rho A_1)} + \rho A_1 \leq A_2 \sqrt{P(X \geq \rho A_1)} + \rho A_1$$

yielding (2.8.25).

Let us fix some sufficiently large k so that (2.8.23) holds true. Applying (2.8.25) with $X = \frac{1}{k}\hat{V}_k$ and $\rho = \frac{1}{2}$ and taking into account (2.8.23) and (2.8.24) we deduce that

$$P_\Theta(\hat{V}_k \geq \frac{1}{2}bk) \geq \varepsilon_1 := \frac{b^2}{4C_1^2} > 0. \qquad (2.8.26)$$

Next, for any $n \geq 0$ set

$$\hat{\Theta}_n = (\Theta_n, \Theta_{n+1}, ..., \Theta_{n+k-1}), \quad \mathcal{Z}_n = (Z_n^{(1)}, Z_{2n}^{(2)}, ..., Z_{(\ell-1)n}^{(\ell-1)})$$

and $\hat{\mathcal{Z}}_n = (\mathcal{Z}_{n+k-1}, ..., \mathcal{Z}_n)$, where $\{Z_m^{(i)} : m \geq 0\}, i = 1, 2, ..., \ell - 1$ are independent copies of the process $\{Z_m : m \geq 0\}$. We note that the process $\{\mathcal{Z}_n : n \geq 0\}$ is the stationary process generated by the map $\hat{T} := T \times T^2 \times \cdots \times T^{\ell-1}$ and the $\hat{T}$-invariant measure $\mu^{\ell-1}$. We assume that our probability space is large enough so that both the process $\{\mathcal{Z}_n, n \geq 0\}$ and other processes below can be constructed on it and we denote everywhere by P the corresponding probability on it. Consider the function $V_k : (\mathcal{X}^{\ell-1})^k \to \mathbb{R}$ given by

$$V_k(y) = \text{Var}\left(\sum_{s=1}^k F(y_s^{(1)}, ..., y_{s(\ell-1)}^{(\ell-1)}, \zeta_{s\ell})\right),$$

$$y = \{(y_s^{(1)}, y_{2s}^{(2)}, ..., y_{(\ell-1)s}^{(\ell-1)}) : 1 \leq s \leq k\} \in (\mathcal{X}^{\ell-1})^k.$$

We observe that for any $n \in \mathbb{N}$ the random vectors $\{\hat{V}_k \circ \vartheta^{jk}\}_{j=0}^n$, $\{V_k(\hat{\Theta}_{jk})\}_{j=0}^n$ and $\{V_k(\hat{\mathcal{Z}}_{(n-j)k})\}_{j=0}^n$ have the same distribution. Indeed, the first two are identically distributed since $(\Omega_\Theta, \mathcal{B}_\Theta, P_\Theta, \vartheta)$ is the invertible MPS corresponding to the processes $\{\Theta_n : n \geq 0\}$, while the second and third random vectors are identically distributed in view of (2.5.11). Therefore, for any $c_2, \varepsilon_0 > 0$,

$$P_\Theta(\Gamma_{n,\varepsilon_0,c_2,k}) = P\{\frac{1}{n}\sum_{j=1}^{n}\mathbb{I}_{\{V_k(\hat{Z}_{(n-j)k})\geq c_2 k\}} < \varepsilon_0\}. \tag{2.8.27}$$

Furthermore, by (2.8.26) and since P_Θ is ϑ-invariant, for any $0 \leq j \leq n$,

$$P\{V_k(\hat{Z}_{(n-j)k}) \geq \frac{1}{2}bk\} = P_\Theta\{\hat{V}_k \circ \vartheta^{jk} \geq \frac{1}{2}bk\} \tag{2.8.28}$$

$$= P_\Theta\{\hat{V}_k \geq \frac{1}{2}bk\} \geq \varepsilon_1.$$

Next, fix some $c_2 < \frac{1}{3}b$ and $\varepsilon_0 < \frac{1}{4}\varepsilon_1$, where $b > 0$ and ε_1 appear in (2.8.26). By considering the product space of the distributions of the processes $\{(Z_{in}, Z_{in,s}) : n, s \geq 0\}$, $i = 1, 2, ..., \ell - 1$ we can always assume that there exist independent copies $\{Z_{in,s}^{(i)} : n, s \geq 0\}$ of $\{Z_{in,s} : n, s \geq 0\}$, $i = 1, 2, ..., \ell - 1$ so that

$$\|\rho(Z_{in}^{(i)}, Z_{in,s}^{(i)})\|_{L^\infty} = \|\rho(Z_{in}, Z_{in,s})\|_{L^\infty}$$

for any $n, s \geq 0$ and $i = 1, 2, ..., \ell - 1$. Notice that

$$V_k \in \mathcal{H}^{\alpha,\xi}(\mathcal{X}, \rho_{\infty,(\ell-1)k})$$

since $F \in \mathcal{H}^{\alpha,\xi}(\mathcal{X}^{\ell-1}, \rho_{\infty,\ell-1})$. Therefore, there exists a constant $c_0 = c_0(k)$, which depends only on k and C_F in (2.5.18), such that for any $s \in \mathbb{N}$, $n \in \mathbb{N}$ and $0 \leq j \leq n$,

$$\|V_k(\hat{Z}_{(n-j)k}) - V_k(\hat{Z}_{(n-j)k,s})\|_{L^\infty} \leq c_0 \beta_{\infty,\alpha,\xi}(s) \tag{2.8.29}$$

where $\hat{Z}_{m,s}, m \in \mathbb{N}$ is defined similarly to $\hat{Z}_m$ but with the $Z_{j,s}^{(i)}$'s in place of the $Z_j^{(i)}$'s. Fix some $s \in \mathbb{N}$ such that $c_0 \beta_{\infty,\alpha,\xi}(s) < \frac{1}{4}c_2 k$ and set

$$W_{s,j,n} = \mathbb{I}_{\{V_k(\hat{Z}_{(n-j)k,s}) > \frac{5}{4}c_2 k\}}, \quad 0 \leq j \leq n.$$

Then

$$P\{\frac{1}{n}\sum_{j=1}^{n}\mathbb{I}_{\{V_k(\hat{Z}_{(n-j)k})\geq c_2 k\}} < \varepsilon_0\} \leq P\{\frac{1}{n}\sum_{j=1}^{n}W_{s,j,n} < \varepsilon_0\}. \tag{2.8.30}$$

In order to estimate the above right-hand side we will estimate first the expectation and variance of the sums $\sum_{j=1}^{n} W_{s,j,n}$. It follows from Lemma 2.8.6 that

$$\text{Cov}(W_{s,j_1,n}, W_{s,j_2,n}) \leq 4(\ell - 1)\alpha\big((j_2 - j_1 - 1)k - 2s + 1\big)$$

for any $1 \leq j_1, j_2 \leq n$ such that $(j_2 - j_1 - 1)k > 2s - 1$. Since $\alpha(m)$ is decreasing in m, we infer that

$$\operatorname{Var}(\sum_{j=1}^{n} W_{s,j,n}) \leq \sum_{1 \leq j_1 \leq j_2 \leq n: j_2 - j_1 \leq 2s} \operatorname{Cov}(W_{s,j_1,n}, W_{s,j_2,n}) \quad (2.8.31)$$

$$+ \sum_{1 \leq j_1 \leq j_2 \leq n: j_2 - j_1 > 2s} 4(\ell - 1)\alpha(j_2 - j_1 - 2s) \leq 2ns + 4(\ell - 1)n \sum_{j=1}^{\infty} \alpha(j)$$

and observe that the last sum converges by Assumption 2.5.10 since $\alpha(n) \leq \phi_n$ for any $n \geq 0$. Next, by (2.8.29), taking into account that $c_0 \beta_{\infty,\alpha,\xi}(s) < \frac{1}{4}c_2 k$ and $c_2 < \frac{b}{3}$, for any $0 \leq j \leq n$ we have

$$\mathbb{E}W_{s,j,n} = P\{V_k(\hat{Z}_{(n-j)k,s}) > \frac{5}{4}c_2 k\} \quad (2.8.32)$$

$$\geq P\{V_k(\hat{Z}_{(n-j)k}) \geq \frac{1}{2}bk\} \geq \varepsilon_1$$

where in the last inequality (2.8.28) is used. We conclude from (2.8.31) and (2.8.32) and the Markov inequality that there exists a constant C_2 which is independent of k such that for any $n \in \mathbb{N}$,

$$P\{\frac{1}{n}\sum_{j=1}^{n} W_{s,j,n} < \varepsilon_0\} \leq P\{\frac{1}{n}|\sum_{j=1}^{n} W_{s,j,n} - \mathbb{E}\sum_{j=1}^{n} W_{s,j,n}| \geq \frac{1}{4}\varepsilon_0\}$$

$$\leq 16 \cdot \frac{2ns + 4(\ell - 1)n \sum_{j=1}^{\infty}\alpha(j)}{\varepsilon_0^2 n^2} \leq \frac{C_2}{n}$$

where in the first inequality we used that $\varepsilon_0 < \frac{1}{4}\varepsilon_1$ and Claim 2.8.5 follows, taking into account (2.8.27). □

2.8.5 Remarks

We conclude this section with several remarks whose purpose is to discuss weaker assumptions under which the above proof of Claim 2.8.5 proceeds similarly, and the possibility of proving this claim without using Assumption 2.5.10, relying on some large deviation argument instead (see Remark 2.8.10). Note that Assumption 2.5.10 is still needed in order to obtain (2.7.8), which is an estimate which requires stronger mixing conditions than just the decay of correlations obtained in Proposition 2.5.8 which was used in the proof of Claim 2.8.4.

Remark 2.8.7. The proof of Claim 2.8.5 proceeds similarly under some restrictions on the L^q-approximation rate

$$\beta_q(s) := \sup_{n \geq 0} \|\rho(Z_n, Z_{n,s})\|_{L^q}$$

for some $q \geq 1$. For instance, in place of (2.8.30) we have

$$P\{\frac{1}{n}\sum_{j=1}^{n}\mathbb{I}_{\{V_k(\hat{Z}_{(n-j)k})\geq c_2 k\}} < \varepsilon_0\} \leq P\{\frac{1}{n}\sum_{j=1}^{n}W_{s,j,n} < \varepsilon_0\}$$

$$+ \sum_{k=1}^{n}P\{|V_k(\hat{Z}_{(n-j)k}) - V_k(\hat{Z}_{(n-j)k,s})| \geq \frac{1}{4}c_2 k\}$$

and a similar inequality holds true in place of (2.8.32). Using the Markov inequality in order to estimate the probabilities appearing in the sum in the above right-hand side, the proof of Claim 2.8.5 will proceed similarly if we assume that there exists a sequence of natural numbers $(s_n)_{n=1}^{\infty}$ so that $n^{\frac{3}{2}}\beta_q^q(s_n)$ and $n^{-\frac{1}{2}}s_n$ converge to 0 as $n \to \infty$. Moreover, an inequality similar to (2.7.8) will hold true under similar conditions, as well and therefore the nonconventional local limit theorem itself holds true under conditions concerning the rate of decay of L^q-approximation and mixing coefficients.

Remark 2.8.8. Another way to deduce (2.8.28) goes as follows. By Theorem 2.3 in [38] the limit $\sigma^2 = \lim_{n\to\infty}\frac{1}{n}\text{Var}_\mu(S_{n,\ell}^\omega \mathbf{F})$ exists P_Θ-a.s., it does not depend on ω and $\sigma^2 > 0$ if and only if $\mathbf{F}$ does not admit a (random) coboundary representation. In our situation this coboundary representation is equivalent to F_ℓ having an appropriate coboundary representation, which is excluded. Thus, for any sufficiently large n we have

$$P_\Theta\{\omega : \text{Var}_\mu(S_{n,\ell}^\omega \mathbf{F}) > \frac{n}{2}\sigma^2\} > \frac{1}{2}.$$

Remark 2.8.9. The proof that Assumption 2.2.2 holds true proceeds similarly without approximating the Z_i's by the $Z_{i,s}$'s, assuming some rate of decay of correlation for bounded observables. For instance, an inequality of the form

$$|\text{Cov}_{\mu^{\ell-1}}(g, h \circ \hat{T}^k)| \leq c\|g\|_\infty\|h\|_\infty k^{-2-\delta}$$

for some $c, \delta > 0$, all $k \geq 1$ and bounded Borel functions g and h is sufficient, where we recall that

$$\hat{T} = T \times T^2 \times \cdots \times T^{\ell-1}.$$

Nevertheless, this condition implies that the process $\{\hat{T}^m : m \geq 0\}$ taken with the initial distribution $\mu^{\ell-1}$ is sufficiently well α-mixing, which, in some sense, is stronger than our assumptions and it does not hold in important examples such as one sided topologically mixing subshifts of finite type and expanding C^2 endomorphisms of a Riemannian manifold.

Remark 2.8.10. A different strategy to prove (2.8.18) is to rely on some large deviation results instead of mixing and approximation conditions. In many dynamical systems such results are available, while the mixing and approximation conditions usually hold true when Markov partitions exist. Still, there are several difficulties in applying such large deviation results, as described below.

Consider the functions $\tilde{V}_k : \mathcal{X}^{\ell-1} \to \mathbb{R}$, $k \in \mathbb{N}$ given by

$$\tilde{V}_k(\bar{x}) = \mathrm{Var}\Big(\sum_{j=1}^{k} F(\hat{T}^{k-j}\bar{x}, \zeta_{\ell j}) \Big) = \mathbb{E}\Big(\sum_{j=1}^{k} F_\ell(\hat{T}^{k-j}\bar{x}, \zeta_{\ell j}) \Big)^2.$$

Then each $\tilde{V}_k$ is a (locally) Hölder continuous and by (2.5.11) when $\bar{x}$ is drawn according to $\mu^{\ell-1}$ then the stationary processes

$$\{\tilde{V}_k \circ \hat{T}^{kn} : n \geq 0\} \text{ and } \{\hat{V}_k \circ \vartheta^{kn} : n \geq 0\}$$

have the same distribution. Relying on the exponential decay of correlations of the system $(\mathcal{X}, \mathcal{B}_\mathcal{X}, \mu, T)$ obtained in Theorem 2.5.7 we derive that the limits

$$\sigma_k^2 = \lim_{n \to \infty} \frac{1}{n} \mathrm{Var}_{\mu^{\ell-1}} \sum_{j=1}^{n} \tilde{V}_k \circ \hat{T}^{kj}, \ k \in \mathbb{N}$$

exist and that $\sigma_k^2 = 0$ if and only if the function $\tilde{V}_k$ admits an appropriate coboundary representations (see [33]). When $\sigma_k^2 > 0$ then we can apply Theorem E (i) in [28] in order to estimate the $\mu^{\ell-1}$-probabilities of sets of the form

$$\Gamma_n = \Big\{ \bar{x} : \sum_{j=1}^{n} \tilde{V}_k(\hat{T}^{jk}\bar{x}) - \int \Big(\sum_{j=1}^{n} \tilde{V}_k(\hat{T}^{jk}\bar{x}) \Big) d\mu^{\ell-1}(\bar{x}) \geq n\varepsilon \Big\}$$

which is sufficient for showing that Assumption 2.2.2 holds true. The disadvantage here is that in order to use such estimates we have to assume that $\sigma_k^2 > 0$ for some unknown sufficiently large k which is an assumption that is hard to verify.

2.9 Decay of characteristic functions for large t's

In this section we will show that Assumption 2.2.1 holds true in the circumstances of Theorem 2.6.3 in the non-lattice case and Theorem 2.6.5 in the lattice case. We will use directly the stationary process $\Theta = \{\Theta_n : n \geq 0\}$ and not its associated MPS $M(\Theta) = (\Omega_\Theta, \mathcal{B}_\Theta, P_\Theta, \vartheta)$ and note that we will not use here Assumption 2.5.10 but only the decay of correlations established in Proposition 2.5.8.

2.9.1   *Basic estimates and strategy of the proof*

For any $n \in \mathbb{N}$, $t \in \mathbb{R}$ and $\bar{v} = (v^{(0)}, v^{(1)}, ..., v^{(n-1)}) \in (\mathcal{X}^{\ell-1})^n$ consider the transfer operator $\mathcal{L}_{it}^{\bar{v},n}$ which acts on function g on $\mathcal{X}$ by the formula

$$\mathcal{L}_{it}^{\bar{v},n} g(x) = \sum_{y \in T^{-n\ell}\{x\}} e^{S_{n\ell}f(y) + it\sum_{k=0}^{n-1} F(v^{(k)}, T^{k\ell}y)} g(y) \qquad (2.9.1)$$

and notice that

$$\mathcal{L}_{it}^{\bar{v},n} = \mathcal{L}_{it}^{v^{(n-1)},1} \circ \mathcal{L}_{it}^{v^{(n-2)},1} \circ \cdots \circ \mathcal{L}_{it}^{v^{(0)},1}. \qquad (2.9.2)$$

Next, by (2.7.9) and the definition of the MPS $M(\Theta)$, for any $t \in \mathbb{R}$ and $N \geq 1$,

$$|\varphi_N(t)| = |\mathbb{E}e^{itS_N}| \leq c_N + \mathbb{E}\|\mathcal{L}_{it}^{(\Theta_{M+1},...,\Theta_N), N-M}\mathbf{1}\|_\infty \qquad (2.9.3)$$

for some sequence $(c_n)_{n=1}^\infty$ which satisfies $\lim_{n\to\infty} \sqrt{n}c_n = 0$. Thus, in order to show that Assumption 2.2.1 holds true it is sufficient to estimate the expectations

$$\mathbb{E}\|\mathcal{L}_{it}^{(\Theta_{M+1},...,\Theta_N), N-M}\mathbf{1}\|_\infty$$

appropriately. Consider again the map

$$\hat{T} = T \times T^2 \times \cdots \times T^{\ell-1} : \mathcal{X}^{\ell-1} \to \mathcal{X}^{\ell-1}.$$

Then by stationarity of Θ and (2.5.11),

$$\mathbb{E}\|\mathcal{L}_{it}^{(\Theta_{M+1},...,\Theta_N), N-M}\mathbf{1}\|_\infty = \mathbb{E}\|\mathcal{L}_{it}^{(\Theta_1,...,\Theta_{N-M}), N-M}\mathbf{1}\|_\infty \qquad (2.9.4)$$

$$= \int \|\mathcal{L}_{it}^{\bar{v}_{\hat{T}}, N-M-1}(\bar{x}), N-M}\mathbf{1}\|_\infty d\mu^{\ell-1}(\bar{x})$$

where for any $\bar{x} = (x_1, .., x_{\ell-1}) \in \mathcal{X}^{\ell-1}$ and $n \geq 0$,

$$\bar{v}_{\hat{T},n}(\bar{x}) = (\hat{T}^n\bar{x}, \hat{T}^{n-1}\bar{x}, ..., \bar{x})$$

is the inversion of the truncated orbit $\mathcal{O}_{0,n}(\bar{x}) = \{\hat{T}^k\bar{x} : 0 \leq k \leq n\}$ of $\bar{x}$. Let $x_0 \in \mathcal{X}$ be a periodic point of T, i.e. $T^{m_0}x_0 = x_0$ for some $m_0 \in \mathbb{N}$, and set

$$\bar{x}_0 = (x_0, x_0, ..., x_0) \in \mathcal{X}^{\ell-1}$$

which satisfies $\hat{T}^{m_0}\bar{x}_0 = \bar{x}_0$.

Next, for each $t \in \mathbb{R}$, consider the transfer operator R_{it} generated by the map $\tau_0 = T^{\ell m_0}$ and the potential $S_{\ell m_0}f + itF_{x_0,m_0}$, which acts on functions $g : \mathcal{X} \to \mathbb{C}$ by the formula

$$R_{it}g(x) = \mathcal{L}_f^{\ell m_0}(e^{itF_{x_0,m_0}}g)(x) \qquad (2.9.5)$$

$$= \sum_{y \in \tau_0^{-1}\{x\}} e^{S_{\ell m_0}f(y) + it\sum_{k=0}^{m_0-1} F(\hat{T}^k\bar{x}_0, T^{\ell k}y)} g(y)$$

where $S_m f = \sum_{k=0}^{m-1} f \circ T^m$ for any $m \in \mathbb{N}$. Set $\mathcal{Y} = \mathcal{X}^{\ell-1}$, let $n_0 \in \mathbb{N}$ and consider the periodic element $u^{\bar{x}_0, n_0}$ of $(\mathcal{Y}^{m_0})^{n_0}$ given by

$$u^{\bar{x}_0, n_0} = \{\bar{x}_0, \hat{T}\bar{x}_0, ..., \hat{T}^{m_0-1}\bar{x}_0, \bar{x}_0, \hat{T}\bar{x}_0, ..., \hat{T}^{m_0-1}\bar{x}_0, ...\} \qquad (2.9.6)$$

which consists of n_0 consecutive copies of $\mathcal{O}_{0,m_0-1}(\bar{x}_0)$. Note that, in fact,

$$u^{\bar{x}_0, n_0} = \mathcal{O}_{0,n_0 m_0 - 1}(\bar{x}_0)$$

and observe that

$$R_{it}^{n_0} = R_{it} \circ R_{it} \circ \cdots \circ R_{it} = \mathcal{L}_{it}^{u^{\bar{x}_0, n_0}, n_0 m_0}. \qquad (2.9.7)$$

The idea behind the proof is to show that for any $N \in \mathbb{N}$, on a set whose $\mu^{\ell-1}$ probability is sufficiently close to 1, the sequence

$$\bar{v}_{\hat{T}, N-M-1}(\bar{x})$$

contains proportional to N number of $\mathcal{Y}$-valued subsequences of length $n_0 m_0$ which are close to $u^{\bar{x}_0, n_0}$ (see Corollary 2.9.3 for the precise formulation). Relying on Section 2.9.4, this will insure that $\mathcal{L}_{it}^{v_{\hat{T}, N-M-1}(\bar{x}), N-M}$ can be represented in the form

$$\mathcal{L}_{it}^{v_{\hat{T}, N-M-1}(\bar{x}), N-M} = \mathcal{A}_1 \circ \mathcal{B}_1 \circ \mathcal{A}_2 \circ \mathcal{B}_2 \circ \cdots \circ \mathcal{B}_k \circ \mathcal{A}_{k+1}$$

with operators $\mathcal{A}_i$ and $\mathcal{B}_i$ depending on t, $k \geq c_1(N - M)$ for some $c_1 > 0$, $\|\mathcal{A}_i\| \leq D(1 + |t|)$ for some constant D and all i's and each one of the $\mathcal{B}_i$'s is sufficiently close to $R_{it}^{n_0}$ in the norm $\|\cdot\|_{\alpha,\xi}$ (see (2.9.27) for the precise formulation). This representation reduces the problem of approximating the norm of the transfer operator inside the integral on the right-hand side of (2.9.4) to approximation of norms of the powers $R_{it}^{n_0}$ of a single transfer operator which is a well studied problem (see Section 2.9.5). Since we can take x's from a set whose probability is sufficiently close to 1 this yields (see Section 2.9.6) appropriate upper bounds on the integral in (2.9.4).

2.9.2 *Probabilities of large number of visits to open sets*

Let $(\mathcal{Y}_i, d_i), i = 1, ..., k$ be metric spaces equipped with the Borel σ-algebras $\mathcal{B}_{\mathcal{Y}_i}$. For each i let $S_i : \mathcal{Y}_i \to \mathcal{Y}_i$ be a Borel measurable map and ν_i be a S_i-invariant Borel probability measure on $\mathcal{Y}_i$ which assigns positive probability to open sets. Suppose that each measure preserving system $(\mathcal{Y}_i, \mathcal{B}_{\mathcal{Y}_i}, S_i, \nu_i), i = 1, 2, ..., k$ satisfies

$$\sum_{m=0}^{\infty} |\mathrm{Cov}_{\nu_i}(g_i, g_i \circ S_i^m)| < \infty \qquad (2.9.8)$$

for any bounded Lipschitz function $g_i : \mathcal{Y}_i \to \mathbb{R}$. Consider the product measure preserving system $(\mathcal{Y}, \mathcal{B}_\mathcal{Y}, \nu, S)$ where $\mathcal{Y} = \mathcal{Y}_1 \times \mathcal{Y}_2 \times \cdots \times \mathcal{Y}_k$, $\nu = \nu_1 \times \nu_2 \times \cdots \times \nu_k$ and $S = S_1 \times S_2 \times \cdots \times S_k$. Let $B \subset \mathcal{Y}$ be an open set and let $n \in \mathbb{N}$. Consider the function $\mathcal{N}_{n,B} : \mathcal{Y} \to \mathbb{R}$ which counts the number of visits to B from time 0 to time $n-1$ given by

$$\mathcal{N}_{n,B}(y) = \sum_{m=0}^{n-1} \mathbb{I}_B(S^m y)$$

where $\mathbb{I}_B$ is the indicator function of the set B. For any $c > 0$ set

$$\mathcal{B}_{n,c}(B) = \{y \in \mathcal{Y} : \mathcal{N}_{n,B}(y) \geq cn\}.$$

Lemma 2.9.1. *There exist constants $b > 0$ and $c \in (0,1)$, which may depend on B, such that for any $n \in \mathbb{N}$,*

$$1 - \nu(\mathcal{B}_{n,c}(B)) \leq \frac{b}{n}. \tag{2.9.9}$$

Proof. The set B is open and thus there exist open balls $\mathcal{B}_i \subset \mathcal{Y}_i$, $i = 1, ..., k$ such that

$$\mathcal{B}_1 \times \mathcal{B}_2 \times \cdots \times \mathcal{B}_k \subset B.$$

Write $\mathcal{B}_i = B(y_i, s_i)$ for some $y_i \in \mathcal{Y}_i$ and $s_i > 0$ and set $r_i = \frac{1}{2}s_i$. For each i, let $g_i : \mathcal{Y}_i \to [0,1]$ be a bounded Lipschitz function such that

$$\mathbb{I}_{B(y_i, s_i)} \geq g_i \geq \mathbb{I}_{B(y_i, r_i)}. \tag{2.9.10}$$

For instance, the function g_i given by

$$g_i(x) = \max\left(0, 1 - \max(0, \frac{d_i(x, y_i) - r_i}{r_i})\right)$$

satisfies the above conditions and has Lipschitz constant r_i^{-1}. Consider the function $g : \mathcal{Y} \to [0,1]$ given by

$$g(y) = \prod_{i=1}^{k} g_i(y_i), \quad y = (y_1, ..., y_k)$$

and set $G_n = \sum_{m=0}^{n-1} g \circ S^m$, $n \in \mathbb{N}$. Then by (2.9.10), for any $y \in \mathcal{Y}$ and $n \in \mathbb{N}$,

$$\mathcal{N}_{n,B}(y) \geq G_n(y)$$

and therefore for any $c > 0$,

$$1 - \nu(\mathcal{B}_{n,c}(B)) \leq 1 - \nu(\mathcal{B}_{n,c}(g)) \tag{2.9.11}$$

where

$$\mathcal{B}_{n,c}(g) = \{y \in \mathcal{Y} : G_n(y) \geq cn\}.$$

By (2.9.10), for each $i = 1, ..., k$ we have

$$\mathbb{E}_{\nu_i} g_i = \int g_i(y_i) d\nu_i(y_i) \geq \nu_i(B(y_i, r_i)) := a_i > 0$$

where a_i is positive since we have assumed that each ν_i assigns positive mass to open sets. Since S_i preserves ν_i, $i = 1, ..., k$ we conclude that

$$\frac{1}{n}\mathbb{E}_\nu G_n = \mathbb{E}_\nu g = \prod_{i=1}^{k} \mathbb{E}_{\nu_i} g_i \geq \prod_{i=1}^{k} a_i := a > 0. \tag{2.9.12}$$

Next, for any $m \geq 0$ we have

$$\mathrm{Cov}_\nu(g, g \circ S^m) = \prod_{i=1}^{k} \int g_i \cdot (g_i \circ S_i^m) d\nu_i - \prod_{i=1}^{k} \left(\int g_i d\nu_i \right)^2.$$

Using the inequality

$$\left| \prod_{i=1}^{k} \alpha_i - \prod_{i=1}^{k} \beta_i \right| \leq \sum_{i=1}^{k} |\alpha_i - \beta_i|$$

which holds true for any $\alpha_i, \beta_i, i = 1, ..., k$ such that $|\alpha_i|, |\beta_i| \leq 1$, we deduce that

$$|\mathrm{Cov}_\nu(g, g \circ S^m)| \leq \sum_{i=1}^{k} |\mathrm{Cov}_{\nu_i}(g_i, g_i \circ S_i^m)|$$

where we have used that each g_i takes values on the interval $[0, 1]$. We conclude from this and (2.9.8) that there exists $M > 0$ which depends only on g such that for any $j \geq 0$,

$$\sum_{m \geq j} |\mathrm{Cov}_\nu(g \circ S^j, g \circ S^m)| = \sum_{m \geq 0} |\mathrm{Cov}_\nu(g, g \circ S^m)|$$

$$\leq \sum_{i=1}^{k} \sum_{m \geq 0} |\mathrm{Cov}_\nu(g_i, g_i \circ S_i^m)| \leq M < \infty$$

and therefore

$$\mathrm{Var}_\nu(G_n) \leq 2nM. \tag{2.9.13}$$

Set $c = \frac{a}{2}$, where a comes from (2.9.12). Then by (2.9.12), (2.9.13) and the Chebyshev inequality,

$$\nu(G_n < nc) \leq \nu(|G_n - \mathbb{E}_\nu G_n| > \frac{1}{2}an) \leq \frac{8Mn}{a^2 n^2} = \frac{8Ma^{-2}}{n}. \tag{2.9.14}$$

The lemma follows with the above c and with $b = 8Ma^{-2}$, taking into account (2.9.11). $\qquad\square$

Remark 2.9.2. In many situations large deviation type results exist for the map S and, in fact, the arguments in the proof above show that we could have used some large deviation results for sums $G_n = \sum_{j=0}^{n-1} g \circ S^j$ generated by a Lipschitz continuous function g and the map S. Still, as in Remark 2.8.10, it is unclear whether the corresponding large deviation upper bound is strictly negative. For instance, in the situation of nonconventional sums consider the map $S = \hat{T} = T \times T^2 \times \cdots \times T^{\ell-1}$. Applying Theorem E (i) in [28] with the function g we will get exponentially fast converges to 0 of probabilities of sets of the form $\{G_n \geq \varepsilon_0 n\}$, $\varepsilon_0 > 0$ assuming that $\lim_{n \to \infty} n^{-1} \mathrm{Var} G_n > 0$, or, equivalently, that g does not admit a coboundary representation with respect to S, which in general is hard to verify.

2.9.3 *Segments of periodic orbits*

Let $(\mathcal{Y}_i, S_i, \nu_i), i = 1, ..., k$ be as in Section 2.9.2. It will be convenient to have the following notation of concatenation. Let $m_1, ..., m_l \in \mathbb{N}$ and $w^{(i)} = \{w_j^{(i)} : 1 \leq j \leq m_i\} \in \mathcal{Y}^{m_i}, i = 1, ..., l$, where we recall that $\mathcal{Y} = \mathcal{Y}_1 \times \mathcal{Y}_2 \times \cdots \times \mathcal{Y}_k$. Set $M_k = \sum_{i=1}^{k} m_i$, $M_0 = 0$ and let the concatenation of the $w^{(i)}$'s

$$w = w^{(1)} w^{(2)} ... w^{(l)} \in \mathcal{Y}^{M_l}$$

be given by

$$w_i = w_{i-M_{j-1}}^{(j)} \quad \text{if } M_{j-1} + 1 \leq i \leq M_j.$$

For notational convenience, we will allow concatenation with the empty symbol, that is when $w_0 = \emptyset$ we write

$$w_0 w = w w_0 = w.$$

Next, suppose that each S_i is continuous. Consider the metric $d = d_\infty$ on $\mathcal{Y}$ given by

$$d(y, z) = \max_{1 \leq i \leq k} d_i(y_i, z_i), \quad y = (y_1, ..., y_k), \quad z = (z_1, ..., z_k)$$

and for any $m \geq 1$ consider the metric $d_{m,\infty}$ on $\mathcal{Y}^m = \mathcal{Y} \times \mathcal{Y} \times \cdots \times \mathcal{Y}$ given by

$$d_{m,\infty}(\bar{y}, \bar{z}) = \max_{1 \leq i \leq m} d(\bar{y}_i, \bar{z}_i)$$

where $\bar{y} = (\bar{y}_1, ..., \bar{y}_m)$ and $\bar{z} = (\bar{z}_1, ..., \bar{z}_m)$. Fix some $y = (y_1, ..., y_k) \in \mathcal{Y}$, $m \geq 1$ and $\delta > 0$ and consider the Bowen ball

$$B_m(y, \delta) = \{z \in \mathcal{Y} : \max_{0 \leq i \leq m-1} d(S^i z, S^i y) < \delta\}.$$

Since each S_i is continuous the set $B_m(y, \delta)$ is open. For any $0 \le a \le b \le \infty$ and $z \in \mathcal{Y}$ consider the orbit segment

$$\mathcal{O}_{a,b}(z) = \{S^i z : a \le i \le b\} \in \mathcal{Y}^{b-a+1}$$

of z from time a to time b. Lemma 2.9.1 provides estimate on the ν-probability of the set of all z's such that the truncated orbit $\mathcal{O}_{0,n-1}(z)$ contains at least nc orbit segments of the form $\mathcal{O}_{a,a+m-1}(z)$, $a \le n - m$ which satisfy

$$d_{m,\infty}(\mathcal{O}_{a,a+m-1}(z), \mathcal{O}_{0,m-1}(y)) < \delta.$$

If z satisfies this condition then there exist numbers $\{a_i : 1 \le i \le L+1\}$, $L \ge \lceil \frac{nc}{m} \rceil$ so that for any $1 \le i \le L$,

$$a_i + m \le a_{i+1} \le n \text{ and } d(S^{a_i+j}z, S^j y) < \delta \text{ for any } 0 \le j \le m-1.$$
$$(2.9.15)$$

Therefore on a set whose probability is not less than $1 - \frac{b}{n}$ we can write

$$\mathcal{O}_{0,n-1}(z) = w^{(1)}v^{(1)}w^{(2)}v^{(2)}...v^{(L)}w^{(L+1)}, \quad L \ge \lceil \frac{nc}{m} \rceil$$

where each $v^{(i)} = (v_0^{(i)}, ..., v_{m-1}^{(i)}) \in \mathcal{Y}^m$ satisfies

$$d_{m,\infty}(v^{(i)}, \mathcal{O}_{0,m-1}(y)) = \max_{0 \le j \le m-1} d(v_j^{(i)}, S^j y) < \delta$$

and some of the $w^{(i)}$'s are allowed to be empty.

Suppose now that y is a periodic point and let $m_y \in \mathbb{N}$ be so that $S^{m_y}y = y$. For any $n_0 \in \mathbb{N}$ consider the orbit segment $u^{y,n_0} = \mathcal{O}_{0,n_0 m_y-1}(y)$. Then

$$u^{y,n_0} = ww...w, \quad w = \mathcal{O}_{0,m_y-1}(y)$$

namely it is periodic and consists of n_0 consecutive copies of $\mathcal{O}_{0,m_y-1}(y)$. Considering m's of the form $m = n_0 m_y$, $k \in \mathbb{N}$ it follows that up to a set whose ν-probability does not exceed $\frac{b}{n}$ we can write

$$\mathcal{O}_{0,n-1}(z) = w^{(1)}v^{(1)}w^{(2)}v^{(2)}w^{(3)}...v^{(L)}w^{(L+1)}, \quad L \ge \lceil \frac{nc}{m} \rceil$$

where each $v^{(i)} = (v_0^{(i)}, ..., v_{n_0 m_y-1}^{(i)}) \in \mathcal{Y}^{n_0 m_y}$ satisfies

$$d_{n_0 m_y,\infty}(v^{(i)}, u^{y,n_0}) < \delta$$

and some of the $w^{(i)}$'s are allowed to be empty.

We return now to our situation of nonconventional sums. In this case we take $S_i = T^i$ and $\nu_i = \mu$ for any $1 \le i \le k = \ell - 1$. Then (2.9.8) holds true by Proposition 2.5.8, taking into account that the norms $\| \cdot \|_{\alpha,\xi}$ and

$\| \cdot \|_{1,\xi}$ are equivalent since the space $\mathcal{X}$ is compact. For any $s \in \mathbb{N}$ and $\bar{u}, \bar{v} \in (\mathcal{X}^{\ell-1})^s$ set

$$\rho_{s,\infty}(\bar{u}, \bar{v}) = \max_{1 \leq i \leq s, \, 1 \leq j \leq \ell-1} \rho(u_{i,j}, v_{i,j})$$

where $\bar{u} = (\bar{u}_1, ..., \bar{u}_s)$, $\bar{v} = (\bar{v}_1, ..., \bar{v}_s)$, $\bar{u}_i = \{u_{i,j} : 1 \leq j \leq \ell - 1\} \in \mathcal{X}^{\ell-1}$ and $\bar{v}_i = \{v_{i,j} : 1 \leq j \leq \ell - 1\} \in \mathcal{X}^{\ell-1}$ for each i. Let x_0 be a periodic point of T and let m_0 be so that $T^{m_0} x_0 = x_0$. Consider the point $\bar{x}_0 = (x_0, ..., x_0) \in \mathcal{X}^{\ell-1}$, which satisfies $\hat{T}^{m_0} \bar{x}_0 = \bar{x}_0$, and for any natural n_0 the periodic sequence

$$u^{\bar{x}_0, n_0} = ww...w, \quad w = \mathcal{O}_{0,m_0-1}(\bar{x}_0)$$

which consists of n_0 consecutive copies of

$$\mathcal{O}_{0,m_0-1}(\bar{x}_0) = \{\bar{x}_0, \hat{T}\bar{x}_0, ..., \hat{T}^{m_0-1}\bar{x}_0\}.$$

We summarize the above discussion in the following corollary.

Corollary 2.9.3. *For any $n_0 \in \mathbb{N}$ and $\delta_0 > 0$ there exist constants $b_1 = b_1(n_0, \delta_0) > 0$ and $c_1 = c_1(n_0, \delta_0) > 0$ and measurable sets $\mathcal{T}_n = \mathcal{T}_{n,n_0,\delta_0} \subset \mathcal{X}^{\ell-1}, n \in \mathbb{N}$ such that for any $n \in \mathbb{N}$,*

$$\mu^{\ell-1}(\mathcal{T}_n) \leq \frac{b_1}{n} \qquad (2.9.16)$$

and for any $\bar{y} = (y_1, ..., y_{\ell-1}) \in \mathcal{X}^{\ell-1} \setminus \mathcal{T}_n$ we can write

$$\{\bar{y}, \hat{T}\bar{y}, ..., \hat{T}^{n-1}\bar{y}\} = w^{(1)}v^{(1)}w^{(2)}v^{(2)}w^{(3)}...v^{(L)}w^{(L+1)}, \quad L \geq c_1 n \quad (2.9.17)$$

where some of the $w^{(i)}$'s can be empty and each $v^{(i)}$ satisfies

$$\rho_{n_0 m_0, \infty}(v^{(i)}, u^{\bar{x}_0, n_0}) < \delta_0.$$

In other words, there exist indexes $a_i, i = 1, ..., L + 1$, $L \geq c_1 n$ such that for any $1 \leq i \leq L$,

$$a_i + km_0 \leq a_{i+1} \leq n$$

and

$$\rho(T^{s(a_i + jm_0 + l)}y_s, T^{sl}x_0) < \delta_0 \qquad (2.9.18)$$

for any $0 \leq j \leq n_0 - 1$, $0 \leq l \leq m_0 - 1$ and $1 \leq s \leq \ell - 1$.

Remark 2.9.4. Consider the situation when $(\mathcal{X}, T)$ is a one sided topologically mixing subshift of finite type (see [10] and Section 2.11). A Bowen ball B is just a cylinder set and its indicator function $\mathbb{I}_B$ is Lipshitz continuous, and it is easy to check directly that the corresponding asymptotic variance is positive. A power and a product of one sided (topologically mixing) subshifts of finite type is a one sided (topologically mixing) subshift of finite type, and in this case as discussed in the end of Remark 2.9.2 we can find sets Γ_n with the properties specified in the above corollary so that $\mu^{\ell-1}(\Gamma_n) \leq b_1 e^{-c_1 n}$ for some $b_1, c_1 > 0$.

Example 2.9.5. Consider the above situation of a one sided topologically mixing subshift of finite type. Let a be a periodic point with period $K \geq 1$ and write $a = \alpha\alpha\alpha...$ where α is a word of length K. Then the above corollary shows that for any k there exist $c_1, b_1 > 0$ such that up to a probability of $\frac{b_1}{n}$ the truncated word $x_0 x_1, ..., x_{n-1}$ of an element $x = x_0 x_1 x_2...$ of $\mathcal{X}$ contains at least $c_1 n$ disjoint blocks of the form $\alpha\alpha...\alpha$ of length K. In fact, according to Remark 2.9.4 this happens on a set whose probability is not less than $1 - b_1 e^{-nc_1}$.

2.9.4 Parametric continuity of transfer operators

Consider the transfer operators $\mathcal{L}_{it}^{\bar{u},n}$ defined in (2.9.1) and recall (2.9.2). Then the arguments in the proof of Lemma 2.7.4 show that there exists a constant $D > 0$ so that for any $n \in \mathbb{N}$, $\bar{u} \in \left(\mathcal{X}^{\ell-1}\right)^n$ and $t \in \mathbb{R}$,

$$\|\mathcal{L}_{it}^{\bar{u},n}\|_{\alpha,\xi} \leq D(1 + |t|) := D_t. \qquad (2.9.19)$$

Next, for any $\bar{v}, \bar{u} \in \left(\mathcal{X}^{\ell-1}\right)^n$ write $\bar{v} = (v^{(0)}, ..., v^{(n-1)})$, $\bar{u} = (u^{(0)}, ..., u^{(n-1)})$ and set

$$\rho_{n,\infty}(\bar{v}, \bar{u}) = \max_{0 \leq i \leq n-1} \rho_{\ell-1,\infty}(v^{(i)}, u^{(i)}).$$

Lemma 2.9.6. *Let $k_0 \in \mathbb{N}$ and $\bar{u} \in \left(\mathcal{X}^{\ell-1}\right)^{k_0}$. Suppose that the function $u \to F_u = F(u, \cdot)$ is continuous at the points $u = u^{(0)}, u^{(1)}, ..., u^{(k_0-1)}$ when considered as a function from $\mathcal{X}^{\ell-1}$ to $\left(\mathcal{H}^{\alpha,\xi}(\mathcal{X}), \|\cdot\|_{\alpha,\xi}\right)$. Then for any $\varepsilon_0 > 0$ there exists $\delta_0 = \delta_0(k_0, \varepsilon_0) > 0$ which may depend only on ε_0 and k_0 (and the function F) such that*

$$\|\mathcal{L}_{it}^{\bar{v},k_0} - \mathcal{L}_{it}^{\bar{u},k_0}\|_{\alpha,\xi} \leq B^2 k_0(|t| + t^2)(1 + |t|)^2 \varepsilon_0$$

for any $t \in \mathbb{R}$ and $\bar{v} \in \left(\mathcal{X}^{\ell-1}\right)^{k_0}$ so that $\rho_{k_0,\infty}(\bar{v}, \bar{u}) < \delta_0$, where $B > 0$ is a constant which does not depend on k_0 $\bar{u}, \bar{v}$, ε_0 and t.

Proof. Let k_0 and $\bar{u}$ be as in the statement of the lemma and let $\varepsilon_0 > 0$ and $t \in \mathbb{R}$. We recall first the decompositions (2.9.2) of $\mathcal{L}_{it}^{\bar{v},k_0}$ and $\mathcal{L}_{it}^{\bar{u},k_0}$. Then, using the convention that $\mathcal{L}^{w,0}$ is the identity map, we can write

$$\mathcal{L}_{it}^{\bar{v},k_0} - \mathcal{L}_{it}^{\bar{u},k_0}$$

$$= \sum_{j=0}^{k_0-1} \mathcal{L}_{it}^{(v^{(k_0-1)},...,v^{(j+1)}),k_0-j-1}\left(\mathcal{L}_{it}^{v^{(j)},1} - \mathcal{L}_{it}^{u^{(j)},1}\right)\mathcal{L}_{it}^{(u^{(j-1)},...,u^{(0)}),j}$$

and therefore by submultiplicativity of the operator norm and (2.9.19),

$$\|\mathcal{L}_{it}^{\bar{v},k_0} - \mathcal{L}_{it}^{\bar{u},k_0}\|_{\alpha,\xi} \leq D^2(1 + |t|)^2 \sum_{j=0}^{k_0-1} \|\mathcal{L}_{it}^{v^{(j)},1} - \mathcal{L}_{it}^{u^{(j)},1}\|_{\alpha,\xi}.$$

Now, let $0 \leq j < k_0$. Then

$$\mathcal{L}_{it}^{v^{(j)},1} - \mathcal{L}_{it}^{u^{(j)},1} = \mathcal{L}_f^\ell\big(e^{itF_{v^{(j)}}} - e^{itF_{u^{(j)}}}\big)$$

and therefore

$$\|\mathcal{L}_{it}^{v^{(j)},1} - \mathcal{L}_{it}^{u^{(j)},1}\|_{\alpha,\xi} \leq \|\mathcal{L}_f^\ell\|_{\alpha,\xi}\|e^{itF_{v^{(j)}}} - e^{itF_{u^{(j)}}}\|_{\alpha,\xi}.$$

In order to estimate the above right-hand side, we apply first the mean value theorem and the Hölder continuity of F (see (2.5.18)) to obtain

$$\|e^{itF_{v^{(j)}}} - e^{itF_{u^{(j)}}}\|_\infty \leq 2|t|\|F_{v^{(j)}} - F_{u^{(j)}}\|_\infty \leq 2C_F|t|\big(\rho_{\ell-1,\infty}(v^{(j)}, u^{(j)})\big)^\alpha$$

assuming that $\rho_{\ell-1,\infty}(v^{(j)}, u^{(j)}) < \xi$. Next, we will estimate

$$v_{\alpha,\xi}\big(e^{itF_{v^{(j)}}} - e^{itF_{u^{(j)}}}\big).$$

Let $x, x' \in \mathcal{X}$ be such that $\rho(x, x') < \xi$, and consider the functions U and V given by

$$V(x) = V_j(x) = F_{v^{(j)}}(x) \quad \text{and} \quad U(x) = U_j(x) = F_{u^{(j)}}(x).$$

Then we can write

$$\big(e^{itF_{v^{(j)}}(x)} - e^{itF_{u^{(j)}}(x)}\big) - \big(e^{itF_{v^{(j)}}(x')} - e^{itF_{u^{(j)}}(x')}\big)$$

$$= e^{itU(x)}\big(e^{it(V(x)-U(x))} - e^{it(V(x')-U(x'))}\big)$$

$$+ \big(e^{it(U(x)-U(x'))} - 1\big)\big(e^{itV(x')} - e^{itU(x')}\big) := I_1 + I_2.$$

By the mean value theorem we have

$$|I_1| \leq 2|t|\big|\big(V(x) - U(x)\big) - \big(V(x') - U(x')\big)\big|$$

$$\text{and} \quad |I_2| \leq 4|t|^2|U(x) - U(x')| \cdot \|V - U\|_\infty.$$

Using now (2.5.18) we derive that $|U(x) - U(x')| \leq C_F\rho^\alpha(x, x')$ and that

$$\|V - U\|_\infty \leq C_F\big(\rho_{\ell-1,\infty}(v^{(j)}, u^{(j)})\big)^\alpha$$

assuming that $\rho_{\ell-1,\infty}(v^{(j)}, u^{(j)}) < \xi$. Next, it is clear from the definition of V and U that

$$\big|\big(V(x) - U(x)\big) - \big(V(x') - U(x')\big)\big| \leq \|F_{v^{(j)}} - F_{u^{(j)}}\|_{\alpha,\xi}\rho^\alpha(x, x').$$

By the continuity assumption in the statement of the lemma, there exists $\delta_1 = \delta_1(k_0, \varepsilon_0) > 0$ so that for any $0 \leq j < k_0$,

$$\|F_{v^{(j)}} - F_{u^{(j)}}\|_{\alpha,\xi} \leq \varepsilon_0$$

for any $v^{(j)} \in \mathcal{X}^{\ell-1}$ such that $\rho_{\ell-1,\infty}(v^{(j)}, u^{(j)}) < \delta_1$. Let $0 < \delta_0 < \min(\delta_1, \xi)$ be so that $(\delta_0)^\alpha < \varepsilon_0$. The lemma follows now (with such δ_0) from the above estimates. $\qquad\square$

2.9.5 *Quasi compactness of R_{it}, lattice and non-lattice cases*

We first claim that for any $t \in \mathbb{N}$ the transfer operator R_{it} satisfies the Lasota-Yorke type inequality with respect to the norm $\| \cdot \|_{\alpha,\xi}$ and the (semi) norm $\| \cdot \|_{\infty}$. More precisely, we assert that there exist constants $c \in (0,1)$ and $A > 0$ such that

$$\|R_{it}^n g\|_{\alpha,\xi} \leq A\big(c^n \|g\|_{\alpha,\xi} + (1 + |t|)\|g\|_{\infty}\big) \qquad (2.9.20)$$

for any $n \in \mathbb{N}$, $t \in \mathbb{R}$ and $g \in \mathcal{H}^{\alpha,\xi} = \mathcal{H}^{\alpha,\xi}(\mathcal{X})$. Indeed, (2.9.20) follows by arguments similar to the ones in the proof of (2.7.5) in Lemma 2.7.4 since in that proof we have only used the pairing property described in Section 2.5.2, the Hölder continuity of F and some elementary inequalities. Observe now that by (2.9.20) the spectral radius $r(it)$ of R_{it} with respect to the norm $\| \cdot \|_{\alpha,\xi}$ does not exceed 1. We conclude from Theorem II.5 in [28] (applied with the (semi) norm $\|| \cdot ||| = \| \cdot \|_{\infty}$) that R_{it} is quasi-compact with respect to the norm $\| \cdot \|_{\alpha,\xi}$ when $r(it) = 1$. Note that conditions $K1$ and $K2$ in Chapter XI of [28] hold true with the map $\tau_0 = T^{\ell m_0}$, the transfer operator $Q = R_0 = \mathcal{L}_f^{\ell m_0}$ and the space $\mathcal{B} = (\mathcal{H}^{\alpha,\xi}, \| \cdot \|_{\alpha,\xi})$. We conclude that all the results from Chapter XI in [28] (cited below) hold true for the family of transfer operators $R_{it}, t \in \mathbb{R}$.

Next, in the non-lattice case let $J \subset \mathbb{R}$ be a compact set not containing 0. We claim that there exist constants $A > 0$ and $c \in (0,1)$, which may depend on J, such that for any $n \in \mathbb{N}$ and $t \in J$,

$$\|R_{it}^n\|_{\alpha,\xi} \leq Ac^n. \qquad (2.9.21)$$

Indeed, since F_{m_0,x_0} is non-arithmetic, Proposition XI.7 in [28] implies that $r(it) < 1$ for any $t \in \mathbb{R} \setminus \{0\}$, and (2.9.21) follows from Corollary III.13 in [28].

Now, in the lattice case, let $J \subset (-\frac{2\pi}{h}, \frac{2\pi}{h})$ be a compact set not containing the origin. As in the non-lattice case, we claim that there exist constants $A > 0$ and $c \in (0,1)$, which may depend on J, such that for any $n \in \mathbb{N}$ and $t \in J$,

$$\|R_{it}^n\|_{\alpha,\xi} \leq Ac^n. \qquad (2.9.22)$$

Indeed, by Proposition XI.7 in [28] for any $t \neq 0$ the spectral radius $r(it)$ equals 1 if and only if there exist $g \in \mathcal{H}^{\alpha,\xi} \setminus \{0\}$ and $\alpha \in \mathbb{R}$ such that

$$ge^{itF_{x_0,m_0}} = e^{i\alpha} g \circ T^{\ell m_0}, \quad \mu\text{-a.s.}$$

where we recall that R_{it} is indeed quasi-compact when $r(it) = 1$. Since h is maximal (see Assumption 2.6.4) the spectral radius $r(it)$ of R_{it} is strictly less than 1 when $t \in (-\frac{2\pi}{h}, \frac{2\pi}{h}) \setminus \{0\}$, and (2.9.22) also follows from Corollary III.13 in [28].

2.9.6　*Norms estimates*

We complete here the proof that Assumption 2.2.1 holds true in both non-lattice and lattice cases. Recall first that by Assumption 2.6.1 the function $u \to F_u$ is continuous at the points

$$u = \hat{T}^k \bar{x}_0, \ 0 \le k < m_0.$$

Set $\mathcal{Y} = \mathcal{X}^{\ell-1}$ and let $n_0 \in \mathbb{N}$ and $\varepsilon_0 > 0$. Consider the periodic element $u^{\bar{x}_0, n_0}$ of $(\mathcal{Y}^{m_0})^{n_0}$ given by (2.9.6). Then the function $u \to F_u$ is continuous at each of the $\mathcal{Y}$-valued components of $u^{\bar{x}_0, n_0}$. Consider the number $\delta_0 = \delta_0(n_0 m_0, \varepsilon_0) > 0$ appearing in Lemma 2.9.6, and let $\bar{v} \in (\mathcal{Y}^{m_0})^{n_0}$ be so that

$$\rho_{n_0 m_0, \infty}(\bar{v}, u^{\bar{x}_0, n_0}) < \delta_0. \tag{2.9.23}$$

Using (2.9.7) and then applying Lemma 2.9.6, we deduce that

$$\|\mathcal{L}_{it}^{\bar{v}, n_0 m_0} - R_{it}^{n_0}\|_{\alpha, \xi} \tag{2.9.24}$$

$$= \|\mathcal{L}_{it}^{\bar{v}, n_0 m_0} - \mathcal{L}_{it}^{u^{\bar{x}_0, n_0}, n_0 m_0}\|_{\alpha, \xi} \le B n_0 m_0 (|t| + t^2)(1 + |t|)^2 \varepsilon_0$$

for any $t \in \mathbb{R}$. Next, let $\delta \in (0, 1)$. In the non-lattice case set

$$J = J_\delta = [-\delta^{-1}, -\delta] \cup [\delta, \delta^{-1}]$$

while in the lattice case we set

$$J = J_\delta = [-\frac{\pi}{h}, \frac{\pi}{h}] \setminus (-\delta, \delta).$$

By (2.9.21) in the non-lattice case and by (2.9.22) in the lattice case there exist constants $A > 0$ and $c \in (0, 1)$, which may depend on δ, such that for any $n \in \mathbb{N}$ and $t \in J = J_\delta$,

$$\|R_{it}^n\|_{\alpha, \xi} \le A c^n. \tag{2.9.25}$$

Set $D_\delta = \max\{D_t : t \in J_\delta\}$ where D_t is defined in (2.9.19). Fix some $n_0 \in \mathbb{N}$ so large such that $A c^{n_0} < \frac{1}{4 D_\delta}$ and then fix some $\varepsilon_0 > 0$ so small such that

$$B n_0 m_0 (|t| + t^2)(1 + |t|)^2 \varepsilon_0 < \frac{1}{4 D_\delta} \quad \text{for any } t \in J.$$

Then by (2.9.7), (2.9.24) and (2.9.25),

$$\|\mathcal{L}_{it}^{\bar{v}, n_0 m_0}\|_{\alpha, \xi} \le \frac{1}{2 D_\delta}$$

for any $\bar{v}$ satisfying (2.9.23) with these fixed n_0 and $\delta_0 = \delta_0(n_0 m_0, \varepsilon_0)$. Consider now the sets $\mathcal{T}_n = \mathcal{T}_{n, n_0, \delta_0}, n \in \mathbb{N}$ and the numbers $b_1 = b_1(n_0, \delta_0) > 0$

and $c_1 = c_1(n_0, \delta_0) > 0$ from Corollary 2.9.3. Then by (2.9.4), for any $N \in \mathbb{N}$ and $t \in \mathbb{R}$,

$$\mathbb{E}\|\mathcal{L}_{it}^{(\Theta_{M+1},\dots,\Theta_N),N-M}1\|_\infty = \mathbb{E}_{\mu^{\ell-1}}\|\mathcal{L}_{it}^{v_{\hat{T},N-M-1}(\cdot),N-M}1\|_\infty \quad (2.9.26)$$

$$\leq \mu^{\ell-1}(\mathcal{T}_{N-M}) + \int \mathbb{I}_{\mathcal{T}_{N-M}^c}(\bar{x})\|\mathcal{L}_{it}^{v_{\hat{T},N-M-1}(\bar{x}),N-M}\|_{\alpha,\xi}\,d\mu^{\ell-1}(\bar{x})$$

where $\mathcal{T}_n^c = \mathcal{X} \setminus \mathcal{T}_n$ for any $n \in \mathbb{N}$. By Corollary 2.9.3, on the set $\mathcal{T}_n^c$ we can write

$$v_{\hat{T},N-M-1}(\bar{x}) = w^{(1)}\mathbf{v}^{(1)}w^{(2)}\mathbf{v}^{(2)}w^{(3)}\dots\mathbf{v}^{(L)}w^{(L+1)}, \ \ L \geq c_1(N-M)$$

where each $\mathbf{v}^{(i)} \in (\mathcal{Y}^{m_0})^{n_0}$ satisfies

$$\rho_{m_0 n_0,\infty}(\mathbf{v}^{(i)}, u^{\bar{x}_0,n_0}) < \delta_0$$

while some of the $w^{(i)}$'s can be empty. We remark that Corollary 2.9.3 guarantees that we can decompose the truncated orbit $\mathcal{O}_{0,N-M-1}(\bar{x})$ of $\bar{x}$ in such a way, but since $u^{\bar{x}_0,n_0}$ is periodic it is invariant under inversion and so the reverse truncated orbit $v_{\hat{T},N-M-1}(\bar{x})$ inherits this decomposition from the truncated orbit. We conclude that for any $\bar{x} \in \mathcal{T}_n$ and $t \in J_\delta$, the random product $\mathcal{L}_{it}^{v_{\hat{T},N-M}(\bar{x}),N-M-1}$ can be represented in the form

$$\mathcal{L}_{it}^{v_{\hat{T},N-M-1}(\bar{x}),N-M} = \mathcal{A}_1 \circ \mathcal{B}_1 \circ \mathcal{A}_2 \circ \mathcal{B}_2 \circ \cdots \circ \mathcal{B}_k \circ \mathcal{A}_{k+1} \quad (2.9.27)$$

where $k \geq c_1(N-M)$, the norms $\|\mathcal{A}_j\|_{\alpha,\xi}$, $1 \leq j \leq k+1$ are bounded from above by D_δ and the norms $\|\mathcal{B}_j\|_{\alpha,\xi}$, $1 \leq j \leq k$ are bounded from above by $\frac{1}{2D_\delta}$. There exists a constant a_0 which depends only on ℓ such that $N - M \geq a_0 N$ for any $N > 3$. We conclude that here exists a constant $d > 0$ such that

$$\|\mathcal{L}_{it}^{v_{\hat{T},N-M-1}(\bar{x}),N-M}\|_{\alpha,\xi} \leq 2^{-Nd}$$

for any $N > 3$ and $\bar{x} \in \mathcal{T}_{N-M}^c$, which together with (2.9.3), (2.9.26) and (2.9.16) completes the proof that Assumption 2.2.1 holds true.

Remark 2.9.7. At first glance it seems that the arguments in the proof that Assumption 2.2.1 holds true can be modified in order to show that Assumption 2.2.2 holds. Indeed, for any $c > 0$ the set $\{\tilde{V}_k > ck\}$ is open and thus contains an open ball, where $\tilde{V}_k$ is defined in Remark 2.8.10. The problem here is that in general we have no control on the radius of these balls and so their $\mu^{\ell-1}$-measures cannot be controlled uniformly over k. As a consequence, it is not possible to choose ε_0 independently of k as we did in the proof of Claim 2.8.5.

2.10   The periodic point approach to quenched and annealed dynamics

The idea behind the proof (in Section 2.9) that Assumption 2.2.1 holds true is not restricted to random transformations arising in the nonconventional setup. In this section we present a scheme to prove a fiberwise (and annealed) local limit theorems using the ideas of Section 2.9, and we refer the readers to Chapter 7 for more details.

Let $(\Omega, \mathcal{B}, P, \tau)$ be an ergodic measure preserving system (not necessarily invertible) on a topological space Ω where $\mathcal{B}$ is the Borel σ-algebra. Let B_ω, $\omega \in \Omega$ be a family of complex Banach spaces.

Assumption 2.10.1. The probability measure P assigns positive measure to open sets, τ is continuous and there exist $\omega_0 \in \Omega$ and $m_0 \in \mathbb{N}$ so that $\tau^{m_0}\omega_0 = \omega_0$. Moreover, the map $\omega \to B_\omega$ is locally constant around the points $\tau^i\omega_0$, $0 \leq i < m_0$.

Let $\mathcal{I} \subset \mathbb{R}$ be an interval around 0 (possibly $\mathcal{I} = \mathbb{R}$) and $\{A_{it}^\omega : \omega \in \Omega\}$, $t \in \mathcal{I}$ be families of continuous linear maps $A_t^\omega : B_\omega \to B_{\theta\omega}$. Consider the random products $A_{it}^{\omega,n} : B_\omega \to B_{\theta^n\omega}$ given by

$$A_{it}^{\omega,n} := A_{it}^{\tau^{n-1}\omega} \circ \cdots \circ A_{it}^{\tau\omega} \circ A_{it}^\omega, \quad t \in \mathcal{I}, \ n \in \mathbb{N}.$$

Assumption 2.10.2. For each compact set $J \subset \mathcal{I}$ the family of maps $\omega \to A_{it}^\omega$, where t ranges over J, is equicontinuous at the points $\omega = \tau^i\omega_0$, $0 \leq i < m_0$ with respect to the operator norm and there exists a constant $B = B(J) \geq 1$, which may depend on J, such that P-a.s.,

$$\|A_{it}^{\omega,n}\| \leq B \tag{2.10.1}$$

for any $n \in \mathbb{N}$ and $t \in J$.

Note that the equicontinuity requirement in Assumption 2.10.2 makes sense when Assumption 2.10.1 is satisfied. Consider the deterministic family of operators $\mathcal{A}_{it}$, $t \in \mathcal{I}$ given by $\mathcal{A}_{it} = A_{it}^{\omega_0,m_0} : B_{\omega_0} \to B_{\omega_0}$, and note that

$$\mathcal{A}_{it}^s = \mathcal{A}_{it} \circ \mathcal{A}_{it} \circ \cdots \circ \mathcal{A}_{it} = A_{it}^{\omega_0,sm_0}$$

for any $s \in \mathbb{N}$ and $t \in \mathcal{I}$.

Assumption 2.10.3. For any compact $J \subset \mathcal{I}$ which does not contain the origin there exist constants $c = c(J) > 0$ and $b = b(J) \in (0,1)$ such that

$$\|\mathcal{A}_{it}^s\| \leq cb^s$$

for any $s \in \mathbb{N}$ and $t \in J$.

This assumption holds true when the map $t \to \mathcal{A}_{it}$ is continuous and the spectral radius of all $\mathcal{A}_{it}$'s is strictly less than 1 except for the spectral radius of $\mathcal{A}_0$.

Lemma 2.10.4. *Under Assumptions 2.10.1, 2.10.2 and 2.10.3, for each compact subset J of $\mathcal{I}$ which does not contain the origin there exist a constant $d = d_J > 0$ and a random variable r_ω such that P-a.s.,*

$$\|A_{it}^{\omega,n}\| \le r_\omega 2^{-nd}$$

for any $n \in \mathbb{N}$ and $t \in J$.

Proof. Fix some compact subset J of $\mathcal{I}$ which does not contain the origin. For P-a.a. ω, let n_0 be sufficiently large so that $cb^{n_0} < \frac{1}{4B}$ where $c = c(J), b = b(J)$ and $B = B(J)$. Relying on Assumptions 2.10.1 and 2.10.2, there exists an open set $U_{n_0} = U_{n_0}(J) \subset \Omega$ containing ω_0 such that for any $\omega \in U_{n_0}$ we have $B_{\tau^i \omega} = B_{\tau^i \omega_0}$, $0 \le i < m_0$ and

$$\|A_{it}^{\omega,n_0 m_0} - \mathcal{A}_{it}^{n_0}\| = \|A_{it}^{\omega,n_0 m_0} - A_{it}^{\omega_0,n_0 m_0}\| < \frac{1}{4B} \qquad (2.10.2)$$

for any $t \in J$. Set $p_0 = P(U_{n_0}) > 0$. Applying the mean ergodic theorem with the map τ and the indicator function of U_{n_0} we see that P-a.s. there exists an infinite strictly increasing sequence $(n_k)_{k=1}^\infty = (n_k(\omega))_{k=1}^\infty$ of natural numbers such that for any $m \ge 1$,

$$\tau^m \omega \in U_{n_0} \quad \text{if and only if} \quad m \in \{n_k : k \ge 1\}$$

and

$$\lim_{k \to \infty} \frac{n_k}{k} = \frac{1}{p_0} > 0.$$

For each $k \in \mathbb{N}$ set $m_k = n_{k m_0 n_0}$ and

$$k_n = k_n(\omega) = \max\{k \in \mathbb{N} : m_k \le n\}.$$

Then

$$\lim_{n \to \infty} \frac{k_n}{n} = \frac{p_0}{m_0 n_0}, \quad P\text{-a.s..} \qquad (2.10.3)$$

Finally, P-a.s. for each n and $t \in J$ we can write

$$A_{it}^{\omega,n} = \mathcal{B}_1 \circ A_{it}^{\tau^{m_1}\omega,n_0 m_0} \circ \mathcal{B}_2 \circ A_{it}^{\tau^{m_2}\omega,n_0 m_0} \circ \cdots \circ \mathcal{B}_l \circ A_{it}^{\tau^{m_{k_n}}\omega,n_0 m_0} \circ \mathcal{B}_{k_n+1}$$

where $\|\mathcal{B}_i\| \le B = B(J)$ for each i. Since $\tau^{m_i}\omega \in U_{n_0}$ for each i, we obtain from (2.10.2) that

$$\|A_{it}^{\tau^{m_i}\omega,n_0 m_0}\| \le \|\mathcal{A}_{it}^{n_0}\| + \frac{1}{4B} \le bc^{n_0} + \frac{1}{4B} \le \frac{1}{2B}$$

for each i, and the lemma follows, taking into account (2.10.3). $\qquad \square$

Note that when, in addition, $\|A_t^\omega\| \leq 1$ for any $t \in \mathcal{I}$, then bounds of the form

$$\|\mathcal{A}_t^s\| \leq ce^{-dt^2}$$

for some $d > 0$ and t's sufficiently close to 0 will yield in a similar way estimates of the form

$$\|A_t^{\omega,n}\| \leq c(\omega)e^{-d_1 t^2}, \ d_1 > 0$$

while in general bounding the random norms $\|A_t^{\omega,n}\|$ for small t's requires a different approach such as the random complex RPF theorem in Part 2, and we refer again to Chapter 7 for details.

When, in addition to the above assumptions, the measure preserving system $(\Omega, \mathcal{B}, P, \tau)$ satisfies some decay of correlations then using similar arguments to the ones from Sections 2.8 and 2.9 we obtain appropriate estimates on the norms of iterates of the operators $A_{it} = \int A_{it}^\omega dP(\omega)$ which will yield a local central limit theorem for a Markov chain whose transition operator is A_0.

2.11 Extensions to dynamical systems

In this section we will explain how to prove a local limit theorem for sums of the form

$$S_N(x) = S_N F(x) = \sum_{n=1}^{N} F(Tx, T^2x, ..., T^\ell x) \qquad (2.11.1)$$

in the case when T is either a topologically mixing two (one) sided subshift of finite type or a C^2 Axiom A diffeomorphism (in particular, Anosov) in a neighborhood of an attractor (see [10]), x is distributed according to a Gibbs measure and F is a Hölder continuous function with additional regularity around the elements of the orbit of one periodic point, where we recall that periodic points are dense in both cases.

2.11.1 *Subshifts of finite type*

Let A be a $0-1$ matrix of size $n \times n$ and set $\mathcal{A} = \{1, 2, ..., n\}$. We assume that there exists $M \in \mathbb{N}$ such that A^M has only positive entries. Set

$$\Sigma_A = \{(x_i)_{i \in \mathbb{Z}} : A_{x_i, x_{i+1}} = 1, \ \forall i \in \mathbb{Z}\} \subset \mathcal{A}^\mathbb{Z},$$

$$\Sigma_A^+ = \{(x_i)_{i \geq 0} : A_{x_i, x_{i+1}} = 1, \ \forall i \geq 0\} \subset \mathcal{A}^\mathbb{N} \ \text{and}$$

$$\Sigma_A^- = \{(x_i)_{i \leq 0} : A_{x_i, x_{i-1}} = 1, \ \forall i \geq 0\} \subset \mathcal{A}^\mathbb{N}.$$

Then Σ_A, Σ_A^+ and Σ_A^- are compact subsets of the product spaces $\mathcal{A}^{\mathbb{Z}}$, $\mathcal{A}^{\mathbb{N}}$ and $\mathcal{A}^{\mathbb{N}}$, respectively. We equip Σ_A with the metric

$$\rho(x, y) = 2^{-n(x,y)} \text{ where } n(x, y) = \min\{n \geq 0 : x_n \neq y_n \text{ or } x_{-n} \neq y_{-n}\}$$

where we use the conventions $\min \emptyset = \infty$ and $2^{-\infty} = 0$. Similarly, we equip Σ_A^+ and Σ_A^- with the metrics ρ_+ and ρ_- which are defined similarly but with

$$n_+(x, y) = \min\{n \geq 0 : x_n \neq y_n\}$$

and

$$n_-(x, y) = \min\{n \geq 0 : x_{-n} \neq y_{-n}\}.$$

in place of $n(x, y)$. Then the corresponding Borel σ-algebras $\mathcal{B}, \mathcal{B}_+$ and $\mathcal{B}_-$ are generated by appropriate cylinder sets.

Next, consider the metric $\rho_{\ell,\infty}$ on $(\Sigma_A)^\ell$ given by

$$\rho_{\ell,\infty}(\bar{x}, \bar{y}) = \max_{1 \leq i \leq \ell} \rho(x^{(i)}, y^{(i)})$$

where $x = (x^{(1)}, ..., x^{(\ell)})$ and $y = (y^{(1)}, ..., y^{(\ell)})$. Let (Σ_A, T) be the corresponding two sided togologically mixing subshift of finite type, where $T : \Sigma_A \to \Sigma_A$ is the left shift given by

$$(Tx)_k = x_{k+1}, \; \forall\, k \in \mathbb{Z}.$$

Denote by $T_+ : \Sigma_A^+ \to \Sigma_A^+$ the one sided left shift and by $T_- : \Sigma_A^- \to \Sigma_A^-$ the one sided right shift. We equip Σ_A with a T-invariant Gibbs measure $\mu = \mu^{(f)}$ (see [10]) associated with some Hölder continuous (potential) function f. Recall that μ is constructed first as a T_+-invariant measure T_+ on Σ_A^+ associated with some Hölder continuous function $f^+ : \Sigma_A^+ \to \mathbb{R}$ (which, when lifted to Σ_A, differs from f by a coboundary term) and then extended to Σ_A. Using the reflection map $\mathcal{R}$ given by $(x_i)_{i \geq 0} \to (x_i)_{i \leq 0}$, the measure μ_+ defines an appropriate T_--invariant measure μ_- on Σ_A^- which is the Gibbs measure corresponding to the function $f^- = f^+ \circ \mathcal{R}^{-1}$.

Note that the maps T_+ and T_- are locally distance expanding in the sense of Section 2.5.1 with the metrics ρ_+ and ρ_-, respectively, while the invertible map T is not locally distance expanding.

2.11.2 *The strategy of the proof*

The proof of the LLT in the conventional situation ($\ell = 1$) when F depends only on the coordinates with nonnegative indexes is a direct consequence of the relation (2.5.12), where the Markov chain $\{\zeta_n : n \geq 0\}$ from there is the one defined in Section 2.5 using the map T_+ and the function f^+ in place of T and f. When F depends also on other coordinates then (see Lemma 1.6 from [10]) it differs by a coboundary term from a function G^+ which depends only on coordinates with nonnegative indexes. This makes it possible to use the above Markov chain and obtain the desired LLT, taking into account that the measurable map $p_+ : \Sigma_A \to \Sigma_A^+$ given by $p_+(x) = (x_i)_{i \geq 0}$ is a factor map between the measure preserving systems $(\Sigma_A, \mathcal{B}, \mu, T)$ and $(\Sigma_A^+, \mathcal{B}_+, \mu_+, T_+)$, namely p_+ is bijective, $p_+ \circ T = T_+ \circ p_+$ and $\mu \circ p_+^{-1} = \mu_+$.

In the nonconventional situation ($\ell > 1$) the distribution of $S_N F(x)$ when $x \overset{d}{\sim} \mu$ is no longer invariant under the reflection map

$$(Tx, T^2 x, ..., T^{N\ell} x) \to (T^{N\ell} x, T^{N\ell - 1} x, ..., Tx)$$

and instead of (2.5.12) the relations (2.5.14) and (2.5.15) from the end of Section 2.5.4 hold true. In view of (2.5.15), in order to obtain the LLT when F depends only on the coordinates with nonpositive indexes we should consider the process $\mathbf{Z}$ generated by the map T^{-1} and the Markov chain $\zeta^- = \{\zeta_n^- : n \geq 0\}$ defined in Section 2.5 using the map T_- and the function f^-. Indeed, the measurable map $p_- : \Sigma_A \to \Sigma_A^-$ given by $p(x) = (x_i)_{i \leq 0}$, where $x = (x_i)_{i \in \mathbb{Z}}$ is a factor map between the measure preserving systems $(\Sigma_A, \mathcal{B}, \mu, T^{-1})$ and $(\Sigma_A^-, \mathcal{B}_-, \mu_-, T_-)$, and thus we derive from (2.5.15) that for any $N \in \mathbb{N}$,

$$\sum_{n=1}^{N} F(T^n x, T^{2n} x, ..., T^{\ell n} x) \overset{d}{=} \sum_{n=1}^{N} F^-(\zeta_n^-, \zeta_{2n}^-, ..., \zeta_{\ell n}^-)$$

when F depends only on the coordinates with nonpositive indexes, where the function F^- is defined by the relation $F = F^- \circ p_-$ and $x \overset{d}{\sim} \mu$. When F depends also on other coordinates then an appropriate version of the above Lemma 1.6 holds true (see Lemma 2.11.2), and so the desired LLT will essentially follow from the arguments in Sections 2.8 and 2.9 applied with the Markov chain ζ^-.

2.11.3 *Local limit theorem: one sided case*

Consider the variable $\bar{x} = (x^{(1)}, ..., x^{(\ell)}) \in (\Sigma_A)^\ell$. Let $F : (\Sigma_A)^\ell \to \mathbb{R}$ be a function which depends only on the coordinates with nonnegative indexes (i.e. on $x_j^{(i)}$, $i = 1, 2, ..., \ell - 1$, $j \geq 0$). Let $\{\zeta_n^+ : n \geq 0\}$ be the associated Markov chain defined in Section 2.5 using the map T_+ and the function f^+.

Then similarly to (2.5.13) (or (2.5.15)) and Section 2.11.2, the sums

$$R_N(x) = \sum_{n=1}^{N} F(T^{-n}x, T^{-2n}x, ..., T^{-\ell n}x)$$

when $x \overset{d}{\sim} \mu$ have the same distribution as the sums $\sum_{n=1}^{N} F(\zeta_n^+, \zeta_{2n}^+, ..., \zeta_{\ell n}^+)$ from Section 2.5, and therefore appropriate LLT's for the sums R_N follow directly from Theorems 2.6.3 and 2.6.5. Replacing T_+ with T_- and Σ_A^+ with Σ_A^- we obtain LLT's for the sums

$$Q_N(x) = \sum_{n=1}^{N} F(T^n x, T^{2n}x, ..., T^{\ell n}x)$$

when F depends only on the coordinates with nonpositive indexes. Consequently, using the symbolic representation via Markov partitions (see Sections 3 and 4 from [10]), we derive appropriate LLT's for sums of the form

$$\tilde{R}_N(x) = \sum_{n=1}^{N} F(g^{-n}x, g^{-2n}x, ..., g^{-\ell n}x)$$

and

$$\tilde{Q}_N(x) = \sum_{n=1}^{N} F(g^n x, g^{2n}x, ..., g^{\ell n}x)$$

when g is a C^2 Axiom A diffeomorphism (in particular, Anosov) in a neighborhood of an attractor, x is distributed according to an appropriate Gibbs measure and in the first case F is constant on the atoms of the partition $\bigcap_{j=0}^{\infty} g^{-j}(\mathcal{M})$ while in the second case it is constant on the atoms of the partition $\bigcap_{j=0}^{\infty} g^j(\mathcal{M})$, where $\mathcal{M}$ is a Markov partition with a sufficiently small diameter (see [10]).

2.11.4 *Two sided case*

The above LLT's can be extended to the case when F depends only on the coordinates in places from $-s$ to ∞ (or from $-\infty$ to s) for some fixed $s > 0$ in the subshift case, and when F is constant on atoms of partitions $\bigcap_{j=-s}^{\infty} g^{-j}(\mathcal{M})$ (or $\bigcap_{j=s}^{\infty} g^j(\mathcal{M})$) in the Axiom A case. Extensions of these LLT's to the case when F is allowed to depend on all the coordinates is done similarly to the conventional case ($\ell = 1$), and for readers' convenience in this section we provide the main details. We first consider the situation of a two sided topologically mixing subshift of finite type and the corresponding extensions in the Axiom A maps case will follow.

Before discussing an LLT, note first that all the results from [45] and [30] hold true for the sequence of random variables $\{X_n : n \geq 0\}$ where $X_n = T^n X_0$ and X_0 is distributed according to μ. Namely, Theorem 2.5.11 holds true for the sums $S_N F$ and the characterization (2.5.19) for positivity of the asymptotic variance holds true. Replacing F with F_ℓ we deduce, in particular, that the limit

$$\sigma_\ell^2 = \lim_{N \to \infty} \frac{1}{N} \mathbb{E}_\mu (S_N F_\ell)^2$$

exists and that it vanishes if and only if (2.5.19) holds. Therefore, the limit $\sigma^2 = \lim_{N \to \infty} N^{-1} \mathrm{Var} S_N F$ is positive when $\sigma_\ell^2 > 0$.

We distinguish here between lattice and non-lattice cases in the same way as in Section 2.5 but with the function $\tilde{F}_{x_0, m_0}$ given by

$$\tilde{F}_{x_0, m_0}(x) = \sum_{k=1}^{m_0} F(\hat{T}^k \bar{x}_0, T^{k\ell} x) \qquad (2.11.2)$$

where $\hat{T} = T \times T^2 \times \cdots \times T^{\ell-1}$, $\bar{x}_0 = (x_0, ..., x_0) \in \Sigma_A^{\ell-1}$ and $x_0 \in \Sigma_A$ is a T-periodic point of period m_0.

Now we can state the main result of this section.

Theorem 2.11.1. *Suppose that F is a Hölder continuous function which satisfies (2.6.1) at any point v. Further assume that $\sigma_\ell^2 > 0$. Then in the non-lattice case the sequence $\{S_N : N \geq 1\}$ satisfies the LLT with the centralizing constant $m = \bar{F}$, the normalizing constant σ and the Lebesgue measure while in the lattice case it satisfies the LLT with m, σ and the measure assigning mass h to each point of the lattice $h\mathbb{Z}$.*

2.11.5   *Reduction to one sided case*

As in the conventional case ($\ell = 1$), the following version of Lemma 1.6 from [10] is the key to deduce limit theorems for two sided subshifts of finite type from the corresponding limit theorems for the one sided subshifts.

Lemma 2.11.2. *Let $F : (\Sigma_A)^\ell \to \mathbb{R}$ be a Hölder continuous function. Then there exist Hölder continuous functions $G^+, H^+ : (\Sigma_A)^\ell \to \mathbb{R}$ such that G^+ depends only on coordinates with nonnegative indexes and*

$$F = G^+ - H^+ + H^+ \circ \bar{T} \qquad (2.11.3)$$

where $\bar{T} = T \times T^2 \times \cdots \times T^\ell$. Moreover, the function G^+ satisfies (2.6.1) at any point v when F does. The same thing holds true with a function G^-

which depends only on coordinates with nonpositive indexes, a function H^- and with $\bar{T}^{-1} = T^{-1} \times T^{-2} \times \cdots \times T^{-\ell}$ in place of $\bar{T}$.

The proof of the first part of the lemma goes essentially as the proof of Lemma 1.6 from [10] with the map $\bar{T}$ in place of T, while the proof of the second part (about G^+ satisfying (2.6.1)) is a direct consequence of the arguments from there.

Next, consider the function G^- from Lemma 2.11.2, namely the function satisfying (2.11.3) with $\bar{T}^{-1}$ in place of $\bar{T}$ which depends only on the coordinates with nonpositive indexes. Consider the random variables $S_N G^-$ (defined on Σ_A) given by

$$S_N G^-(x) = \sum_{n=1}^{N} G^-(T^n x, T^{2n} x, ..., T^{\ell n} x), \quad N \in \mathbb{N}$$

where x is drawn according to μ. Let $F^- : (\Sigma_A^-)^\ell \to \mathbb{R}$ be the unique function satisfying $G^- = F^- \circ p_-$ and consider the Markov chain $\{\zeta_n^- : n \geq 0\}$ generated as in Section 2.5 by the map T_- and the Hölder continuous (potential) function $f^- : \Sigma_A^- \to \mathbb{R}$ (corresponding to the measure μ_-). Define the function G_ℓ^- similarly to (2.5.20) but with G^- in place of F and the function F_ℓ^- similarly to (2.5.20) but with F^- in place of F and μ_- in place of μ. Set

$$\sigma_\ell^2(F_\ell^-) := \lim_{N \to \infty} \frac{1}{N} \mathbb{E}_{\mu_-} \Big(\sum_{n=1}^{N} F_\ell^-(\zeta_n^-, \zeta_{2n}^-, ..., \zeta_{\ell n}^-) \Big)^2$$

where we recall that this limit exists by Theorem 2.5.11. In order to make the reduction to the one sided case we need first the following.

Lemma 2.11.3. *The inequality $\sigma_\ell^2 > 0$ implies that $\sigma_\ell^2(F_\ell^-) > 0$.*

Proof. Suppose that $\sigma_\ell^2 > 0$. Then F_ℓ does not admit a coboundary representation of the form (2.5.19). Since G^- and F differ by the coboundary term $H^- - H^- \circ \bar{T}^{-1}$ the functions G_ℓ^- and F_ℓ differ by a coboundary term, as well and so the function G_ℓ^- also cannot admit a coboundary representation of this form. Using the map p_- defined in Section 2.11.2, we conclude that the function F_ℓ^- cannot admit a coboundary representation of the form (2.5.19) with T_-, and therefore Theorem 2.5.11 implies that $\sigma_\ell^2(F_\ell^-) > 0$. $\square$

We also need the following.

Lemma 2.11.4. *Suppose that the non-lattice (lattice) condition holds true with some T-periodic point $x_0 \in \Sigma_A$. Then the function F^- satisfies the non-lattice (lattice) condition from Section 2.5 with the map T_- and the (T_--periodic) point $p_-(x_0)$.*

Proof. Let x_0 be a periodic point of period m_0 as in the statement of the lemma. First, since G^- and F differ only by the coboundary term $H^- - H^- \circ \bar{T}^{-1}$ then G^- satisfies the non-lattice (lattice) condition with the point x_0, as well. Secondly, any solution of (2.6.3) (or (2.6.5)) with the function $F^-_{p_-(x_0),m_0}$ and the map T_- extends (after pulling it back with the factor map p_- and applying the map $T^{-m_0\ell}$) to a solution of (2.6.3) (or (2.6.5)) with the function $\tilde{G}^-_{x_0,m_0}$ defined similarly to (2.11.2) with G^- in place of F, where we also used that T preserves μ and that (since T_- is ergodic) any solution g of the equation (2.6.3) (or (2.6.5)) must have a constant nonzero absolute value, implying that $1/g$ is Hölder continuous. $\qquad\square$

We show now that Assumptions 2.2.1 and 2.2.2 hold true in both lattice and non-lattice cases considered above. First observe that

$$S_N(x) := S_N F(x) = \sum_{n=1}^{N} F(T^n x, T^{2n} x, ..., T^{\ell n} x) = S_N G(x)$$

$$+ H(x, ..., x) - H \circ \bar{T}^N(x, ..., x),$$

where we abbreviate $G = G^-$ and $H = H^-$. For any $r \geq 0$, let H_r be a function which depends only on the coordinates with indexes from $-\infty$ to r such that

$$\|H_r\|_{\kappa,1} \leq \|H\|_{\kappa,1} \text{ and } \|H - H_r\|_{\infty} \leq c2^{-\kappa r}$$

where κ is the Hölder exponent of H and $c > 0$ is a constant which depends only on H. For instance, we can consider the function

$$H_r(\bar{x}) = \left(\mu^\ell(E_r(\bar{x}))\right)^{-1} \int_{E_r(\bar{x})} H(\bar{y}) d\mu^\ell(\bar{y})$$

where with $\bar{x} = (x^{(1)}, ..., x^{(\ell)})$ and $\bar{y} = (y^{(1)}, ..., y^{(\ell)})$,

$$E_r(\bar{x}) = \bigcap_{j=1}^{\ell} \{\bar{y} : y_i^{(j)} = x_i^{(j)} \text{ for all } |i| \leq r\} = \{\bar{y} : \rho_{\ell,\infty}(\bar{x}, \bar{y}) \leq 2^{-r}\}$$

which, in fact, depends only on the coordinates with indexes from $-r$ to r. It follows that

$$\left|S_N(x) - S_N G(x) - \left(H_r \circ \bar{T}^{-1}(x, x, ..., x) - H_r \circ \bar{T}^{N-1}(x, ..., x)\right)\right| \leq 2c2^{-\kappa r}$$
$$\tag{2.11.4}$$

for any $\bar{x}$. Set $u_r = H_r \circ (T^{-r} \times \cdots \times T^{-r})$. Then u_r depends only on the coordinates in places from $-\infty$ to 0 and for any $\bar{x}$,

$$\begin{aligned}
\big|S_N(x) - S_N G(x) - \big(u_r(T^r x, T^r x, ..., T^r x) \quad\quad (2.11.5)\\
-u_r(T^{N+r} x, T^{2N+r} x, ..., T^{\ell N+r} x)\big)\big| \leq 2c2^{-\kappa r}.
\end{aligned}$$

Since μ is T^{-1} invariant we can replace x with $T^{-\ell N - r}x$ without changing the distributions. Recall that ζ^- is the Markov chain whose transition probabilities are defined by the transfer operator generated by the map T_-, and the Hölder continuous function $f^- : \Sigma_A^- \to \mathbb{R}$. This together with (2.5.11) and the discussion in Section 2.11.2 yields that

$$\begin{aligned}
S_N G(x) &+ u_r(T^r x, T^r x, ..., T^r x) \\
&- u_r(T^{N+r} x, T^{2N+r} x, ..., T^{\ell N+r} x) \\
\overset{d}{=} \sum_{n=1}^{N} F^- &(T_-^{N\ell + r - n} x^-, T_-^{N\ell + r - 2n} x^-, ..., T_-^{N\ell + r - n\ell} x^-) \\
+ u_r \circ p_- (T_-^{N\ell} x^-, ..., &T_-^{N\ell} x^-) - u_r \circ p_- (T_-^{(\ell-1)N} x^-, T_-^{(\ell-2)N} x^-, ..., x^-) \\
\overset{d}{=} \sum_{n=1}^{N} F^- (\zeta_n^-, &\zeta_{2n}^-, ..., \zeta_{\ell n}^-) + u_r \circ p_- (\zeta_r^-, ..., \zeta_r^-) - u_r \circ p_- (\zeta_{N+r}^-, ... \zeta_{\ell N + r}^-)
\end{aligned}$$

when $x \overset{d}{\sim} \mu$ and $x^- \overset{d}{\sim} \mu_-$.

Next, let $r < \frac{N}{4}$. Repeating the arguments from Section 2.5 we obtain the upper bound

$$\begin{aligned}
|\mathbb{E} e^{it S_N}| &\leq \mathbb{E} |\mathbb{E} [e^{it \sum_{k=M+1}^{N} F^-(\zeta_k^-, \zeta_{2k}^-, ..., \zeta_{\ell k}^-)} \quad\quad (2.11.6)\\
&\times u_r \circ p_- (\zeta_{N+r}^-, \zeta_{2N+r}^-, ..., \zeta_{\ell N + r}^-) | \zeta_1^-, ..., \zeta_{\ell M}^-] |.
\end{aligned}$$

Now, using the Markov property and the identity

$$\mathbb{E}[h_1(\zeta_k^-, ..., \zeta_{k+l}^-) h_2(\zeta_{k+l+s}^-) | \zeta_{k-1}^-] = \mathbb{E}[h_1(\zeta_k^-, ..., \zeta_{k+l}^-) \mathcal{L}_{f^-}^s h_2(\zeta_{k+l}^-) | \zeta_{k-1}^-]$$

which holds true for any $k \in \mathbb{Z}$, $l, s > 0$ and bounded measurable functions h_1 and h_2, we derive that the expression inside the absolute value from the right-hand side of (2.11.6) equals

$$\left(\prod_{k=M+1}^{N} \mathcal{L}_{f^- + it F^-(\zeta_k^-, \zeta_{2k}^-, ..., \zeta_{(\ell-1)k}^-, \cdot)}^{\ell} \right) \mathbf{v}_{r,N}$$

where the random function $\mathbf{v}_{r,N}$ is given by

$$\mathbf{v}_{r,N}(x) = \mathcal{L}_{f^-}^{\ell r} e^{it u_r \circ p_-(\zeta_{N+r}^-, ..., \zeta_{(\ell-1)N+r}^-, \cdot)}(x).$$

Next, the map T^{-r} can, at most, expand distance by a factor of 2^r. Therefore, $u_r \circ p_-$ is Hölder continuous with exponent κ and a constant $v_{\kappa,1}(u_r)$ which does not exceed $2^{\kappa r} v_{\kappa,1}(H_r)$. Since Σ_A^- is compact the norms $\|\cdot\|_{\kappa,\xi}$ and $\|\cdot\|_{\kappa,1}$ are equivalent and since $\|\mathcal{L}_{f-}^s\|_{\kappa,\xi}$ is bounded in s then $\|\mathcal{L}_{f-}^s\|_{\kappa,1}$ is bounded in s, as well. Let $r = r_N = C \ln N$ for an appropriate constant C, so that $2^{-\kappa r_N} \sqrt{N}$ converges to 0 as $N \to \infty$. Then 2^{r_N} grows polynomially fast in N, and so the norms $\|u_{r_N}\|_{\alpha,1}$ and the random norms $\|\mathbf{v}_{r_N,N}(\cdot)\|_{\kappa,1}$ grow polynomially fast in N, as well. Moreover, for this choice of r_N the estimate (2.11.5), which comes from replacing H with H_{r_N}, does not affect the validity of Assumptions 2.2.1 and 2.2.2. Furthermore, when x is fixed then the function $\bar{y} \to \mathcal{L}_{f-}^{\ell r} e^{itu_{r_N}(\bar{y},\cdot)}(x)$ is Hölder continuous with exponent κ and a constant which does not exceed $v_{\kappa,1}(u_{r_N})$, where we used that $\mathcal{L}_{f-} 1 = 1$. Relying on this together with the above estimates of $\|\mathbf{v}_{r_N,N}(\cdot)\|_{\kappa,1}$, the proof of a corresponding version of (2.7.8) proceeds similarly since in the situation of a topologically mixing subshift of finite type the approximation and mixing coefficients defined in Section 2.5.5 decay exponentially fast to 0 (see [10]). Using the above estimates, similar arguments to the ones in Sections 2.8 and 2.9, but with the random function $\mathbf{F}_\omega^+ = F^+(p_0(\omega), \cdot)$ in place of $\mathbf{F}$, show that Assumptions 2.2.1 and 2.2.2 hold true in our situation, and a nonconventional LLT theorem for the sums $S_N = S_N F$ follows.

Finally, employing symbolic representations via Markov partitions (see Sections 3 and 4 from [10]) we derive this limit theorem for the sums $\sum_{n=1}^N F(T^n x, T^{2n} x, ..., T^{\ell n} x)$ in the Axiom A case. $\square$

Chapter 3

Limit theorems for nonconventional arrays

We extend the nonconventional law of large numbers, the central limit theorem and Poisson limit theorems from [42], [45] and [43] to sums of the form $\sum_{n=1}^{N} F\left(\xi_{p_1 n + q_1 N}, \xi_{p_2 n + q_2 N}, \ldots, \xi_{p_\ell n + q_\ell N}\right)$ where $\{\xi_n, \, n \geq 0\}$ is a sufficiently fast mixing stochastic process with some moment conditions and stationarity properties while F is a continuous function with polynomial growth and certain regularity properties. We discuss also the crucial question on positivity of the variance in this central limit theorem.

3.1 Introduction

In this chapter we study limit theorems for nonconventional sums of the form

$$S_N = \sum_{N \geq n \geq 1} F(\xi_{q_1(n,N)}, \ldots, \xi_{q_\ell(n,N)}) \tag{3.1.1}$$

where ξ_m, $m = 0, 1, \ldots$ is a sequence of random variables with sufficiently weak dependence, $q_i(n, N)$, $i = 1, \ldots, \ell$ are functions taking on integer values on integers and F is a Borel measurable function which depending on conditions imposed on ξ_m's will also be assumed to be sufficiently regular, for instance, Hölder continuous. Since the summands in (3.1.1) depend on the number of summands N it is natural according to the probabilistic terminology to call such expressions by the name (triangular) arrays. To avoid extensive technicalities we will restrict ourselves to the case of linear functions $q_i(n, N) = p_i n + q_i N$, $i = 1, \ldots, \ell$ which includes, in particular, the symmetric case $\ell = 2k$, $q_i(n, N) = (k - i + 1)(N - n)$, $i = 1, \ldots, k$ and $q_i(n, N) = in$, $i = k + 1, \ldots, 2k$.

Thus, we will be dealing here with expressions of the form

$$S_N = \sum_{N \geq n \geq 1} F(\xi_{p_1 n + q_1 N}, \xi_{p_2 n + q_2 N}, \ldots, \xi_{p_\ell n + q_\ell N}) \tag{3.1.2}$$

137

where p_i, q_i, $i = 1, ..., \ell$ are integers. In Section 3.2 we will show that under appropriate weak dependence conditions on the sequence ξ_m, $m = 0, 1, ...$ the strong law of large numbers for normalized sums (3.1.2) holds true, i.e. with probability one

$$\lim_{N \to \infty} \frac{1}{N} S_N = \int F(x_1, x_2, ..., x_\ell) d\mu(x_1) d\mu(x_2) \cdots d\mu(x_\ell) \qquad (3.1.3)$$

where μ is the distribution of $\xi(m)$ which is supposed to be the same for all m. Unlike in the nonconventional sums case considered in [42] and [45], where $q_j = 0$ for all j, in the nonconventional arrays (3.1.2) some $p_i n + q_i N$ and $p_j n + q_j N$, $i \neq j$ can be close to each other (and even coincide) for large n and N, and so the corresponding entries $\xi_{p_i n + q_i N}$ and $\xi_{p_j n + q_j N}$ can be strongly dependent which complicates the use of the weak dependence condition imposed on the sequence ξ_m, $m \geq 0$. When the stochastic sequence ξ_m, $m = 0, 1, 2, ...$ is generated by a dynamical system, i.e. $\xi_m = T^m g$, for some measure preserving transformation T and a function g, then for $F(x_1, ..., x_\ell) = f_1(x_1) \cdots f_\ell(x_\ell)$ the L^2 convergence in (3.1.3) can be proved under weak mixing condition on T (see [48]) but the almost sure convergence obtained here is new for such limits.

In Section 3.3 we will study the central limit theorem type results for expressions of the form (3.1.2). It is clear that if $N^{-1/2}(S_N - \mathbb{E}S_N)$ converges in distribution as $N \to \infty$ to the normal distribution then the normalized variance $N^{-1} \text{Var} S_N$ must converge as $N \to \infty$. It turns out that the latter convergence cannot be ensured for general coefficients p_i, q_i, $i = 1, ..., \ell$, as above, but we will prove this here under certain assumption (see Assumption 3.3.1) which holds true, for instance, in the symmetric case

$$S_N = \sum_{N \geq n \geq 1} F(\xi_{k(N-n)}, \xi_{(k-1)(N-n)}, ..., \xi_{N-n}, \xi_n, \xi_{2n}, ..., \xi_{kn}). \qquad (3.1.4)$$

In fact, the martingale approximation technique will work here for any coefficients p_i, q_i assuming appropriate weak dependence in the stochastic sequence ξ_m, $m = 0, 1, ...$, and so the central limit theorem can be proved here similarly to [45] whenever convergence of variances holds true. We will give also conditions which ensure positivity of the limiting variance since without knowing this central limit theorem is not very meaningful. The conditions imposed in Sections 3.2 and 3.3 are valid for sufficiently fast mixing stochastic processes, in particular, Markov chains satisfying some form of the Doeblin condition while on the dynamical systems side our results can be applied to topologically mixing subshifts of finite type taken with a Gibbs measure constructed by a Hölder continuous function, and so we can deal also with expanding transformations and Axiom A diffeomorphisms.

In Section 3.4 we will study Poisson type limit theorems for nonconventional arrays. The classical Poisson limit theorem can be expressed in the following way. Let $\{\xi_n, n \geq 0\}$ be independent identically distributed (i.i.d.) random variables having a distribution μ and Γ_N, $N \geq 1$ be a sequence of measurable sets of real numbers such that $\lim_{N \to \infty} N\mu(\Gamma_N) = \lambda > 0$. Then the sum $S_N = \sum_{n=1}^N \mathbb{I}_{\Gamma_N}(\xi_n)$, where $\mathbb{I}_\Gamma$ is the indicator of a set Γ, converges in distribution to a Poisson random variable with the parameter λ. An extension of such type of limit theorems to the nonconventional setup was obtained in [43] and [47]. We will extend here some of the results from [43] and [47] to the arrays case. Namely, we will consider nonconventional arrays of the form

$$S_N = \sum_{n=1}^N \prod_{j=1}^\ell \mathbb{I}_{\Gamma_N}(\xi_{p_i n + q_i N}) \qquad (3.1.5)$$

where ξ_m, $m = 0, 1, \ldots$ is a sequence of random variables, Γ_N is a sequence of shrinking with N sets and $\mathbb{I}_\Gamma$ is the indicator of a set Γ. We will show that if ξ_m, $m = 0, 1, \ldots$ is a ψ-mixing stationary sequence then under certain conditions on p_i's the sums S_N weakly converge to a Poisson random variable with a parameter λ provided

$$\lim_{N \to \infty} N(\mu(\Gamma_N))^\ell = \lambda.$$

We will also obtain more involved results which concern arrivals to shrinking cylinder sets by subshifts. Namely, we will obtain Poisson type limit theorems for expressions of the form

$$S_{N_m} = \sum_{n=1}^{N_m} \prod_{j=1}^\ell \mathbb{I}_{B_m} \circ T^{p_i n + q_i N} \qquad (3.1.6)$$

where B_m is a cylinder set of length m built by a fixed nonperiodic sequence from a space of sequences Ω, T is the left shift, P is a T-invariant ψ-mixing measure and the sequence $\{N_m : m \geq 1\}$ satisfies

$$\lim_{m \to \infty} N_m(P(B_m))^\ell = \lambda. \qquad (3.1.7)$$

3.2 Strong law of large numbers for nonconventional arrays

3.2.1 *Setup and the main result*

Our setup consists of a $\wp$-dimensional stochastic process $\{\xi_n, n = 0, 1, \ldots\}$ on a probability space $(\Omega, \mathcal{F}, P)$ and of a family of σ-algebras

$\mathcal{F}_{kl} \subset \mathcal{F}$, $-\infty \le k \le l \le \infty$ such that $\mathcal{F}_{kl} \subset \mathcal{F}_{k'l'}$ if $k' \le k$ and $l' \ge l$. It is often convenient to measure the dependence between two sub-σ-algebras $\mathcal{G}, \mathcal{H} \subset \mathcal{F}$ via the quantities

$$\varpi_{q,p}(\mathcal{G}, \mathcal{H}) = \sup\{\|\mathbb{E}[g|\mathcal{G}] - \mathbb{E}[g]\|_p : g \text{ is } \mathcal{H}\text{-measurable and } \|g\|_q \le 1\}, \tag{3.2.1}$$

where the supremum is taken over real functions and $\|\cdot\|_r$, $r \in [1, \infty)$ is the $L^r(\Omega, \mathcal{F}, P)$-norm. Then more familiar α, ρ, ϕ and ψ-mixing (dependence) coefficients can be expressed via the formulas (see [11], Ch. 4),

$$\alpha(\mathcal{G}, \mathcal{H}) = \tfrac{1}{4}\varpi_{\infty,1}(\mathcal{G}, \mathcal{H}), \ \rho(\mathcal{G}, \mathcal{H}) = \varpi_{2,2}(\mathcal{G}, \mathcal{H})$$
$$\phi(\mathcal{G}, \mathcal{H}) = \tfrac{1}{2}\varpi_{\infty,\infty}(\mathcal{G}, \mathcal{H}) \text{ and } \psi(\mathcal{G}, \mathcal{H}) = \varpi_{1,\infty}(\mathcal{G}, \mathcal{H}).$$

We set also

$$\varpi_{q,p}(n) = \sup_{k \ge 0} \varpi_{q,p}(\mathcal{F}_{-\infty,k}, \mathcal{F}_{k+n,\infty}) \tag{3.2.2}$$

and accordingly

$$\alpha(n) = \frac{1}{4}\varpi_{\infty,1}(n), \ \rho(n) = \varpi_{2,2}(n), \ \phi(n) = \frac{1}{2}\varpi_{\infty,\infty}(n), \ \psi(n) = \varpi_{1,\infty}(n).$$

Our setup also includes conditions on the approximation rate

$$\beta(p,r) = \sup_{r' \ge r} \sup_{k \ge 0} \|\xi_k - \mathbb{E}[\xi_k|\mathcal{F}_{k-r',k+r'}]\|_p. \tag{3.2.3}$$

If $p = 2$ the exterior supremum here is not needed but for general p it is taken to obtain $\beta(p,r)$ nonincreasing in r.

In what follows we can always extend the definitions of $\mathcal{F}_{kl}$ given only for $k, l \ge 0$ to negative k by defining $\mathcal{F}_{kl} = \mathcal{F}_{0l}$ for $k < 0$ and $l \ge 0$. Furthermore, we do not require stationarity of the process $\{\xi_n, n \ge 0\}$ assuming only that the distribution of ξ_n does not depend on n and the joint distribution of $\{\xi_n, \xi_{n'}\}$ depends only on $n - n'$ which we write for further references by

$$\xi_n \overset{d}{\sim} \mu \text{ and } (\xi_n, \xi_{n'}) \overset{d}{\sim} \mu_{n-n'} \text{ for all } n, n' \tag{3.2.4}$$

where $Y \overset{d}{\sim} \mu$ means that Y has μ for its distribution.

Next, let $F = F(x_1, ..., x_\ell)$, $x_j \in \mathbb{R}^\wp$ be a function on $\mathbb{R}^{\wp\ell}$ such that for some $\iota, K > 0, \kappa \in (0, 1]$ and all $x_i, y_i \in \mathbb{R}^\wp, i = 1, ..., \ell$, we have

$$|F(x_1, ..., x_\ell) - F(y_1, ..., y_\ell)| \le K\Big(1 + \sum_{j=1}^{\ell} |x_j|^\iota + \sum_{j=1}^{\ell} |y_j|^\iota\Big) \sum_{j=1}^{\ell} |x_j - y_j|^\kappa \tag{3.2.5}$$

and

$$|F(x_1, ..., x_\ell)| \leq K\left(1 + \sum_{j=1}^{\ell} |x_j|^\iota\right). \qquad (3.2.6)$$

It will be clear from the proof that if ξ_n is $\mathcal{F}_{n-k,n+k}$-measurable for a fixed k and all n then the regularity conditions (3.2.5) and (3.2.6) are not needed and it suffices to assume that F is a Borel measurable and bounded. To simplify formulas we assume a centering condition

$$\bar{F} = \int F(x_1, ..., x_\ell) \, d\mu(x_1) \cdots d\mu(x_\ell) = 0 \qquad (3.2.7)$$

which is not really a restriction since we can always replace F by $F - \bar{F}$.

Our main result relies on

Assumption 3.2.1. With $d = (\ell - 1)\wp$ there exist $\infty > p, q \geq 1$ and $\delta, m > 0$ with $\delta < \kappa - \frac{d}{p}$ satisfying

$$\sum_{n=0}^{\infty} \left(\varpi_{q,p}(n) + (\beta(q, r))^\delta\right) < \infty, \qquad (3.2.8)$$

$$\gamma_m + \gamma_{2q(\iota+2)} < \infty \text{ and } \frac{1}{4} \geq \frac{3}{p} + \frac{\iota+2}{m} + \frac{\delta}{q} \qquad (3.2.9)$$

where $\gamma_\theta^\theta = \|\xi_n\|_\theta^\theta = \mathbb{E}|\xi_n|^\theta = \int |x|^\theta d\mu$.

Our strong law of large numbers for nonconventional arrays is represented by the following result.

Theorem 3.2.2. *Suppose that (3.2.7) and Assumption 3.2.1 hold true, the integers $p_1, p_2, ..., p_\ell$ in the sum S_N defined by (3.1.2) are nonzero, distinct and $p_i n + q_i N \geq 0$ for all $i, n, N \geq 1$. Then with probability one,*

$$\lim_{N \to \infty} \frac{1}{N} S_N = 0. \qquad (3.2.10)$$

3.2.2 *Auxiliary lemmas*

We will rely on the following general result which appears in [45] as Corollary 3.6.

Lemma 3.2.3. *Let $\mathcal{G}$ and $\mathcal{H}_1 \subset \mathcal{H}_2$ be σ-subalgebras on a probability space $(\Omega, \mathcal{F}, P)$, X and Y be d-dimensional random vectors and $f_i = f_i(x, \omega)$, $i = 1, 2$ be collections of random variables that are continuously (or separable)* *dependent on* $x \in \mathbb{R}^d$ *for almost all* ω, *measurable with respect to* $\mathcal{H}_i$, $i = 1, 2$, *respectively, and satisfy*

$$\|f_i(x, \omega) - f_i(y, \omega)\|_q \leq C_1(1 + |x|^\iota + |y|^\iota)|x - y|^\kappa \qquad (3.2.11)$$

$$\text{and } \|f_i(x, \omega)\|_q \leq C_2(1 + |x|^\iota).$$

Set $\tilde{f}_i(x, \omega) = \mathbb{E}[f_i(x, \cdot)|\mathcal{G}](\omega)$ *and* $g_i(x) = \mathbb{E}[f_i(x, \omega)]$.

(i) *Assume that* $q \geq p$, $1 \geq \kappa > \theta > \frac{d}{p}$ *and* $\frac{1}{a} \geq \frac{1}{p} + \frac{\iota+1}{m}$. *Then for* $i = 1, 2$,

$$\|\tilde{f}_i(X(\omega), \omega) - g_i(X)\|_a \leq c\, \varpi_{q,p}(\mathcal{G}, \mathcal{H}_i)(C_1 + C_2)^{\frac{d}{p\theta}} C_2^{1 - \frac{d}{p\theta}}(1 + \|X\|_m^{\iota+1}),$$
$$(3.2.12)$$

where $c = c(\iota, \kappa, \theta, p, q, a, \delta, d) > 0$ *depends only on the parameters in brackets.*

(ii) *Next, assume that* $\frac{1}{a} \geq \frac{1}{p} + \frac{\iota+2}{m} + \frac{\delta}{q}$. *Then for* $i = 1, 2$,

$$\|\mathbb{E}[f_i(X, \cdot)|\mathcal{G}] - g_i(X)\|_a \leq R + 2c(C_1 + C_2)(1 + 2\|X\|_m^{\iota+2})\|X - \mathbb{E}[X|\mathcal{G}]\|_q^\delta$$
$$(3.2.13)$$

where R *denotes the right-hand side of (3.2.12).*

(iii) *Furthermore, let* $x = (v, z)$ *and* $X = (V, Z)$, *where* V *and* Z *are* d_1 *and* $d - d_1$-*dimensional random vectors, respectively, and let* $f_i(x, \omega) = f_i(v, z, \omega)$ *satisfy (3.2.11) in* $x = (v, z)$. *Set* $\tilde{g}_i(v) = \mathbb{E}f_i(v, Z(\omega), \omega)$. *Then for* $i = 1, 2$,

$$\|\mathbb{E}[f_i[V, Z, \cdot]|\mathcal{G}] - \tilde{g}_i(V)\|_a \leq c(1 + \|X\|_m^{\iota'}) \qquad (3.2.14)$$

$$\times \left(\varpi_{q,p}(\mathcal{G}, \mathcal{H}_i)(C_1 + C_2)^{\frac{d_1}{p\theta}} C_2^{1 - \frac{d_1}{p\theta}} + \|V - \mathbb{E}[V|\mathcal{G}]\|_q^\delta + \|Z - \mathbb{E}[Z|\mathcal{H}_i]\|_q^\delta\right).$$

(iv) *Finally, for* $a, p, q, \iota, m, \delta$ *satisfying conditions of (ii),*

$$\|\tilde{f}_1(X(\omega), \omega) - \tilde{f}_2(Y(\omega), \omega) - g_1(X) + g_2(Y)\|_a \qquad (3.2.15)$$

$$\leq c\, \varpi_{q,p}(\mathcal{G}, \mathcal{H}_2)\left(1 + \|X\|_m^{\iota+2} + \|Y\|_m^{\iota+2}\right)\|X - Y\|_q^\delta$$

where $c = c(\iota, \kappa, \theta, p, q, a, \delta, d) > 0$ *depends only on the parameters in brackets.*

We will also need the following lemmas from Section 6 of the unpublished version of [45] (see arXiv:1012.2223v2).

Lemma 3.2.4. *Suppose that* $\{a_n : n \geq 1\}$ *is a sequence of nonnegative numbers such that for some integer* $l \geq 1$ *and any integer* $n \geq 1$,

$$a_{n+1} \leq c \sum_{j=1}^{n} \sum_{r=2}^{2l} C^r a_j^{\frac{2l-r}{2l}}.$$

Then

$$a_n \leq A n^l$$

with $A = \max\{2^l c^l C^{2l}, C^{2l}, a_1\}$.

Proof. We derive the above inequality by induction. It is clearly valid for $n = 1$. Assume it is valid for $j = 1, 2, \ldots, n$. Then

$$a_{n+1} \leq c \sum_{j=1}^{n} \sum_{r=2}^{2l} C^r (A j^l)^{\frac{2l-r}{2l}}$$

$$\leq c \, C^2 \, A^{1-\frac{1}{l}} \sum_{r=2}^{2l} C^{r-2} A^{-\frac{r-2}{2l}} \sum_{j=1}^{n} j^{l-1} \leq A' \frac{(n+1)^l}{l}$$

where

$$A' = c \, C^2 \, A^{1-\frac{1}{l}} \sum_{r=0}^{2l-2} C^r A^{-\frac{r}{2l}}$$

and we need to pick A so that $\frac{A'}{l} \leq A$. In particular, $A = \max\{2^l c^l C^{2l}, C^{2l}, a_1\}$ will do because $C A^{-\frac{1}{2l}} \leq 1$, $2 c C^2 A^{-\frac{1}{l}} \leq 1$ and

$$c \, C^2 \, A^{1-\frac{1}{l}} \sum_{r=0}^{2l-2} C^r A^{-\frac{r}{2l}} \leq c \, C^2 \, A^{1-\frac{1}{l}} (2l-1) \leq c \, C^2 \, A^{1-\frac{1}{l}} 2l \leq l \, A.$$

$\square$

Now let $(\Omega, \mathcal{F}, P)$ be a probability space with a filtration of σ-algebras $\mathcal{G}_j$, $j \geq 1$. Suppose that random variables X_j, $j \geq 1$ are $\mathcal{G}_j$ measurable and for each $1 \leq p < \infty$ set

$$A_p = \sup_i \sum_{j \geq i} \|\mathbb{E}(X_j | \mathcal{G}_i)\|_p \geq \sup_j \|X_j\|_p := \gamma_p. \tag{3.2.16}$$

Lemma 3.2.5. *Let the sequence $\{X_j : i \geq j\}$ of random variables be as above and $S_n = \sum_{j=1}^{n} X_j$. Then for each $l \geq 1$ there is a constant c_l depending only on l such that*

$$\mathbb{E} S_n^{2l} \leq c_l A_{2l}^{2l} n^l.$$

Proof. Without loss of generality we assume that $A_{2l} < \infty$ since otherwise the above inequality is trivial. We begin by expanding $S_{j+1}^{2l} = (S_j + X_{j+1})^{2l}$ by the binomial theorem,

$$S_{j+1}^{2l} = S_j^{2l} + 2l S_j^{2l-1} X_{j+1} + \sum_{r=2}^{2l} \binom{2l}{r} S_j^{2l-r} X_{j+1}^r$$

and expressing

$$S_j^{2l-1} = \sum_{i=1}^{j} (S_i^{2l-1} - S_{i-1}^{2l-1}) = \sum_{1 \leq i \leq j} X_i \sum_{r=0}^{2l-2} S_i^r S_{i-1}^{2l-2-r}$$

where $S_0 = 0$. This enables us to rewrite

$$S_{j+1}^{2l} = S_j^{2l} + 2l \sum_{1 \le i \le j} Z_i X_{j+1} + \sum_{r=2}^{2l} \binom{2l}{r} S_j^{2l-r} X_{j+1}^r$$

where $Z_i = X_i \sum_{r=0}^{2l-2} S_i^r S_{i-1}^{2l-2-r}$. Then,

$$\mathbb{E}S_{n+1}^{2l} = \mathbb{E}X_1^{2l} + 2l \sum_{1 \le i \le j \le n} \mathbb{E}(Z_i X_{j+1}) + \sum_{j=1}^{n} \sum_{r=2}^{2l} \binom{2l}{r} \mathbb{E}(S_j^{2l-r} X_{j+1}^r)$$

$$= 2l \sum_{1 \le i \le n} \mathbb{E}(Z_i W_i) + \sum_{j=1}^{n} \sum_{r=2}^{2l} \binom{2l}{r} \mathbb{E}(S_j^{2l-r} X_{j+1}^r)$$

where $W_i = \sum_{j=i}^{n} \mathbb{E}(X_{j+1} | \mathcal{F}_i)$. We note that $\|X_i\|_{2l} \le \gamma_{2l} \le A_{2l}$ and $\|W_i\|_{2l} \le A_{2l}$. Taking into account that $|S_i^r S_{i-1}^{2l-2-r}| \le \max(S_i^{2l-2}, S_{i-1}^{2l-2})$ together with the Cauchy–Schwarz inequality we obtain

$$\mathbb{E}|Z_i W_i| \le \| \sum_{r=0}^{2l-2} S_i^r S_{i-1}^{2l-2-r} \|_{\frac{l}{l-1}} \|X_i\|_{2l} \|W_i\|_{2l}$$

$$\le (2l-1) A_{2l}^2 \big((\mathbb{E}S_i^{2l})^{\frac{l-1}{l}} + (\mathbb{E}S_{i-1}^{2l})^{\frac{l-1}{l}} \big).$$

Next, for $r \ge 2$,

$$|\mathbb{E}(S_j^{2l-r} X_{j+1}^r)| \le \|S_j\|_{2l}^{2l-r} \|X_{j+1}\|_{2l}^r \le A_{2l}^r \|S_j\|_{2l}^{2l-r}.$$

It follows that

$$\mathbb{E}[S_{n+1}^{2l}] \le (2l)! \sum_{j=1}^{n} \Big(\sum_{r=2}^{2l} A_{2l}^r \|S_j\|_{2l}^{2l-r} + A_{2l}^2 \|S_j\|_{2l}^{2l-2} + A_{2l}^2 \|S_{j-1}\|_{2l}^{2l-2} \Big)$$

$$\le c_l \sum_{j=1}^{n} \sum_{r=2}^{2l} A_{2l}^r \|S_j\|_{2l}^{2l-r}$$

where $c_l = 3(2l)!$. The sequence $a_n = \mathbb{E}[S_n^{2l}]$ satisfies the condition of Lemma 3.2.4 with $c = c_l$, $C = A_{2l}$ and $a_1 \le \gamma_{2l}^{2l}$ and the result follows. $\square$

3.2.3 *Ordering and decompositions*

It will be important for our method to order the numbers $p_i n + q_i N$, $i = 1, ..., \ell$ but, in general, the same ordering is not possible for all n between 1 and N and we will have to consider several stretches of n's with different ordering.

Without loss of generality we assume that $p_1 < p_2 < ... < p_\ell$ and since $p_i n + q_i N \geq 0$ always then $q_i \geq 0$ for $i = 1, ..., \ell$. Let $\mathcal{N}_N$ be the set of $n \in \{1, 2, ..., N\}$ such that all $p_i n + q_i N$, $i = 1, ..., \ell$ are distinct. For each $n \in \mathcal{N}_N$ we define distinct integers $\varepsilon_i(n, N)$, $i = 1, 2, ..., \ell$ such that

$$p_{\varepsilon_i(n,N)} n + q_{\varepsilon_i(n,N)} N < p_{\varepsilon_{i+1}(n,N)} n + q_{\varepsilon_{i+1}(n,N)} N \text{ for all } i = 1, 2, ..., \ell - 1.$$
$$(3.2.17)$$

Let $\mathcal{E}_\ell$ be the set of all permutations of $(1, 2, ..., \ell)$. For each $\varepsilon = (\varepsilon_1, ..., \varepsilon_\ell) \in \mathcal{E}_\ell$ set

$$\mathcal{N}_{\varepsilon,N} = \{n \in \{1, 2, ..., N\} : \varepsilon_j(n, N) = \varepsilon_j \text{ for each } j = 1, ..., \ell\}.$$

Some of the sets $\mathcal{N}_{\varepsilon,N}$ can be empty and for each $n \in \mathcal{N}_{\varepsilon,N}$,

$$p_{\varepsilon_i} n + q_{\varepsilon_i} N < p_{\varepsilon_{i+1}} n + q_{\varepsilon_{i+1}} N, \text{ i.e.}$$

$$n > (q_{\varepsilon_i} - q_{\varepsilon_{i+1}})(p_{\varepsilon_{i+1}} - p_{\varepsilon_i})^{-1} N \text{ if } \varepsilon_{i+1} > \varepsilon_i \text{ and}$$

$$n < (q_{\varepsilon_i} - q_{\varepsilon_{i+1}})(p_{\varepsilon_{i+1}} - p_{\varepsilon_i})^{-1} N \text{ if } \varepsilon_{i+1} < \varepsilon_i.$$

Hence,

$$\mathcal{N}_{\varepsilon,N} = \{n : a_\varepsilon N < n < b_\varepsilon N\}$$

for some (not unique) $a_\varepsilon \geq 0$ and $b_\varepsilon \leq 1$, $\mathcal{N}_{\varepsilon,N}$ are disjoint for different $\varepsilon \in \mathcal{E}_\ell$ and, clearly,

$$\mathcal{N}_N = \cup_{\varepsilon \in \mathcal{E}_\ell} \mathcal{N}_{\varepsilon,N}. \qquad (3.2.18)$$

There is always $\varepsilon = (\varepsilon_1, ..., \varepsilon_\ell)$ with $a_\varepsilon = 0$ and then $\varepsilon_1 = \min\{i : q_i = \min_{1 \leq j \leq \ell} q_j\}$ and $\varepsilon_\ell = \max\{i : q_i = \max_{1 \leq j \leq \ell} q_j\}$. Set $\hat{\mathcal{N}}_N = \{1, 2, ..., N\} \setminus \mathcal{N}_N$. Then

$$\hat{\mathcal{N}}_N \subset \{(q_i - q_j)(p_j - p_i)^{-1} N : \text{ for some } i, j = 1, ..., \ell, i \neq j\},$$

and so the cardinality of $\hat{\mathcal{N}}_N$ does not exceed ℓ^2.

Lemma 3.2.6. *If* $a_\varepsilon N < n < b_\varepsilon N$ *then*

$$(p_{\varepsilon_{i+1}} - p_{\varepsilon_i})n + (q_{\varepsilon_{i+1}} - q_{\varepsilon_i})N \geq \min(n - a_\varepsilon N - 1, b_\varepsilon N - n - 1). \quad (3.2.19)$$

Proof. Let integers $m_1, m_2 \geq 1$ satisfy $a_\varepsilon N < n - m_1$ and $n + m_2 < b_\varepsilon N$. Then

$$p_{\varepsilon_i}(n - m_1) + q_{\varepsilon_i} N < p_{\varepsilon_{i+1}}(n - m_1) + q_{\varepsilon_{i+1}} N$$

and

$$p_{\varepsilon_i}(n + m_2) + q_{\varepsilon_i} N < p_{\varepsilon_{i+1}}(n + m_2) + q_{\varepsilon_{i+1}} N.$$

Hence,

$$(p_{\varepsilon_{i+1}} - p_{\varepsilon_i})n + (q_{\varepsilon_{i+1}} - q_{\varepsilon_i})N > m_1(p_{\varepsilon_{i+1}} - p_{\varepsilon_i}) \geq m_1$$

if $\varepsilon_{i+1} > \varepsilon_i$ and if $\varepsilon_i > \varepsilon_{i+1}$ then

$$(p_{\varepsilon_{i+1}} - p_{\varepsilon_i})n + (q_{\varepsilon_{i+1}} - q_{\varepsilon_i})N > m_2(p_{\varepsilon_i} - p_{\varepsilon_{i+1}}) \geq m_2,$$

and so the assertion follows. $\qquad \square$

Next, for each $\varepsilon = (\varepsilon_1, ..., \varepsilon_\ell) \in \mathcal{E}_\ell$ we define

$$F_{\ell,\varepsilon}(x_{\varepsilon_1}, ..., x_{\varepsilon_\ell}) = F(x_1, ..., x_\ell) - \int F(x_1, ..., x_\ell) d\mu(x_{\varepsilon_\ell})$$

and for all $j = \ell - 1, \ell - 2, ..., 1$,

$$F_{j,\varepsilon}(x_{\varepsilon_1}, ..., x_{\varepsilon_j}) = \int F(x_1, ..., x_\ell) d\mu(x_{\varepsilon_\ell}) d\mu(x_{\varepsilon_{\ell-1}}) \cdots d\mu(x_{\varepsilon_{j+1}})$$
$$- \int F(x_1, ..., x_\ell) d\mu(x_{\varepsilon_\ell}) \cdots d\mu(x_{\varepsilon_j}).$$

Observe that $\mathbb{E}F_{j,\varepsilon}(x_{\varepsilon_1}, ..., x_{\varepsilon_{j-1}}, \xi_n) = 0$. For $j = 1, ..., \ell$ set

$$S_{j,\varepsilon}(N) = \sum_{n \in \mathcal{N}_{\varepsilon,N}} F_{j,\varepsilon}\big(\xi_{p_{\varepsilon_1} n + q_{\varepsilon_1} N}, \xi_{p_{\varepsilon_2} n + q_{\varepsilon_2} N}, ..., \xi_{p_{\varepsilon_j} n + q_{\varepsilon_j} N}\big).$$

Then

$$S_N = \sum_{N \geq n \geq 1} F(\xi_{p_1 n + q_1 N}, ..., \xi_{p_\ell n + q_\ell N}) = \sum_{\varepsilon \in \mathcal{E}_\ell} \sum_{j=1}^{\ell} S_{j,\varepsilon}(N) + \hat{S}_N \quad (3.2.20)$$

where

$$\hat{S}_N = \sum_{n \in \hat{\mathcal{N}}_N} F(\xi_{p_1 n + q_1 N}, ..., \xi_{p_\ell n + q_\ell N}).$$

3.2.4   *Proof of Theorem 3.2.2*

For any $k, n, m, N \in \mathbb{N}$, $r \geq 0$ and $1 \leq j \leq \ell$ set $\xi_{k,r} = \mathbb{E}[\xi_k | \mathcal{F}_{k-r,k+r}]$, $\rho_{\varepsilon_j}(n, N) = p_{\varepsilon_j} n + q_{\varepsilon_j} N$, $Y_{j,\varepsilon,\rho_{\varepsilon_j}}(n, N) = F_{j,\varepsilon}(\xi_{\rho_{\varepsilon_1}(n,N)}, ..., \xi_{\rho_{\varepsilon_j}(n,N)})$ and $Y_{j,\varepsilon,m} = 0$ if $m \neq \rho_{\varepsilon_j}(n, N)$,

$$F_{j,\varepsilon,n,r} = F_{j,\varepsilon,n,r}(x_1, ..., x_{j-1}, \omega) = \mathbb{E}[F_{j,\varepsilon}(x_1, ..., x_{j-1}, \xi_n) | \mathcal{F}_{n-r,n+r}]$$

and $Y_{j,\varepsilon,\rho_{\varepsilon_j}(n,N),r} = F_{j,\varepsilon,\rho_{\varepsilon_j}(n,N),r}\big(\xi_{\rho_{\varepsilon_1}(n,N),r}, ..., \xi_{\rho_{\varepsilon_{j-1}}(n,N),r}, \omega\big)$.

Recall that by the construction,

$$\mathbb{E}F_{j,\varepsilon,n,r}(x_1, ..., x_{j-1}, \omega) = 0. \quad (3.2.21)$$

Hence, for any $r \geq r'$ and $l + r \leq \rho_{\varepsilon_{j-1}}(n, N) + r \leq \rho_{\varepsilon_j}(n, N) - r$ we obtain from Lemma 3.2.3(iv) with $\mathcal{G} = \mathcal{F}_{-\infty, \rho_{\varepsilon_{j-1}}(n,N)-r}$, $\mathcal{H}_1 = \mathcal{F}_{\rho_{\varepsilon_j}(n,N)-r', \rho_{\varepsilon_j}(n,N)+r'}$, $\mathcal{H}_1 = \mathcal{F}_{\rho_{\varepsilon_j}(n,N)-r, \rho_{\varepsilon_j}(n,N)+r}$, $f_1 = F_{j,\varepsilon,\rho_{\varepsilon_j}(n,N),r'}$ and $f_2 = F_{j,\varepsilon,\rho_{\varepsilon_j}(n,N),r}$ together with the contraction property of conditional expectations that

$$\big\|\mathbb{E}\big[Y_{j,\varepsilon,\rho_{\varepsilon_j}(n,N),r'} - Y_{j,\varepsilon,\rho_{\varepsilon_j}(n,N),r} \big| \mathcal{F}_{-\infty,l+r}\big]\big\|_4 \quad (3.2.22)$$
$$\leq \big\|\mathbb{E}\big[Y_{j,\varepsilon,\rho_{\varepsilon_j}(n,N),r'} - Y_{j,\varepsilon,\rho_{\varepsilon_j}(n,N),r} \big| \mathcal{F}_{-\infty,\rho_{\varepsilon_{j-1}}(n,N)+r}\big]\big\|_4$$
$$\leq C_1 \varpi_{q,p}(\rho_{\varepsilon_j}(n, N) - \rho_{\varepsilon_{j-1}}(n, N) - 2r)(\beta(q, r'))^\delta$$

where $C_1 > 0$ depends on parameters in Assumption 3.2.1 but it is independent of n, N, r, r', ε and j. Similarly, if $\rho_{\varepsilon_{j-1}}(n, N) + r \leq l + r \leq \rho_{\varepsilon_j}(n, N) - r$ then Lemma 3.2.3(iv) yields

$$\left\| \mathbb{E}\left[Y_{j,\varepsilon,\rho_{\varepsilon_j}(n,N),r'} - Y_{j,\varepsilon,\rho_{\varepsilon_j}(n,N),r} \big| \mathcal{F}_{-\infty,l+r} \right] \right\|_4 \qquad (3.2.23)$$
$$\leq C_2 \varpi_{q,p}(\rho_{\varepsilon_j}(n, N) - l - 2r)(\beta(q, r'))^\delta$$

where, again, $C_2 > 0$ does not depend on n, N, r, r', ε and j. Thus, if $\rho_{\varepsilon_j}(n, N) \geq l + 2r$ then one of estimates (3.2.22) or (3.2.23) is valid. Observe that there are at most $2r$ numbers n with $l \leq \rho_{\varepsilon_j}(n, N) < l + 2r$ for which we estimate the left-hand sides in (3.2.22) and (3.2.23) just relying on the contraction property of conditional expectations bounding them by

$$\left\| Y_{j,\varepsilon,\rho_{\varepsilon_j}(n,N),r'} - Y_{j,\varepsilon,\rho_{\varepsilon_j}(n,N),r} \right\|_4 \leq C_3(\beta(q, r'))^\delta \qquad (3.2.24)$$

where we used Lemma 3.2.3(iv) with $\mathcal{G} = \mathcal{F}$ (on our probability space $(\Omega, \mathcal{F}, P)$) and $C_3 > 0$ does not depend on n, N, r, r', ε and j.

Next, replacing r by 2^{r+1} and r' by 2^r we obtain from Assumption 3.2.1, Lemma 3.2.6 and (3.2.22)-(3.2.24) that

$$\sup_l \sum_{n \in \mathcal{N}_{\varepsilon,N}, \, \rho_{\varepsilon_j}(n,N) \geq l} \left\| \mathbb{E}\left[Y_{j,\varepsilon,\rho_{\varepsilon_j}(n,N),2^r} \right. \right. \qquad (3.2.25)$$
$$\left. \left. - Y_{j,\varepsilon,\rho_{\varepsilon_j}(n,N),2^{r+1}} \big| \mathcal{F}_{l+2^{r+1}} \right] \right\|_4 \leq C_4 2^r (\beta(q, 2^r))^\delta$$

where $C_4 > 0$ does not depend on ε, j and $r, N \geq 1$. Since $2^r(\beta(q, 2^r))^\delta$ is bounded in view of Assumption 3.2.1 we can apply Lemma 3.2.5 to obtain that

$$\mathbb{E}(S_{j,\varepsilon,2^r}(N) - S_{j,\varepsilon,2^{r+1}}(N))^4 \leq C_5 N^2 2^{4r} (\beta(q, 2^r))^{4\delta} \qquad (3.2.26)$$

where

$$S_{j,\varepsilon,m}(N) = \sum_{n \in \mathcal{N}_{\varepsilon,N}} Y_{j,\varepsilon,\rho_{\varepsilon_j}(n,N),m}$$

and $C_5 > 0$ does not depend on ε, j and $r, N \geq 1$.

Observe that by similar estimates employing Lemma 3.2.3(i) in place of Lemma 3.2.3(iv) together with Lemma 3.2.5 we obtain that

$$\mathbb{E}(S_{j,\varepsilon,1}(N))^4 \leq C_6 N^2 \qquad (3.2.27)$$

where $C_6 > 0$ does not depend on ε, j and $r, N \geq 1$. It follows from Assumption 3.2.1 that

$$\sum_{r \geq 0} 2^r (\beta(q, 2^r))^\delta < \infty,$$

and so we obtain from (3.2.26) and (3.2.27) that

$$\|S_{j,\varepsilon}(N)\|_4 \leq \|S_{j,\varepsilon,1}(N)\| + \sum_{r \geq 0} \|S_{j,\varepsilon,2^r}(N) - S_{j,\varepsilon,2^{r+1}}(N)\|_4 \leq C_7\sqrt{N}$$

$$(3.2.28)$$

where $C_7 > 0$ does not depend on ε, j and $N \geq 1$.

By Chebyshev's inequality

$$P\{\frac{1}{N}|S_{j,\varepsilon}(N)| \geq N^{-1/8}\} \leq \frac{\|S_{j,\varepsilon}(N)\|_4^4}{N^{7/2}} \leq C_7^4 N^{-3/2} \qquad (3.2.29)$$

which together with the Borel–Cantelli lemma yields that with probability one,

$$\lim_{N \to \infty} \frac{1}{N} S_{j,\varepsilon}(N) = 0. \qquad (3.2.30)$$

Observe also that for any $n \leq N$ it follows from Assumption 3.2.1 and (3.2.26) that

$$\|F_{j,\varepsilon}(\xi_{p_{\varepsilon_1}n+q_{\varepsilon_1}N}, ..., \xi_{p_{\varepsilon_j}n+q_{\varepsilon_j}N})\|_4 \qquad (3.2.31)$$
$$\leq C_8\big(1 + \sum_{i=1}^{\ell}(\mathbb{E}|\xi_{p_{\varepsilon_i}n+q_{\varepsilon_i}N}|^{4\iota})^{1/4}\big) < \infty$$

where $C_8 > 0$ does not depend on ε, j and $N \geq 1$. Since the sum $\hat{S}_N$ in (3.2.20) contains no more than ℓ^2 terms we see that

$$\|\hat{S}_N\|_4 \leq C_9 \qquad (3.2.32)$$

for some $C_9 > 0$ independent of $N \geq 1$. Applying again the Chebyshev inequality and the Borel–Cantelli lemma we obtain that with probability one

$$\lim_{N \to \infty} \frac{1}{N} \hat{S}_N = 0$$

which together with (3.2.20) and (3.2.30) completes the proof of Theorem 3.2.2. $\qquad \square$

Remark 3.2.7. It is possible to provide a proof of Theorem 3.2.2 relying on the strong law of large numbers for mixingale arrays from [34] verifying mixingale conditions by means of Lemma 3.2.3 (cf. [42]) though it is not clear whether this approach will yield Theorem 3.2.2 under substantially weaker conditions.

3.3 Central limit theorem

3.3.1 *I.i.d. case*

We start with considering sums S_N in (3.1.2) with a sequence of independent identically distributed (i.i.d) random variables $\{\xi_n,\, n \geq 0\}$. Clearly, if $N^{-1/2}(S_N - \mathbb{E}S_N)$ converges in distribution to a normal random variable then $n^{-1}\mathrm{Var}S_N$ must converge as $N \to \infty$. It turns out that this is not the case, in general.

Consider the sum

$$S_N = \sum_{n=1}^{N} F(\xi_{2n+N}, \xi_{2N-2n}) \tag{3.3.1}$$

where $\xi_n,\, n \geq 0$ are i.i.d. random variables each having a distribution μ and, say, F is a bounded measurable function on $\mathbb{R}^2$. In notations of (3.1.2) this corresponds to $\ell = 2$, $p_1 = -p_2 = 2, q_1 = 1$ and $q_2 = 2$. Observe also that in the notations of Section 3.2.3 we have here $\mathcal{E}_2$ consisting of two points $\varepsilon^{(1)}$ and $\varepsilon^{(2)}$ with $\mathcal{N}_{\varepsilon^{(1)},N} = \{n : 0 < n < N/4\}$ and $\mathcal{N}_{\varepsilon^{(2)},N} = \{n : N/4 < n < N\}$. Assume as before that

$$\bar{F} = \int F(x, y)d\mu(x)d\mu(y) = 0. \tag{3.3.2}$$

Then

$$\mathbb{E}F(\xi_{2n+N}, \xi_{2N-2n}) = 0 \text{ provided } n \neq N/4. \tag{3.3.3}$$

Set

$$\tilde{S}_N = \sum_{1 \leq n \leq N,\, n \neq N/4} F(\xi_{2n+N},\, \xi_{2N-2n})$$

then the limiting behavior as $N \to \infty$ of $N^{-1/2}S_N$ and of $N^{-1/2}\tilde{S}_N$, as well as of their variances $N^{-1}\mathrm{Var}S_N$ and $N^{-1}\mathrm{Var}\tilde{S}_N$, is the same.

Since $\mathbb{E}\tilde{S}_N = 0$ we have

$$\mathrm{Var}\tilde{S}_N = \mathbb{E}\tilde{S}_N^2 = \sum_{1 \leq n,m \leq N; n,m \neq N/4} \mathbb{E}\big(F(\xi_{2n+N}, \xi_{2N-2n}) \tag{3.3.4}$$

$$\times F(\xi_{2m+N}, \xi_{2N-2m})\big) = U_N^{(1)} + U_N^{(2)} + U_N^{(3)} + U_N^{(4)}.$$

Here

$$U_N^{(1)} := \sum_{1 \leq n \leq N, n \neq N/4} \mathbb{E}F^2(\xi_{2n+N}, \xi_{2N-2n}) = D(N) \int F^2(x, y)d\mu(x)d\mu(y)$$

where $D(N) = N - 1$ if N is divisible by 4 and $D(N) = N$ if not. Next,

$$U_N^{(2)} := \sum_{1 \le n,m < N/4; n \ne m} \mathbb{E}\big(F(\xi_{2n+N}, \xi_{2N-2n})F(\xi_{2m+N}, \xi_{2N-2m})\big) \quad (3.3.5)$$

$$= \sum_{1 \le n,m < N/4; n \ne m} \mathbb{E}F(\xi_{2n+N}, \xi_{2N-2n})\mathbb{E}F(\xi_{2m+N}, \xi_{2N-2m}) = 0$$

since when $1 \le n, m < N/4$ then the numbers $2n+N, 2N-2n, 2m+N$ and $2N - 2m$ are all different, and so we have an expectation of a product of independent random variables which is the product of expectations while the last equality in (3.3.5) follows from (3.3.2). The same is true when $N/4 < n, m \le N$, and so

$$U_N^{(3)} := \sum_{N/4 \le n,m \le N; n \ne m} \mathbb{E}\big(F(\xi_{2n+N}, \xi_{2N-2n})F(\xi_{2m+N}, \xi_{2N-2m})\big) = 0.$$

Finally,

$$U_N^{(4)} := 2 \sum_{1 \le n < N/4 < m \le N} \mathbb{E}\big(F(\xi_{2n+N}, \xi_{2N-2n})F(\xi_{2m+N}, \xi_{2N-2m})\big) \quad (3.3.6)$$

$$= 2V_N^{(1)} + 2V_N^{(2)}.$$

Here

$$V_N^{(1)} = \sum_{1 \le n < N/4 < m \le N, n+m \ne N/2} \mathbb{E}\big(F(\xi_{2n+N}, \xi_{2N-2n}) \quad (3.3.7)$$

$$\times F(\xi_{2m+N}, \xi_{2N-2m})\big) = 0$$

since when $0 \le n < N/4 < m \le N$ and $n+m \ne N/2$ then the numbers $2n+N, 2N-2n, 2m+N$ and $2N-2m$ are different and the above expectation of the product of independent random variables is the product of expectations. Now, if N is odd then the sum $V^{(2)}$ is empty and we obtain

$$\lim_{N \to \infty, N\text{odd}} \frac{1}{N}\mathbb{E}\tilde{S}_N^2 = \int F^2(x, y)d\mu(x)d\mu(y). \quad (3.3.8)$$

On the other hand, if N is even then in the sum $V_N^{(2)}$ we have the terms with $1 \le n < N/4 < m \le N$, $n + m = N/2$, and so

$$V_N^{(2)} := \sum_{1 \le n < N/4} \mathbb{E}\big(F(\xi_{2n+N}, \xi_{2N-2n})F(\xi_{2m+N}, \xi_{2N-2m})\big)$$

$$= ([N/4] - \delta_N) \int F(x, y)F(y, x)d\mu(x)d\mu(y)$$

where $\delta_N = 1$ if N is divisible by 4 and $\delta_N = 0$, for otherwise. Hence,

$$\lim_{N \to \infty, N\text{even}} \frac{1}{N}\mathbb{E}\tilde{S}_N^2 = \int (F^2(x, y) + F(x, y)F(y, x))d\mu(x)d\mu(y). \quad (3.3.9)$$

In general, $\int F(x,y)F(y,x)d\mu(x)d\mu(y) \neq 0$, for instance, if $F(x,y) = F(y,x)$ is symmetric and F is not μ-trivial, i.e. $\int F^2(x,y)d\mu(x)d\mu(y) \neq 0$. In the latter case we obtain that

$$\lim_{N\to\infty,\, N_{\text{odd}}} \frac{1}{N}\mathbb{E}\tilde{S}_N^2 \neq \lim_{N\to\infty,\, N_{\text{even}}} \frac{1}{N}\mathbb{E}\tilde{S}_N^2, \tag{3.3.10}$$

i.e. there is no convergence of normalized variances of $\tilde{S}_N$, and so the central limit theorem does not hold true in this case. Of course, we can provide a large class of similar counterexamples and the main reason for their existence are arithmetic divisibility considerations concerning N and the coefficients $p_j, q_j, j = 1, ..., \ell$ in the sum (3.1.2).

Nevertheless, when $\xi_n, n \geq 0$ are i.i.d. and (3.2.7) holds true then the sum S_N can be represented as a martingale, and so when convergence of variances can be ensured then usually the central limit theorem for martingales is applicable and we obtain the same for $N^{-1/2}S_N$, as well. Using $\varepsilon_j(n, N)$ defined at the beginning of Section 3.2.3 we set $\varepsilon(n, N) = (\varepsilon_1(n, N), ..., \varepsilon_\ell(n, N))$ and

$$M_{n,N}^{(j)} = \sum_{m\in\mathcal{M}_{j,N}(n)} F_{j,\varepsilon(m,N)}\left(\xi_{p_{\varepsilon_1(m,N)}m+q_{\varepsilon_1(m,N)}N}, ..., \xi_{p_{\varepsilon_j(m,N)}m+q_{\varepsilon_j(m,N)}N}\right)$$

$$\tag{3.3.11}$$

where

$$\mathcal{M}_{j,N}(n) = \{m \in \mathbb{N} : m \leq N, p_{\varepsilon_j(m,N)}m + q_{\varepsilon_j(m,N)}N \leq n\}.$$

Then, when j and N fixed, $M_{n,N}^{(j)}, n = 1, 2, ..., N$ is a martingale with respect to the filtration $\{\mathcal{F}_{0,n}, n \geq 0\}$ where $\mathcal{F}_{m,n} = \sigma\{\xi_m, \xi_{m+1}, ..., \xi_n\}$ for any $m \leq n$. Their sum $M_{n,N} = \sum_{j=1}^{\ell} M_{n,N}^{(j)}, n = 1, 2, ..., N$ is also a martingale or, more precisely, a martingale array since the construction depends on N. If we set

$$K_N = \max_{1\leq m\leq N,\, 1\leq j\leq \ell}(p_{\varepsilon_j(m,N)}m + q_{\varepsilon_j(m,N)}N)$$

then $S_N = M_{K_N,N} + \hat{S}_N$ where the sum $\hat{S}_N$ is the same as in (3.2.20) and it contains no more than ℓ^2 terms. Then relying on the central limit theorem for martingale arrays (see, for instance, Section VIII.I in [55] and §3c of Ch. VIII in [35]) we will obtain that $N^{-1/2}S_N$ converges in distribution to a normal random variable provided convergence of variances hold true since the other, Lindeberg type, condition is always satisfied in our case in view of Assumption 3.2.1.

3.3.2   *Convergence of covariances*

We saw above that, in general, $N^{-1}\mathrm{Var}S_N$ does not converge as $N \to \infty$ for nonconventional arrays S_N. Still, we will show here that the convergence takes place under the following additional assumption.

Assumption 3.3.1. For any $i, j = 1, ..., \ell$ the difference $q_i - q_j$ is divisible by the greatest common divisor of p_i and p_j where p_i's and q_i's are the same as in (3.1.2).

Observe that this assumption is satisfied in the symmetric case (3.1.4) since then for each i either $q_i = -p_i$ or $q_i = 0$. On the other hand, this assumption does not hold true for the example of Section 3.3.1 since then p_1 and p_2 are both even while $q_1 - q_2$ is odd.

Using notations of Section 3.2.4 we will study the asymptotic behavior as $N \to \infty$ of the covariances

$$D_{i,j,\varepsilon,\bar{\varepsilon}}(N) = \frac{1}{N}\mathbb{E}S_{i,\varepsilon}(N)S_{j,\bar{\varepsilon}}(N) = \frac{1}{N} \sum_{m\in\mathcal{N}_{\varepsilon,N}, n\in\mathcal{N}_{\bar{\varepsilon},N}} b_{i,j,\varepsilon,\bar{\varepsilon}}(N,m,n)$$

where

$$b_{i,j,\varepsilon,\bar{\varepsilon}}(N,m,n) = \mathbb{E}\big(Y_{i,\varepsilon,\rho_{\varepsilon_i}}(m,n)Y_{i,\bar{\varepsilon},\rho_{\bar{\varepsilon}_j}}(m,n)\big).$$

We will show that the limit

$$\lim_{N\to\infty} D_{i,j,\varepsilon,\bar{\varepsilon}}(N) = \lim_{N\to\infty} \frac{1}{N} \sum_{a_\varepsilon N < m < b_\varepsilon N, a_{\bar{\varepsilon}} N < n < b_{\bar{\varepsilon}} N} b_{i,j,\varepsilon,\bar{\varepsilon}}(N,m,n)$$

(3.3.12)

exists and will compute it. Then in view of (3.2.20) and Theorem 3.2.2,

$$\lim_{N\to\infty} \frac{1}{N}\mathrm{Var}S_N = \lim_{N\to\infty} \frac{1}{N}\mathbb{E}S_N^2 = \sum_{i,j,\varepsilon,\bar{\varepsilon}} \lim_{N\to\infty} D_{i,j,\varepsilon,\bar{\varepsilon}}.$$

In fact, we will show that the limit

$$\lim_{N\to\infty} \frac{1}{N} \sum_{\substack{a_\varepsilon N < m < b_\varepsilon N, \, a_{\bar{\varepsilon}} N < n < b_{\bar{\varepsilon}} N \\ \rho_{\varepsilon_i}(m,N) - \rho_{\bar{\varepsilon}_j}(n,N) = u}} b_{i,j,\varepsilon,\bar{\varepsilon}}(N,m,n) \qquad (3.3.13)$$

exists for each integer u (where the sum over the empty set is considered to be zero) and then, in order to obtain the limit in (3.3.12) we will just sum in u.

We will show first that if $N \to \infty$, $m - a_\varepsilon N \to \infty$, $b_\varepsilon N - m \to \infty$, $n - a_{\bar{\varepsilon}} N \to \infty$, $b_{\bar{\varepsilon}} N - n \to \infty$ and $\rho_{\varepsilon_i}(m,N) - \rho_{\bar{\varepsilon}_j}(n,N) = p_{\varepsilon_i} m - p_{\bar{\varepsilon}_j} n +$

$N(q_{\varepsilon_i} - q_{\bar{\varepsilon}_j}) = u$ then $b_{i,j,\varepsilon,\bar{\varepsilon}}(N, m, n)$ converges to a limit. After that we will estimate the number of solutions of the equation

$$p_{\varepsilon_i} m - p_{\bar{\varepsilon}_j} n + N(q_{\varepsilon_i} - q_{\bar{\varepsilon}_j}) = u \qquad (3.3.14)$$

in $m \in \mathcal{N}_{\varepsilon, N}$ and $n \in \mathcal{N}_{\bar{\varepsilon}, N}$, divide this number by N and show that this ratio converges to a limit which will give the limit in (3.3.13).

The proof will be based on simple number theoretic considerations and on the following result from [45] appearing there as Lemma 4.3 which requires few preliminaries. Let $H(x_1, x_2, \ldots, x_d)$ be a function on $(R^\nu)^d$ that is continuous and satisfies the growth condition $|H(x_1, x_2, \ldots, x_d)| \leq 1 + \sum_i \|x_i\|^\iota$ for some $\iota \geq 1$. Suppose that $\{Y(n) : n \geq 1\}$ is a stochastic process with values in R^ν and there exists an integer $m \geq 1$ such that for any $l \leq m$ the distribution of $\{Y(n_1), Y(n_2), \cdots Y(n_l)\}$ depends only on the spacings $\{n_i - n_{i-1}\}$, $i = 2, \ldots, l$ between them. For $l \geq 2$, we denote this distribution by μ_S where S is a set of $l - 1$ positive numbers describing the spacings between the l integers. We assume that all $\{Y(n), n \geq 1\}$ have a common distribution μ and that the integrability condition $\int \|x\|^\iota d\mu < \infty$ holds true. For some $p, q \geq 1$ and a nested family of sub σ-fields $\mathcal{F}_{m,n}$ as in Section 3.2.1 assume the mixing condition

$$\varpi_{q,p}(l) = \sup_{n-m \geq l} \varpi_{q,p}(\mathcal{F}_{-\infty,m}, \mathcal{F}_{n,\infty}) \to 0 \text{ as } l \to \infty,$$

and the localization condition

$$\lim_{r \to \infty} \sup_n \|Y(n) - \mathbb{E}(Y(n)|\mathcal{F}_{n-r,n+r})\|_{L_1(P)} = 0.$$

Let $n_1 < n_2 < \ldots < n_d$ be a sequence of integers that tend to ∞ with some of gaps $\{n_{i+1} - n_i\}$ tending to infinity while others are kept fixed. This splits the set of integers $1, 2, \ldots, d$ into a partition $\mathcal{P}$ consisting of blocks B_j of different sizes. The pairwise distances between integers in each block B_j remain fixed (so it can be viewed as rigid) while the distances between different blocks tend to ∞. We assume that each block B_j consists of at most m integers. Let m_j denotes the number of integers in a block B_j and S_j denotes the set of spacings in B_j, i.e. sequence of $m_j - 1$ positive integers representing pairwise distances between successive integers in S_j. Let the distribution $\mu_\mathcal{P}$ on $(R^l)^d$ be the product measure

$$\mu_\mathcal{P} = \Pi_j \mu_{S_j}$$

over successive blocks.

Lemma 3.3.2. *Assume that $\{n_j\}$ tends to infinity with rigid blocks determined by $\mathcal{P}$. Then*

$$\lim_{n_1,\ldots,n_d\to\infty} \mathbb{E}(H(Y(n_1),\ldots,Y(n_d))) = \int H(x_1,\ldots,x_d)d\mu_{\mathcal{P}}$$

where the limit is taken so that the sets S_j of spacings in each block B_j remain fixed while the gaps between different blocks tend to infinity.

It follows from this result that special care should be taken of distributions within blocks since the total distribution is just the product of them. In our case (as in [45]) all blocks will be of size 1 or 2.

In order to study the limiting behavior of $b_{i,j,\varepsilon,\tilde{\varepsilon}}(N,m,n)$ under the condition (3.3.14) we set $p = p_{\varepsilon_i}$, $\tilde{p} = p_{\tilde{\varepsilon}_j}$, $q = q_{\varepsilon_i}$, $\tilde{q} = q_{\tilde{\varepsilon}_j}$ and denote by v the greatest common divisor of p and $\tilde{p}$. Then $q - \tilde{q}$ is divisible by v in view of Assumption 3.3.1, and so either u in (3.3.14) is also divisible by v or there are no solutions in m and n for the equation (3.3.14). In the former case we can rewrite (3.3.14) in the form

$$\alpha m - \beta n + \gamma N = d \qquad (3.3.15)$$

where α and β are coprime, $p = v\alpha$, $\tilde{p} = v\beta$, $q - \tilde{q} = v\gamma$ and $u = vd$ for some integers γ and d. Since we do not exclude the case $p = \tilde{p}$ then $\alpha = \beta = 1$ is allowed. In this case also $q = \tilde{q}$ and $u = \gamma = d = 0$.

Next, assume that an integer triple r, $\tilde{r}$ and w satisfies Assumption 3.3.1 in the sense that w is divisible by the greatest common divisor of r and $\tilde{r}$ but $(r,\tilde{r},w)$ differs from triples $(k\alpha, k\beta, k\gamma)$ for any integer k. Then we claim that

$$|rm - \tilde{r}n + wN| \to \infty \quad \text{as } m,n,N \to \infty \qquad (3.3.16)$$

unless m,n belong to a set of pairs $\mathcal{A}_N$ whose cardinality is o(N), and so it can be disregarded in computation of the limit (3.3.13).

Indeed, consider first the case when $r = k\alpha$ and $\tilde{r} = k\beta$. Then by the assumption $w = k\tilde{\gamma}$ with $\tilde{\gamma} \neq \gamma$. Hence, by (3.3.15),

$$rm - \tilde{r}n + wN = k(\alpha m - \beta n + \gamma N) + (\tilde{\gamma} - \gamma)N = kd + (\tilde{\gamma} - \gamma)N$$

which tends to ∞ or to $-\infty$ as $N \to \infty$ depending on the sign of $\tilde{\gamma} - \gamma$. Now assume that the pair r, $\tilde{r}$ differs from pairs $k\alpha$, $k\beta$ for any integer k. If for $m_l, n_l, N_l \to \infty$ as $l \to \infty$ we have

$$|rm_l - \tilde{r}n_l + wN_l| \leq C < \infty \text{ and } \alpha m_l - \beta n_l + \gamma N_l = d \qquad (3.3.17)$$

then

$$\alpha \frac{m_l}{N_l} - \beta \frac{n_l}{N_l} \to -\gamma \text{ and } r \frac{m_l}{N_l} - \tilde{r} \frac{n_l}{N_l} \to -w \text{ as } l \to \infty.$$

Observe that the set of ratios above is contained in the interval $[0,1]$. For any subsequence, which we denote by the same letters, such that

$$\frac{m_l}{N_l} \to x \text{ and } \frac{n_l}{N_l} \to y \text{ as } l \to \infty$$

we obtain

$$\alpha x - \beta y = -\gamma \text{ and } rx - \tilde{r}y = -w. \tag{3.3.18}$$

Since the matrix $\left(\begin{smallmatrix} \alpha & \beta \\ r & \tilde{r} \end{smallmatrix}\right)$ has rank 2 the system of equations (3.3.17) has a unique solution (x_0, y_0). This together with (3.3.17) yields that there exists $\delta_N \to 0$ as $N \to \infty$ such that (3.3.16) holds true unless the pair m, n belongs to the set

$$\mathcal{A}_N = \left\{ (m,n) : (x - \delta_N)N \le m < (x + \delta_N)N \text{ and } n = \frac{\alpha}{\beta}m + \frac{\gamma}{\beta}N - \frac{d}{\beta} \right\}$$

whose cardinality does not exceed $2\delta_N N$, and so the pairs m, n belonging to $\mathcal{A}_N$ can be disregarded in computation of the limit (3.3.13).

In view of Lemmas 3.2.6 and 3.3.2 we conclude from the above that the limit

$$\hat{a}_{i,j,\varepsilon,\tilde{\varepsilon}}(u) = \lim_{\substack{N, m-a_\varepsilon N, b_\varepsilon N-m, n-a_{\tilde{\varepsilon}} N, b_{\tilde{\varepsilon}} N - n \to \infty \\ \rho_{\varepsilon_i}(m,N) - \rho_{\tilde{\varepsilon}_j}(n,N) = u}} b_{i,j,\varepsilon,\tilde{\varepsilon}}(N, m, n) \tag{3.3.19}$$

exists and it can be described in the following way. Let $\{(\varepsilon_{\kappa_1}, \tilde{\varepsilon}_{\tilde{\kappa}_1}), (\varepsilon_{\kappa_2}, \tilde{\varepsilon}_{\tilde{\kappa}_2}), ..., (\varepsilon_{\kappa_k}, \tilde{\varepsilon}_{\tilde{\kappa}_k})\}$ be all pairs of distinct indexes such that

$$\frac{p_{\varepsilon_{\kappa_\sigma}}}{\alpha} = \frac{p_{\tilde{\varepsilon}_{\tilde{\kappa}_\sigma}}}{\beta} = \frac{q_{\varepsilon_{\kappa_\sigma}} - q_{\tilde{\varepsilon}_{\tilde{\kappa}_\sigma}}}{\gamma} = \eta_\sigma, \ \sigma = 1, ..., k$$

where η_σ, $\sigma = 1, ..., k$ are distinct integers, $p_{\varepsilon_i} = \upsilon\alpha$, $p_{\tilde{\varepsilon}_j} = \upsilon\beta$, $q_{\varepsilon_i} - q_{\tilde{\varepsilon}_j} = \upsilon\gamma$ and α, β are coprime. Observe that $\kappa_k = i$ and $\tilde{\kappa}_k = j$ and for these indexes the above ratio equals the greatest common divisor υ of p_{ε_i} and $p_{\tilde{\varepsilon}_j}$. Set

$$A_{i,j,\varepsilon,\tilde{\varepsilon}}(x_{\varepsilon_{\kappa_1}}, ..., x_{\varepsilon_{\kappa_k}}, y_{\tilde{\varepsilon}_{\tilde{\kappa}_1}}, ..., y_{\tilde{\varepsilon}_{\tilde{\kappa}_k}}) \tag{3.3.20}$$

$$= \int F_{i,\varepsilon}(x_{\varepsilon_1}, ..., x_{\varepsilon_i}) F_{j,\tilde{\varepsilon}}(y_{\tilde{\varepsilon}_1}, ..., y_{\tilde{\varepsilon}_j})$$

$$\times \prod_{\sigma \notin \{\varepsilon_{\kappa_1}, ..., \varepsilon_{\kappa_k}\}} d\mu(x_\sigma) \prod_{\tilde{\sigma} \notin \{\tilde{\varepsilon}_{\tilde{\kappa}_1}, ..., \tilde{\varepsilon}_{\tilde{\kappa}_k}\}} d\mu(y_{\tilde{\sigma}})$$

and

$$a_{i,j,\varepsilon,\tilde{\varepsilon}}(n_1, n_2, ..., n_k) \tag{3.3.21}$$

$$= \int A_{i,j,\varepsilon,\tilde{\varepsilon}}(x_{\varepsilon_{\kappa_1}}, ..., x_{\varepsilon_{\kappa_k}}, y_{\tilde{\varepsilon}_{\tilde{\kappa}_1}} ..., y_{\tilde{\varepsilon}_{\tilde{\kappa}_k}}) \prod_{\zeta=1}^{k} d\mu_{n_\zeta}(x_{\varepsilon_{\kappa_\zeta}}, y_{\tilde{\varepsilon}_{\tilde{\kappa}_\zeta}})$$

observing that $\int f(x,y)d\mu_0(x,y) = \int f(x,x)d\mu(x)$. Then, relying on Lemmas 3.2.6 and 3.3.2 together with the above arguments similarly to [45] we conclude that the limit (3.3.19) has the form

$$\hat{a}_{i,j,\varepsilon,\tilde{\varepsilon}}(u) = a_{i,j,\varepsilon,\tilde{\varepsilon}}\left(\frac{u}{\upsilon}\eta_1, \frac{u}{\upsilon}\eta_2, ..., \frac{u}{\upsilon}\eta_k\right) \tag{3.3.22}$$

provided u is divisible by υ, otherwise $\hat{a}_{i,j,\varepsilon,\tilde{\varepsilon}}(u) = 0$.

3.3.3   *The number of solutions*

If u is not divisible by the greatest common divisor v of p_{ε_i} and $p_{\bar\varepsilon_j}$, then (3.3.14) does not have solutions in view of Assumption 3.3.1. If u is divisible by v then (3.3.14) is equivalent to (3.3.15). The set of solutions of (3.3.15) which we have to take into account in computing (3.3.13) is obtained by adding a particular solution (m_0, n_0) of (3.3.15) to a general solution of the homogeneous equation

$$\alpha m - \beta n = 0 \qquad (3.3.23)$$

so that $(m_0 + m, n_0 + n)$ satisfy the constraints

$$a_\varepsilon N < m_0 + m < b_\varepsilon N \quad \text{and} \quad a_{\bar\varepsilon} N < n_0 + n < b_{\bar\varepsilon} N. \qquad (3.3.24)$$

The solutions of the homogeneous equation (3.3.23) are pairs (m, n) such that m is divisible by β and n is divisible by α. Thus $m = k\beta$ and $n = k'\alpha$ and then (3.3.23) implies $k = k'$. Thus the solutions of (3.3.23) have the form

$$(m, n) = k(\beta, \alpha) \quad \text{for any integer} \quad k. \qquad (3.3.25)$$

Since α and β are coprime a solution (m_0, n_0) of the inhomogeneous equation (3.3.15) (without the constraints (3.3.24)) always exist. Thus we have to count the number of k's such that $a_\varepsilon N < m_0 + k\beta < b_\varepsilon N$ and $a_{\bar\varepsilon} N < n_0 + k\alpha < b_{\bar\varepsilon} N$. These inequalities mean that

$$\frac{a_\varepsilon}{\beta} N - \frac{m_0}{\beta} < k < \frac{b_\varepsilon}{\beta} N - \frac{m_0}{\beta} \quad \text{and} \quad \frac{a_{\bar\varepsilon}}{\alpha} N - \frac{n_0}{\alpha} < k < \frac{b_{\bar\varepsilon}}{\alpha} N - \frac{n_0}{\alpha}. \qquad (3.3.26)$$

Thus, both inequalities in (3.3.26) can hold true for large N only if $\frac{b_{\bar\varepsilon}}{\alpha} \geq \frac{a_\varepsilon}{\beta}$ and $\frac{b_\varepsilon}{\beta} \geq \frac{a_{\bar\varepsilon}}{\alpha}$. If $\frac{b_{\bar\varepsilon}}{\alpha} = \frac{a_\varepsilon}{\beta}$ or $\frac{b_\varepsilon}{\beta} = \frac{a_{\bar\varepsilon}}{\alpha}$ then there are no more than $\frac{m_0}{\beta} + \frac{n_0}{\alpha} + 1$ solutions for k in (3.3.26), and so dividing by N we will obtain that the limit in (3.3.13) is zero.

Now assume that

$$\frac{b_{\bar\varepsilon}}{\alpha} > \frac{a_\varepsilon}{\beta} \quad \text{and} \quad \frac{b_\varepsilon}{\beta} > \frac{a_{\bar\varepsilon}}{\alpha}. \qquad (3.3.27)$$

Then up to the uniform bound $\frac{m_0}{\beta} + \frac{n_0}{\alpha} + 1$ there are

$$[N \min(|\frac{b_{\bar\varepsilon}}{\alpha} - \frac{a_\varepsilon}{\beta}|, |\frac{b_\varepsilon}{\beta} - \frac{a_{\bar\varepsilon}}{\alpha}|)]$$

of k's satisfying (3.3.26). Thus the limit (3.3.13) will be equal in this case to

$$\hat a_{i,j,\varepsilon,\bar\varepsilon}(u) \min(|\frac{b_{\bar\varepsilon}}{\alpha} - \frac{a_\varepsilon}{\beta}|, |\frac{b_\varepsilon}{\beta} - \frac{a_{\bar\varepsilon}}{\alpha}|)$$

and the limit (3.3.12) will be equal to

$$\min(|\frac{b_{\bar\varepsilon}}{\alpha} - \frac{a_\varepsilon}{\beta}|, |\frac{b_\varepsilon}{\beta} - \frac{a_{\bar\varepsilon}}{\alpha}|) \sum_{u=-\infty}^{\infty} \hat a_{i,j,\varepsilon,\bar\varepsilon}(u).$$

3.3.4 Martingale approximation

Using notations of Section 3.2.4 set

$$R_{j,\varepsilon,m,r} = \sum_{n \geq m-r} \mathbb{E}[Y_{j,\varepsilon,n,r}|\mathcal{F}_{-\infty,m}]$$

and

$$W_{j,\varepsilon,n,r} = Y_{j,\varepsilon,n-r,r} + R_{j,\varepsilon,n,r} - R_{j,\varepsilon,n-1,r} \qquad (3.3.28)$$

which turns out to be a uniformly square integrable martingale difference sequence. Indeed, clearly $W_{j,\varepsilon,n,r}$ is $\mathcal{F}_{-\infty,n}$-measurable, the martingale difference relation with respect to the filtration $\hat{\mathcal{F}}_n = \mathcal{F}_{-\infty,n}$ is obvious and the uniform square integrability follows from the estimates (3.2.22) and (3.2.23).

Set

$$K_{j,\varepsilon,N} = \max_{a_\varepsilon N < n < b_\varepsilon N} (p_{\varepsilon_j} n + q_{\varepsilon_j} N),$$

$$K_{\varepsilon,N} = \max_{1 \leq j \leq \ell} K_{j,\varepsilon,N} \quad \text{and} \quad K_N = \max_{\varepsilon \in \mathcal{E}_\ell} K_{\varepsilon,N}.$$

Then when r and N are fixed

$$M_{N,r}(n) = \sum_{1 \leq m \leq n,\, 1 \leq j \leq \ell,\, \varepsilon \in \mathcal{E}_\ell} W_{j,\varepsilon,m,r}, \quad n \leq K_N, \qquad (3.3.29)$$

is a martingale which serves as a martingale approximation of the sums

$$S_{N,r}(n) = \sum_{1 \leq m \leq n,\, 1 \leq j \leq \ell,\, \varepsilon \in \mathcal{E}_\ell} Y_{j,\varepsilon,m,r}. \qquad (3.3.30)$$

This means that in view of estimates (3.2.22)–(3.2.24) the expressions $N^{-1/2} M_{K_N,r}$ and $N^{-1/2} S_{N,r}(K_N)$ have the same distributional limits. Moreover, from the estimates of Section 3.2.4 it follows in the same way as in Section 3.3.2 that in the symmetric case the normalized variances $N^{-1}\mathbb{E}(S_{N,r}(K_N))^2$ converge as $N \to \infty$ and

$$\lim_{N \to \infty} N^{-1}\mathbb{E}(S_{N,r}(K_n))^2 = \lim_{N \to \infty} N^{-1}\mathbb{E}(M_{K_N,r})^2. \qquad (3.3.31)$$

In fact, in view of estimates in Section 3.2.2 and Assumption 3.2.1 (which yield the Lindeberg type condition) convergence in (3.3.31) is the only condition required to apply the martingal central limit theorem from [55] which gives that $N^{-1/2} S_{N,r}(K_N)$ and $N^{-1/2} M_{K_N,r}$ converge in distribution to the same normal random variable.

Next, we represent the whole sum in the form

$$S_N = \hat{S}_N + S_{N,1}(K_N) + \sum_{r \geq 0}(S_{N,2^r}(K_N) - S_{N,2^{r+1}}(K_N)) \qquad (3.3.32)$$

and set also

$$S_N^{(u)} = \hat{S}_N + S_{N,1}(K_N) + \sum_{0 \le r \le u} (S_{N,2^r}(K_N) - S_{N,2^{r+1}}(K_N)).$$

The random variables $N^{-1/2}S_N^{(u)}$ converge in distribution as $N \to \infty$ for each fixed u to normal random variables. Estimating tails of the series (3.3.32) using (3.2.26) and Assumption 3.2.1 we conclude that $N^{-1/2}S_N$ converges in distribution as $N \to \infty$ to a normal random variable (cf. [45]).

Thus we obtain the following

Theorem 3.3.3. *Suppose that the conditions of Theorem 3.2.2 are satisfied and $\lim_{N\to\infty} N^{-1} VarS_N$ exists which holds true under Assumption 3.3.1 (as established in Section 3.3.2). Then $N^{-1/2}S_N$ converges in distribution as $N \to \infty$ to a normal random variable.*

After proving the central limit theorem it is natural to enquire when the variance of the limiting normal random variable (limiting variance) is nonzero since otherwise the result is not very substantial. Similarly to [30] and relying on estimates of Section 3.2.4 we can prove the following result.

Theorem 3.3.4. *Let $\xi^{(i)} = \{\xi_n^{(i)} : n \ge 0\}$, $i = 1, 2, ..., \ell$ be independent copies of the random sequence $\xi = \{\xi_n, m \ge 0\}$. Suppose that the conditions of Theorem 3.2.2 hold true and $\sigma^2 = \lim_{N\to\infty} N^{-1}\mathbb{E}(S_N)^2$ exists. Set*

$$U_n = F(\xi_{p_1 n + q_1 N}^{(1)}, \xi_{p_2 n + q_2 N}^{(2)}, ..., \xi_{p_\ell n + q_\ell N}^{(\ell)})$$

$\Sigma_N = \sum_{N \ge n \ge 1} U_n$ *and*

$$s^2 = \lim_{N\to\infty} N^{-1}\mathbb{E}(\Sigma_N)^2.$$

Then, the last limit exists and $\sigma^2 > 0$ if and only if $s^2 > 0$. The latter condition holds true if and only if there exists no representation of the form

$$U_n = Z_{n+1} - Z_n, \quad n = 0, 1, 2, ...$$

where $\{Z_n\}_{n=0}^{\infty}$ is a square integrable weakly (i.e. in the wide sense) stationary process. Furthermore, $s^2 = 0$ if and only if $Var\Sigma_N$ is bounded.

Since each $\xi^{(i)}$ satisfies (3.2.4) the above result reduces the positivity problem to the case of special weakly stationary arrays. In fact, since $\xi^{(1)}, \xi^{(2)}, ..., \xi^{(\ell)}$ are independent then in case that ξ is stationary we have

$$\mathbb{E}U_n U_m = \mathbb{E}\big(F(\xi_{p_1(n-m)}^{(1)}, ..., \xi_{p_\ell(n-m)}^{(\ell)})F(\xi_0^{(1)}, ..., \xi_0^{(\ell)})\big),$$

and so

$$Var\Sigma_N = Var\hat{\Sigma}_N$$

where

$$\hat{\Sigma}_N = \sum_{1 \leq n \leq N} \hat{U}_n \text{ and } \hat{U}_n = F(\xi_{p_1 n}^{(1)}, ..., \xi_{p_\ell n}^{(\ell)}).$$

Now we reduced the problem to the case of sums of stationary sequences and not only arrays and for this case the characterization of positivity of $\lim_{N \to \infty} \mathrm{Var} \hat{\Sigma}_N$ is well known (see [33]) which is used in Theorem 3.3.4.

Remark 3.3.5. Considering sums

$$S_N(t) = \sum_{Nt \geq n \geq 0} F(\xi_{p_1 n + q_1 N}, \xi_{p_2 n + q_2 N}, ..., \xi_{p_\ell n + q_\ell N}), \, t \in [0, 1] \quad (3.3.33)$$

we can derive a functional central limit theorem provided convergence of covariances can be ensured which as we saw holds true in the symmetric case. Moreover, similarly to Chapter 1, considering the sets

$$A_{n,l,N} = \{1 \leq m \leq N : \min_{1 \leq i, j \leq \ell} |(p_i n + q_i N) - (p_j N + q_j N)| \leq l\}$$

the (functional) nonconventional CLT for the case of nonconventional arrays follows by Stein's method, assuming that the limit $\sigma^2 = \lim_{N \to \infty} \frac{1}{N} \mathbb{E}(S_N)^2$ (or the corresponding limiting covariances in the functional case) exist, and when σ^2 is positive the rate of convergence towards the weak limit can be estimated.

3.4 Poisson limit theorems for nonconventional arrays

3.4.1 *Preliminaries and main results*

In this section we will consider two related types of Poisson limit theorems for nonconventional arrays. The first type involves a sequence of random variables $\{\xi_n, n \geq 0\}$ having a common distribution μ and a sequence of sets Γ_N, $N \geq 0$ of real numbers such that

$$\lim_{N \to \infty} N(\mu(\Gamma_N))^\ell = \lambda > 0. \quad (3.4.1)$$

In this case our goal will be to show that under suitable conditions the sum

$$S_N = \sum_{n=1}^{N} \prod_{j=1}^{\ell} \mathbb{I}_{\Gamma_N}(\xi_{p_i n + q_i N}) \quad (3.4.2)$$

converges in distribution to a Poisson random variable with a parameter λ.

The second type of theorems concerns counting arrivals of a dynamical system at shrinking sets. More specifically, let $T : \Omega \to \Omega$ be a left shift on the sequence space $\Omega = \mathcal{A}^{\mathbb{N}} = \{\omega = (\omega_i)_{i \geq 0}, \omega_i \in \mathcal{A}\}$ acting by $(T\omega)_i = \omega_{i+1}$ and $B_m^a = \{\omega \in \Omega : \omega_i = a_i, i = 0, 1, ..., m-1\}$ be cylinder sets of length m built for each $a = (a_0, a_1, ...) \in \Omega$. The alphabet $\mathcal{A}$ is supposed to be finite or countable. Our goal will be to study the limiting behavior of distributions of the sums

$$S_{N_m}^a = \sum_{n=1}^{N_m} \prod_{j=1}^{\ell} \mathbb{I}_{B_m^a} \circ T^{p_i n + q_i N} \qquad (3.4.3)$$

assuming that $N_m = N_m^a$ satisfies

$$\lim_{m \to \infty} N_m (P(B_m^a))^{\ell} = \lambda > 0 \qquad (3.4.4)$$

where P is a shift invariant ψ-mixing probability measure.

First, observe that unlike [43] and [47], in general, sums (3.4.2) do not converge weakly to a Poisson random variable without additional assumptions on p_i's. Indeed, consider the special case

$$S_N = \sum_{n=1}^{N} \mathbb{I}_{\Gamma_N}(\xi_n) \mathbb{I}_{\Gamma_N}(\xi_{N-n}) \qquad (3.4.5)$$

$$= 2 \sum_{n=1}^{[N/2]} \mathbb{I}_{\Gamma_N}(\xi_n) \mathbb{I}_{\Gamma_N}(\xi_{N-n}) + \mathbb{I}_{\Gamma_N}(\xi_0) \mathbb{I}_{\Gamma_N}(\xi_N) - a_N$$

where $a_N = \mathbb{I}_{\Gamma_N}(\xi_{[N/2]})$ if N is even and $a_N = 0$ if N is odd. Assuming that ξ_n's have the same distribution μ and $\mu(\Gamma_N) \to 0$ as $N \to \infty$ we see that both $\mathbb{I}_{\Gamma_N}(\xi_0) \mathbb{I}_{\Gamma_N}(\xi_N)$ and a_N tend to zero in probability as $N \to \infty$. Thus, any limit in distribution of S_N as $N \to \infty$ could only be a random variable taking on even values with probability one, and so it cannot have a Poisson distribution. In fact, if ξ_n, $n \geq 0$ is an i.i.d. sequence of random variables and (3.4.1) holds true with $\ell = 2$ then S_N in (3.4.5) converges weakly to $2Z$ where Z has the Poisson distribution with the parameter λ. We will not investigate here possible weak limits of S_N in (3.4.2) for general p_i's but instead impose a condition which will ensure that the limit is a Poisson random variable.

Assumption 3.4.1. For any nontrivial permutation ζ of ℓ numbers $(1, 2, ..., \ell)$ the matrix $\left(\begin{smallmatrix} p_1 & p_2 & \cdots & p_\ell \\ p_{\zeta(1)} & p_{\zeta(2)} & \cdots & p_{\zeta(\ell)} \end{smallmatrix} \right)$ has rank 2.

Observe that this assumption is satisfied, for instance, when $p_i > 0$ for all $i = 1, ..., \ell$. Indeed, assume as before without loss of generality that $0 < p_1 < p_2 < ... < p_\ell$. If the above matrix has rank 1 for some permutation ζ then $p_{\zeta(i)} = a p_i$, $i = 1, ..., \ell$ for some $a > 0$. But then $p_{\zeta(1)} < p_{\zeta(2)} < ... < p_{\zeta(\ell)}$, and so $p_{\zeta(i)} = p_i$, $i = 1, ..., \ell$, i.e. ζ is a trivial permutation.

Next, we consider the case where $\xi = \{\xi_n, n \geq 0\}$ is a stationary sequence of random variables. Define σ-algebras $\mathcal{F}_{mn} =$

$\sigma\{\xi_m, \xi_{m+1}, ..., \xi_n\}$, $m \leq n$, which is the minimal σ-algebra such that $\xi_m, \xi_{m+1}, ..., \xi_n$ are all measurable with respect to it. Then the ψ-dependence (or mixing) coefficient is defined by

$$\psi(n) = \sup_{m \geq 0}\{|\frac{P(A \cap B)}{P(A)P(B)} - 1| : A \in \mathcal{F}_{0,m}, B \in \mathcal{F}_{m+n,\infty}, P(A)P(B) \neq 0\}$$

(3.4.6)

where $\mathcal{F}_{k,\infty}$ is the minimal σ-algebra containing all $\mathcal{F}_{k,n}$, $n \geq k$. The sequence ξ is called ψ-mixing if

$$\psi(1) < \infty \quad \text{and} \quad \lim_{n \to \infty} \psi(n) = 0. \qquad (3.4.7)$$

Theorem 3.4.2. *Let* $\xi = \{\xi_n, n \geq 0\}$ *be a* ψ-*mixing stationary sequence of random variables such that each* ξ_n *has a distribution* μ. *Suppose that (3.4.1) and Assumption 3.4.1 hold true. Then* S_N *defined by (3.4.2) converges in distribution as* $N \to \infty$ *to a Poisson random variable with the parameter* λ.

Next, we consider another setup with the probability space $(\Omega, \mathcal{F}, P)$ where $\Omega = \mathcal{A}^{\mathbb{N}}$ is the sequence space $\Omega = \{\omega = (\omega_0, \omega_1, ...), \omega_i \in \mathcal{A} \, \forall i\}$ built by a finite or countable alphabet $\mathcal{A}$ which is not a singleton, $\mathcal{F}$ is the σ-algebra generated by all cylinder sets and P is the left shift invariant ψ-mixing probability measure. Recall that the left shift T acts by $(T\omega)_i = \omega_{i+1}$, and so $P(T^{-1}\Gamma) = P(\Gamma)$ for any $\Gamma \in \mathcal{F}$. For each $a \in \mathcal{A}$ denote by $[a] = \{\omega = (\omega_0, \omega_1, ...) : \omega_0 = a\} \subset \Omega$ the 1-cylinder set and set $\mathcal{A}_0 = \{a \in \mathcal{A} : P([a]) > 0\}$. Since P is left shift invariant then it is easy to see that $P(\mathcal{A}_0^{\mathbb{N}}) = 1$, and so without loss of generality we will assume that $\mathcal{A} = \mathcal{A}_0$, i.e. that any 1-cylinder has positive probability. Since $\mathcal{A}$ is not a singleton it follows that $\sup_{a \in \mathcal{A}} P([a]) < 1$. Now, the ψ-mixing defined by (3.4.6) and (3.4.7) is considered here with respect to the σ-algebras $\mathcal{F}_{kn}$ generated by all cylinder sets of the form $\{\omega = (\omega_0, \omega_1, ...) \in \Omega : \omega_i = a_i, i = k, k+1, ..., n\}$ where $a_k, a_{k+1}, ..., a_n \in \mathcal{A}$ is a fixed sequence.

For any $a = (a_0, a_1, ...) \in \Omega$ set $B_n^a = \{\omega = (\omega_0, \omega_1, ...) \in \Omega : \omega_i = a_i, i = 0, 1, ..., n-1\}$ which is the cylinder set of length n built by a and denote by Ω_P the set of $\omega \in \Omega$ such that $P(B_n^\omega) > 0$ for all $n \geq 1$. Clearly, $T\Omega_P \subset \Omega_P$ and $P(\Omega_P) = 1$ since Ω_P is the complement in Ω of the union of all cylinder sets B such that $P(B) = 0$ and the number of such cylinders is countable. Moreover, $\Omega_P = \text{supp}P$ if we take the discrete topology on $\mathcal{A}$ and the corresponding product topology on Ω since cylinders form an open base of this topology.

Theorem 3.4.3. *Suppose that Assumption 3.4.1 holds true and P is a ψ-mixing left shift invariant probability measure on the space $(\Omega, \mathcal{F})$ defined above. Then for any nonperiodic $a \in \Omega$ the sum $S^a_{N_m}$ defined by (3.4.3) converges in distribution as $m \to \infty$ to a Poisson random variable with the parameter λ provided (3.4.4) holds true.*

Employing a more advanced technique from [47] which is based on a version of the Stein-Chen method it is possible to obtain here additional results, in particular, to estimate speed of convergence in Theorem 3.4.3 with respect to the total variation distance between distributions.

3.4.2 Stationary sequences

In order to prove Theorem 3.4.3 we will rely on the following result from [59].

Lemma 3.4.4. *Let $\eta_1^{(N)}, ..., \eta_N^{(N)}$, $N = 1, 2, ...$ be an array of 0–1 random variables, $\mathcal{J}_r(N)$, $r \le N$ be the family of all r-tuples $(i_1, i_2, ..., i_r)$ of mutually distinct indices between 1 and N and for any $(i_1, ..., i_r) \in \mathcal{J}_r(N)$ set*

$$b_{i_1,...,i_r}^{(N)} = P\{\eta_{i_1}^{(N)} = ... = \eta_{i_r}^{(N)} = 1\}.$$

Assume that

$$\lim_{N\to\infty} \max_{1\le i\le N} b_i^{(N)} = 0, \quad \lim_{N\to\infty} \sum_{i=1}^{N} b_i^{(N)} = \lambda > 0, \qquad (3.4.8)$$

for $N = 1, 2, ...$ there exist "rare" sets $I_r(N) \subset \mathcal{J}_r(N)$ such that

$$\lim_{N\to\infty} \sum_{(i_1,...,i_r)\in I_r(N)} b_{i_1...i_r}^{(N)} = \lim_{N\to\infty} \sum_{(i_1,...,i_r)\in I_r(N)} b_{i_1}^{(N)} \cdots b_{i_r}^{(N)} = 0 \quad (3.4.9)$$

and uniformly in $(i_1, ..., i_r) \in \mathcal{J}_r(N) \setminus I_r(N)$,

$$\lim_{N\to\infty} \frac{b_{i_1...i_r}^{(N)}}{b_{i_1}^{(N)} \cdots b_{i_r}^{(N)}} = 1. \qquad (3.4.10)$$

Then for $S_N = \sum_{i=1}^{N} \eta_i^{(N)}$,

$$\lim_{N\to\infty} P\{S_N = k\} = \frac{\lambda^k e^{-\lambda}}{k!}, \quad k = 0, 1, 2, ... \qquad (3.4.11)$$

We will use Lemma 3.4.4 defining

$$\eta_n^{(N)} = \prod_{j=1}^{\ell} \mathbb{I}_{\Gamma_N}(\xi_{p_j n + q_j N}), \quad n = 1, 2, ..., N.$$

By (3.4.1),

$$b_n^{(N)} = P\{\eta_n^{(N)} = 1\} \le P\{\xi_0 \in \Gamma_N\} = \mu(\Gamma_N) \to 0 \text{ as } N \to \infty. \quad (3.4.12)$$

In order to complete proving (3.4.8) we observe first that if $A_i \in \mathcal{F}_{m_i,n_i}, i = 1, 2, ..., l$ and $0 \le m_1 \le n_1 < m_2 \le n_2 < ... < m_\ell \le n_\ell$ then it follows by induction from (3.4.6) that

$$\mathbb{E}(\prod_{i=1}^l \mathbb{I}_{A_i}) \le (1 + \psi(m_l - n_{l-1}))\mathbb{E}\mathbb{I}_{A_l}\mathbb{E}(\prod_{i=1}^{l-1} \mathbb{I}_{A_i}) \quad (3.4.13)$$
$$\le \prod_{i=2}^l (1 + \psi(m_i - n_{i-1})) \prod_{i=1}^l \mathbb{E}\mathbb{I}_{A_i}$$

and setting $\prod_{i=1}^0 = 1$ we can also write

$$|\mathbb{E} \prod_{i=1}^l \mathbb{I}_{A_i} - \prod_{i=1}^l \mathbb{E}\mathbb{I}_{A_i}| \le \sum_{k=1}^{l-1} |(\prod_{i=1}^{k-1} \mathbb{E}\mathbb{I}_{A_i})\mathbb{E}(\prod_{i=k}^l \mathbb{I}_{A_i}) \quad (3.4.14)$$
$$-(\prod_{i=1}^k \mathbb{E}\mathbb{I}_{A_i})\mathbb{E}(\prod_{i=k+1}^l \mathbb{I}_{A_i})| = \sum_{k=1}^{l-1} (\prod_{i=1}^{k-1} \mathbb{E}\mathbb{I}_{A_i})|\mathbb{E}(\prod_{i=k}^l \mathbb{I}_{A_i})$$
$$-\mathbb{E}\mathbb{I}_{A_k}\mathbb{E}(\prod_{i=k+1}^l \mathbb{I}_{A_i})| \le \sum_{k=1}^{l-1} (\prod_{i=1}^k \mathbb{E}\mathbb{I}_{A_i})\psi(m_{k+1} - n_k)\mathbb{E}(\prod_{i=k+1}^l \mathbb{I}_{A_i})$$
$$\le \sum_{k=1}^{l-1} \psi(m_{k+1} - n_k) \prod_{i=k+2}^l (1 + \psi(m_i - n_{i+1})) \prod_{i=1}^l \mathbb{E}\mathbb{I}_{A_i}.$$

Using again the notation $\rho_{\varepsilon_j}(n, N) = p_{\varepsilon_j} n + q_{\varepsilon_j} N, \varepsilon \in \mathcal{E}_\ell$ (see Section 3.2.3) we obtain from (3.4.14) by Lemma 3.2.6 that whenever $a_\varepsilon N < n < b_\varepsilon N$,

$$|b_n^{(N)} - (\mu(\Gamma_N))^\ell| \le (\mu(\Gamma_N))^\ell (1 + \psi(0))^\ell \sum_{k=1}^{\ell-1} \psi(\min(n - a_\varepsilon N - 1, b_\varepsilon N - n - 1)). \quad (3.4.15)$$

Now, it follows easily from (3.4.1), (3.4.7) and (3.4.15) that

$$\lim_{N \to \infty} \sum_{n=1}^N b_n^{(N)} = \lambda \quad (3.4.16)$$

completing the proof of (3.4.8).

Next, we deal with (3.4.9). For any two positive integers $l, \tilde{l}$ set

$$d(l, \tilde{l}) = \min_{1 \le i,j \le \ell} |p_i l - p_j \tilde{l} + (q_i - q_j)N|.$$

Denote also $a(N) = |\ln \mu(\Gamma_N)|$. We recall that there exists no more than ℓ^2 of n's between 1 and N which do not belong to $\mathcal{N}_N = \cup_{\varepsilon \in \mathcal{E}_\ell} \mathcal{N}_{\varepsilon,N}$ (see Section 3.2.3), and so such n's can be ignored in the study of weak convergence in Theorem 3.4.2 since every summand converge to zero in probability.

A sequence $J = \{j_1, j_2, ..., j_l\}$ of distinct positive integers from $\mathcal{N}_N$ will be called an N-cluster here if for any $j, \tilde{j} \in J$ there exists a chain $j = j_{i_1}, j_{i_2}, ..., j_{i_m-1}, j_{i_m} = \tilde{j}$ of integers from J such that

$$d(j_{i_k}, j_{i_{k+1}}) \le a(N) \quad \forall k = 1, 2, ..., m - 1. \quad (3.4.17)$$

We shall say that J is a maximal N-cluster in another finite sequence $\tilde{J}$ of distinct positive integers if $J \cup \{\tilde{j}\}$ is already not an N-cluster for any $\tilde{j} \in \tilde{J} \setminus J$. Let $\mathcal{J}_r(N)$ be the set of all r-tuples of mutually distinct indices between 1 and N. Define rare sets $I_r(N)$ as collections of r-tuples $J = (i_1, i_2, ..., i_r)$ which either contains a cluster with more than one element or

$$i_{\min}(J) = \min_{\varepsilon \in \mathcal{E}_\ell} \min_{1 \leq l \leq r, \, a_\varepsilon N < i_l < b_\varepsilon N} (i_l - a_\varepsilon N, b_\varepsilon N - i_l) \leq a(N). \quad (3.4.18)$$

Next, we represent the rare sets $I_r(n)$ in the form

$$I_r(N) = (\cup_{1 \leq k \leq r} I_r^{(k,1)}(N)) \cup (\cup_{1 \leq k < r} I_r^{(k,0)}(N)) \quad (3.4.19)$$

where $I_r^{(k,1)}(N)$ and $I_r^{(k,0)}(N)$ are the sets of r-tuples from $\mathcal{J}_r(N)$, $r \leq N$ which contain exactly k maximal N-clusters and each r-tuple from $I_r^{(k,1)}(N)$ contains an N-cluster J with $i_{\min}(J) \leq a(N)$ while r-tuples from $I_r^{(k,0)}(N)$ do not contain such N-clusters. Since the cardinality of $\mathcal{E}_\ell$ equals $\ell!$ no r-tuple J from $I_r(N)$ may contain more than $2\ell! + 1$ maximal N-clusters with $i_{\min}(J) \leq a(N)$. Observe that $I_r^{(r,l)}(N)$, $l = 0, 1$ contains only singleton N-clusters, and so it can be a subset of $I_r(N)$ only if $l = 1$. On the other hand, $I_r^{(k,l)}(N)$ with $k < r$ contains at least one N-cluster which is not a singleton.

In order to estimate the cardinality of rare sets, observe that if l, i, j, N and k are fixed, there can only be one solution m of the equation

$$p_i l - p_j m + (q_i - q_j)N = k.$$

Hence, if l and N are fixed then there exists no more than $2\ell^2 a(N)$ possibilities for m to satisfy the inequality $d(l, m) \leq a(N)$. It follows that there are no more than $Nr(2a(N))^r \ell^{2r}$ possibilities for the choice of numbers in any N-cluster $J \in \mathcal{J}_r(N)$ if $i_{\min}(J) > a(N)$. If $i_{\min}(J) \leq a(N)$ then there are no more than $2r(2a(N))^r \ell! \ell^{2r}$ such choices. Hence

$$\#(I_r^{(k,l)}(N)) \leq C_r(a(N))^{rk} N^{k-l}, \quad k = 1, 2, ..., r; \ l = 0, 1 \quad (3.4.20)$$

for some $C_r > 0$ depending only on r but not on N.

Let $(i_1, ..., i_l)$ be a sequence of distinct positive integers such that for some pairs $(m_1, i_{j_1}), ..., (m_k, i_{j_k})$,

$$\rho_{m_1}(i_{j_1}, N) < \rho_{m_2}(i_{j_2}, N) < ... < \rho_{m_k}(i_{j_k}, N) \quad (3.4.21)$$

where we recall that $\rho_j(i, N) = p_j i + q_j N$, and the pairs above are different but either i's or m's may repeat themselves. Then by (3.4.13),

$$b_{i_1...i_l}^{(N)} = \mathbb{E} \prod_{j=1}^{l} \eta_{ij}^{(N)} \leq \mathbb{E} \prod_{a=1}^{k} \mathbb{I}_{\Gamma_N}(\xi_{\rho_{m_a}(i_{j_a}, N)}) \quad (3.4.22)$$

$$\leq (\mu(\Gamma_N))^k \prod_{a=2}^{k} \left(1 + \psi(\rho_{m_a}(i_{j_a}, N) - \rho_{m_{a-1}}(i_{j_{a-1}}, N))\right).$$

Let $J = (i_1, ..., i_l)$ be an N-cluster with $l \geq 2$. If $m, n \in J$, $m \neq n$, $m \in \mathcal{N}_{\varepsilon,N}$ and $n \in \mathcal{N}_{\tilde{\varepsilon},N}$ then

$$\rho_{\varepsilon_1}(m, N) < \rho_{\varepsilon_2}(m, N) < ... < \rho_{\varepsilon_\ell}(m, N) \text{ and} \qquad (3.4.23)$$
$$\rho_{\tilde{\varepsilon}_1}(m, N) < \rho_{\tilde{\varepsilon}_2}(m, N) < ... < \rho_{\tilde{\varepsilon}_\ell}(m, N).$$

The first set of inequalities in (3.4.23) gives ℓ pairs (ε_i, m), $i = 1, ..., \ell$ satisfying (3.4.21). If there is $\rho_{\tilde{\varepsilon}_i}(n, N)$ which is different from any $\rho_{\varepsilon_j}(m, N)$ then $(\tilde{\varepsilon}_i, n)$ can be taken as the $(\ell + 1)$-th pair and so J yields $\ell + 1$ pairs satisfying (3.4.21). If there is no such $\rho_{\tilde{\varepsilon}_i}(n, N)$ then we must have $\rho_{\varepsilon_i}(m, N) = \rho_{\tilde{\varepsilon}_i}(n, N)$ for all $i = 1, ..., \ell$, i.e.,

$$p_{\varepsilon_i} m - p_{\tilde{\varepsilon}_i} n = (q_{\tilde{\varepsilon}_i} - q_{\varepsilon_i})N, \; i = 1, ..., \ell. \qquad (3.4.24)$$

By Assumption 3.4.1 the matrices $\left(\begin{smallmatrix} p_1 & p_2 & \cdots & p_\ell \\ p_{\zeta(1)} & p_{\zeta(2)} & \cdots & p_{\zeta(\ell)} \end{smallmatrix} \right)$ have rank 2, and so for any distinct $\varepsilon, \tilde{\varepsilon} \in \mathcal{E}_\ell$ and N there exists no more than one solution pair $(m, n)_{\varepsilon, \tilde{\varepsilon}, N}$ of (3.4.24).

Let $\mathcal{M}_N$ be the set of all $m \leq N$ such that there exists $n \leq N$ with (3.4.24) being satisfied for some $\varepsilon, \tilde{\varepsilon} \in \mathcal{E}_\ell$. Then the cardinality of $\mathcal{M}_N$ does not exceed $(\ell!)^2$. If all numbers i_j, $j = 1, ..., l$ in J belong to $\mathcal{M}_N$ then J belongs to a set of N-clusters $\tilde{\mathcal{J}}_\ell$ of cardinality not exceeding $(\ell!)^{2l}$. If J does not belong to $\tilde{\mathcal{J}}_\ell$ then there is a pair $m, n \in J$ such that (3.4.24) cannot hold true, and so J yields $\ell + 1$ pairs satisfying (3.4.21). These together with (3.4.20) and (3.4.22) imply

$$\sum_{(i_1, ..., i_r) \in I_r(N)} b_{i_1...i_r}^{(N)} \qquad (3.4.25)$$
$$\leq \sum_{k=1}^{r} \sum_{(i_1, ..., i_r) \in I_r^{(k,1)}(N)} b_{i_1...i_r}^{(N)} + \sum_{k=1}^{r-1} \sum_{(i_1, ..., i_r) \in I_r^{(k,0)}(N)} b_{i_1...i_r}^{(N)}$$
$$\leq \tilde{C}_r (a(N))^{r^2} \left(\sum_{k=1}^{r} N^{k-1} (\mu(\Gamma_N))^{k\ell} + \sum_{k=1}^{r-1} N^k (\mu(\Gamma_N))^{k\ell+1} \right)$$
$$\leq \tilde{C}_r (a(N))^{r^2} ((\mu(\Gamma_N))^\ell + \mu(\Gamma_N)) \sum_{k=0}^{r-1} (\lambda_N)^k \to 0 \text{ as } N \to \infty$$

for some $\tilde{C}_r > 0$ depending only on r, where $\lambda_N = N(\mu(\Gamma_N))^\ell \to \lambda$ as $N \to \infty$.

Recall that if $m \in \mathcal{N}_{\varepsilon,N}$ then we have the first set inequalities in (3.4.23) which means that the pairs (ε_i, m), $i = 1, ..., \ell$ satisfy (3.4.21). Hence, by (3.4.22) for each $m \in \mathcal{N}_{\varepsilon,N}$,

$$b_m^{(N)} \leq C(\mu(\Gamma_N))^\ell. \qquad (3.4.26)$$

It follows that if $i_j \in \mathcal{N}_{\varepsilon^{(j)},N}$, $j = 1, ..., r$ then

$$\sum_{(i_1,...,i_r) \in I_r(N)} b_{i_1}^{(N)} \cdots b_{i_r}^{(N)} \leq \sum_{k=1}^{r} \sum_{(i_1,...,i_r) \in I_r^{(k,1)}(N)} b_{i_1}^{(N)} \cdots b_{i_r}^{(N)} \quad (3.4.27)$$

$$+ \sum_{k=1}^{r-1} \sum_{(i_1,...,i_r) \in I_r^{(k,0)}(N)} b_{i_1}^{(N)} \cdots b_{i_r}^{(N)} \leq \tilde{\tilde{C}}_r (a(N))^{r^2} N^{r-1} (\mu(\Gamma_N))^{r\ell}$$

$$= \tilde{\tilde{C}}_r (a(N))^{r^2} (\lambda_N)^{r-1} (\mu(\Gamma_N))^{\ell} \to 0 \text{ as } N \to \infty$$

for some $\tilde{\tilde{C}}_r > 0$ depending only on r, which completes the proof of (3.4.9).

In order to derive (3.4.10) let $(i_1, ..., i_r) \in \mathcal{J}_r(N) \setminus I_r(N)$ and $i_j \in \mathcal{N}_{\varepsilon^{(j)},N}$, $j = 1, ..., r$. Then all N-clusters in $(i_1, ..., i_r)$ are singletons and we conclude that there exist pairs $(m_1, i_{j_1}), (m_2, i_{j_2}), ..., (m_{r\ell}, i_{j_{r\ell}})$ such that

$$i_{\min}(i_1, ..., i_r) \geq a(N) \text{ and } |\rho_{m_{l+1}}(i_{j_{l+1}}, N) - \rho_{m_l}(i_{j_l}, N)| \geq a(N) \quad (3.4.28)$$

for $l = 1, 2, ..., r\ell - 1$. Then it follows from (3.4.14) and Lemma 3.2.6 similarly to (3.4.15) that

$$|b_{i_1...i_r}^{(N)} - (\mu(\Gamma_N))^{r\ell}| \leq \tilde{\tilde{C}} (\mu(\Gamma_N))^{r\ell} \psi(a(N) - 1)$$

and

$$|b_{i_j}^{(N)} - (\mu(\Gamma_N))^{\ell}| \leq \tilde{\tilde{C}} (\mu(\Gamma_N))^{\ell} \psi(a(N) - 1)$$

for some $\tilde{\tilde{C}} > 0$. These together with (3.4.7) yields (3.4.10). We proved (3.4.8)–(3.4.10) for integers i_j belonging to some $\mathcal{N}_{\varepsilon^{(j)},N}$ but as we observed it earlier the sums $\sum_{n=1}^{N} \prod_{j=1}^{\ell} \mathbb{I}_{\Gamma_N}(\xi_{p_j n + q_j N})$ and $\sum_{n \in \mathcal{N}_N} \prod_{j=1}^{\ell} \mathbb{I}_{\Gamma_N}(\xi_{p_j n + q_j N})$ have the same weak limits since the sum of no more than ℓ^2 terms corresponding to integers from $\{1, 2, ..., N\} \setminus \mathcal{N}_N$ converge to zero in probability. Now, the proof of Theorem 3.4.2 is complete. $\qquad\square$

3.4.3   *Poisson limits for subshifts*

We start with the following simple result.

Lemma 3.4.5. *Suppose that P is ψ-mixing. Then there exists a constant $\gamma > 0$ such that for any cylinder $B_n^a = \{\omega = (\omega_0, \omega_1, ...) \in \Omega : \omega_i = a_i, i = 0, 1, ..., n - 1\}$ of length n,*

$$P(B_n^a) \leq e^{-\gamma n}. \quad (3.4.29)$$

Proof. Let $\delta = \sup_{b \in \mathcal{A}} P([b])$. Since $\mathcal{A}$ is not a singleton and without loss of generality we assumed that $P([b]) > 0$ for any $b \in \mathcal{A}$ then $\delta < 1$. Hence, by ψ-mixing there exists $m > 0$ such that

$$\psi(m) < \delta^{-1} - 1. \tag{3.4.30}$$

Then, by ψ-mixing (cf. (3.4.13)) for any $a = (a_0, a_1, ...)$,

$$P(B_n^a) \leq P([a_0] \cap T^{-m}[a_m] \cap T^{-2m}[a_{2m}] \cap ... \cap T^{l_n m}[a_{l_n m}]) \tag{3.4.31}$$
$$\leq (1 + \psi(m))^{l_n + 1} \delta^{l_n}$$

where l_n is the integral part of $n/(m+1)$. Now (3.4.29) follows from (3.4.30). $\qquad\square$

Set

$$\eta_n^{(N_m)} = \prod_{j=1}^{\ell} \mathbb{I}_{B_m^a} \circ T^{p_j n + q_j N_m}, \quad n = 1, 2, ..., N_m$$

where $N_m = N_m^a$ satisfies (3.4.4). Let $n \in \mathcal{N}_\varepsilon$ such that

$$\min(n - a_\varepsilon N - 1, b_\varepsilon N - n - 1) \geq 2m. \tag{3.4.32}$$

For such n's by (3.4.14) and Lemmas 3.2.6 and 3.4.5, $b_n^{(N_m)} = P\{\eta_n^{(N_m)} = 1\}$ satisfies

$$|b_n^{(N_m)} - (P(B_m^a))^\ell| \leq \ell \psi(m)(1 + \psi(m))^\ell (P(B_m^a))^\ell. \tag{3.4.33}$$

Clearly,

$$\max_{1 \leq n \leq N_m} b_n^{(N_m)} \leq P(B_m^a) \to 0 \text{ as } m \to \infty \tag{3.4.34}$$

and by (3.4.4) and (3.4.33),

$$\sum_{n=1}^{N_m} |b_n^{(N_m)} - N_m (P(B_m^a))^\ell| \tag{3.4.35}$$

$$\leq 4m P(B_m^a) + N_m (P(B_m^a))^\ell \ell \psi(m)(1 + \psi(m))^\ell \to 0 \text{ as } m \to \infty$$

proving (3.4.8) (with N_m in place of N).

Now, let $(i_1, ..., i_l)$ be a sequence of distinct positive integers such that for some pairs $(n_1, i_{j_1}), ..., (n_k, i_{j_k})$,

$$\rho_{n_1}(i_{j_1}, N_m) < \rho_{n_2}(i_{j_2}, N_m) - 2m < \rho_{n_3}(i_{j_3}, N_m) - 4m < ... \tag{3.4.36}$$
$$< \rho_{n_k}(i_{j_k}, N_m) - 2(k-1)m$$

where the pairs are different but either i's or n's may repeat themselves. Then, by (3.4.13),

$$b^{(N_m)}_{i_1 \ldots i_l} = \mathbb{E} \prod_{j=1}^{l} \eta^{(N_m)}_{i_j} \le \mathbb{E} \prod_{\alpha=1}^{k} \mathbb{I}_{B^a_m} \circ T^{\rho_{n_\alpha}(i_{j_\alpha}, N_m)} \le (P(B^a_m))^k (1 + \psi(m))^k.$$

(3.4.37)

Next, we introduce again the notion of an N_m-cluster which is a sequence $J = \{j_1, j_2, \ldots, j_l\}$ of distinct positive integers such that for any $j, \tilde{j} \in J$ there exists a chain $j = j_{i_1}, j_{i_2}, \ldots, j_{i_k} = \tilde{j}$ of integers from J with

$$d(j_{i_\alpha}, j_{i_{\alpha+1}}) \le 2m \quad \forall \alpha = 1, 2, \ldots, k - 1.$$

We set $a(N_m) = 2m$ and define maximal N_m-clusters and rare sets $I_r(N_m)$ in the same way as in Section 3.4.2. Now, we derive (3.4.10) assuming that $(i_1, \ldots, i_r) \in \mathcal{J}_r(N_m) \setminus I_r(N_m)$ and $i_j \in \mathcal{N}_{\varepsilon(j), N}$, $j = 1, \ldots, r$. Then all N_m-clusters in $(i_1, \ldots, i_r)$ are singletons and we conclude that there exist pairs $(m_1, i_{j_1}), (m_2, i_{j_2}), \ldots, (m_{r\ell}, i_{j_{r\ell}})$ such that (3.4.28) holds true with $a(N) = a(N_m) = 2m$ where $i_{\min}(i_1, \ldots, i_r)$ is defined by (3.4.18) with $a(N) = a(N_m)$. It follows now from (3.4.14) and Lemma 3.2.6 similarly to (3.4.15) that

$$|b^{(N_m)}_{i_j} - (P(B^a_m))^\ell| \le C(P(B^a_m))^\ell \psi(m), \quad j = 1, \ldots, r$$

and

$$|b^{(N_m)}_{i_1 \ldots i_r} - (P(B^a_m))^{r\ell}| \le C(P(B^a_m))^{r\ell} \psi(m)$$

for some $C > 0$ which does not depend on N and m. Thus (3.4.10) follows from here and (3.4.7).

It remains to establish (3.4.9). Now we represent $I_r(N_m)$ in the form

$$I_r(N) = (\cup_{1 \le k \le r} \cup_{1 \le l \le k} \tilde{I}^{(k,l)}_r(N_m)) \cup (\cup_{1 \le k < r} \cup_{1 \le l \le k} I^{(k,l)}_r(N_m)) \quad (3.4.38)$$

where $\tilde{I}^{(k,l)}_r(N_m)$ and $I^{(k,l)}_r(N_m)$ are sets of r-tuples from $\mathcal{J}_r(N_m)$, $r \le N_m$ which contain exactly k-maximal N_m-clusters. Furthermore, each r-tuple from $\tilde{I}^{(k,l)}_r(N_m)$ contains exactly $l \le k$ maximal N_m-clusters $\tilde{J}_1, \ldots, \tilde{J}_l$ with $i_{\min}(\tilde{J}_j) \le 2m$ while r-tuples from $I^{(k,l)}_r(N_m)$ do not contain such N_n-clusters but it contains exactly l maximal N_m-clusters which are not singletons.

It is easy to see that the number of r-tuples in $\tilde{I}^{(k,l)}_r(N_m)$ can be estimated by

$$\#(\tilde{I}^{(k,l)}_r(N_m)) \le C_r N^{k-l}_m (2m)^{r+l} \quad (3.4.39)$$

where $C_r > 0$ does not depend on m. Choose $n_1, n_2, \ldots, n_k$ from different maximal N_m-clusters in $(i_1, \ldots, i_r) \in \tilde{I}^{(k,l)}_r(N_m)$ so that $n_1, \ldots, n_l$ belong to maximal N_m-clusters $\tilde{J}_1, ..., \tilde{J}_l$ with $i_{\min}(\tilde{J}_j) \leq 2m$ while $n_{l+1}, ..., n_k$ belong to remaining maximal N_m-clusters $\tilde{J}$ with $i_{\min}(\tilde{J}) > 2m$. Then, by (3.4.13) and Lemma 3.2.6,

$$b_{i_1...i_r}^{(N_m)} = \mathbb{E}\prod_{j=1}^r \eta_{i_j}^{(N_m)} \leq \mathbb{E}((\prod_{\alpha=1}^l \mathbb{I}_{B_m^a} \circ T^{\rho_{j_\alpha}(i_{n_\alpha}, N_m)}) \quad (3.4.40)$$

$$\times (\prod_{j=l+1}^k \eta_{n_j}^{(N_m)})) \leq (1 + \psi(m))^k (P(B_m^a))^{(k-l)\ell+l}$$

where $j_\alpha \in \{1, ..., \ell\}$ are chosen arbitrarily. Similarly,

$$\prod_{j=1}^r b_{i_j}^{(N_m)} = \prod_{j=1}^r \mathbb{E}\eta_{i_j}^{(N_m)} \quad (3.4.41)$$

$$\leq (P(B_m^a))^l \prod_{j=l+1}^k \mathbb{E}\eta_{n_j}^{(N_m)} \leq (1 + \psi(m))^k (P(B_m^a))^{(k-l)\ell+l}.$$

By (3.4.39)–(3.4.41) we conclude that as $m \to \infty$,

$$\sum_{(i_1,...,i_r)\in\cup_{1\leq k\leq r}\cup_{1\leq l\leq k}\tilde{I}_r^{(k,l)}(N_m)} (b_{i_1...i_r}^{(N_m)} + \prod_{j=1}^r b_{i_j}^{(N_m)}) \to 0. \quad (3.4.42)$$

It remains to treat the r-tuples from $I_r^{(k,l)}(N_m)$. Let $J = (i_1, ..., i_s)$ be a maximal N_m-cluster with $1 < s \leq r$. Set

$$i_{\max}(J) = \max_{1\leq j, \tilde{j}\leq l, j\neq \tilde{j}} d(i_j, i_{\tilde{j}})$$

with the distance d defined in Section 3.4.2. Set $d = i_{\max}(J)$. Then if one of the numbers i_j is fixed we conclude in view of Assumption 3.4.1 that there exist no more than $(2d)^{s-1}$ possibilities for other i_α's, $\alpha \neq j$. It follows that there exist no more than $N_m(2d)^s$ possibilities for numbers in such a maximal N_m-cluster. Now let $d(i_j, i_{\tilde{j}}) = i_{\max}(J) = d$, $1 \leq j, \tilde{j} \leq l$, $j \neq \tilde{j}$. Since $i_{\min}(J) \geq 2m$ we obtain from (3.4.13) and Lemmas 3.2.6 and 3.4.5 that

$$\mathbb{E}\eta_{i_j}^{(N_m)}\eta_{i_{\tilde{j}}}^{(N_m)} \leq \mathbb{E}\eta_{i_j}^{(N_m)}\mathbb{I}_{B_m^a} \circ T^{\rho_1(i_{\tilde{j}}, N_m)} \quad (3.4.43)$$

$$\leq (1 + \psi(m))^{\ell-1}(P(B_m^a))^{\ell-1}P(D_m^a) \leq (1 + \psi(m))^\ell (P(B_m^a))^\ell e^{-\gamma d}$$

where D_m^a is a cylinder of length $m + d$ having either the form $[\tilde{a}_0, \tilde{a}_1, ..., \tilde{a}_{d-1}, a_0, a_1, ..., a_{m-1}]$ or the form

$$[a_0, a_1, ..., a_{m-1}, \tilde{a}_0, \tilde{a}_1, ..., \tilde{a}_{d-1}].$$

Now, for each $a \in \Omega$ set

$$L_a(m) = \min\{l : B_m^a \cap T^{-l}B_m^a \neq \emptyset\}.$$

Observe that if a is a nonperiodic sequence then

$$\lim_{m\to\infty} L_a(m) = \infty. \quad (3.4.44)$$

Indeed, since, clearly, $L_a(m)$ is nondecreasing in m then $\lim_{m \to \infty} L_a(m) = M$ exists. If $M < \infty$ then a is an M-periodic sequence which contradicts our assumption, and so $M = \infty$. By the definition of $L_a(m)$,

$$B_m^a \cap T^{-l} B_m^a = \emptyset \text{ for all } l = 1, 2, ..., L_a(m) - 1. \qquad (3.4.45)$$

It follows that for any r-tuple $(i_1, ..., i_r)$ if $d(i_j, i_{\tilde{j}}) < L_a(m)$ for some $1 \leq j, \tilde{j} \leq r$, $j \neq \tilde{j}$ then $b_{i_1...i_r}^{(N_m)} = 0$.

Now, let $(i_1, ..., i_r) \in I_r^{(k,l)}(N_m)$ and $J_1, ..., J_k$ be maximal N_m-clusters in $(i_1, ..., i_r)$ such that $J_{l+1}, ..., J_k$ are singletons while $J_1, ..., J_l$ are not. Then we obtain from (3.4.13), (3.4.43) and Lemma 3.2.6 that

$$b_{i_1...i_r}^{(N_m)} \leq (1 + \psi(m))^{l\ell+k}(P(B_m^a))^{k\ell} e^{-\gamma \sum_{j=l}^{l} d_j} \qquad (3.4.46)$$

where $d_j = i_{\max}(J_j) \geq L_a(m)$. Since $\min(i_j - a_{\varepsilon_j} N, b_{\varepsilon_j} N - i_j) > 2m$ here for some ε_j's we deduce from (3.4.37) and Lemma 3.2.6 that

$$\prod_{j=1}^{r} b_{i_j}^{(N_m)} = \prod_{j=1}^{r} \mathbb{E} \eta_{i_j}^{(N_m)} \leq (1 + \psi(m))^{r\ell}(P(B_m^a))^{r\ell}. \qquad (3.4.47)$$

Finally, we conclude from (3.4.4), (3.4.46) and (3.4.47) together with the counting argument that as $m \to \infty$,

$$\sum_{(i_1,...,i_r) \in \cup_{1 \leq k < r} \cup_{1 \leq l \leq k} I_r^{(k,l)}(N_m)} b_{i_1...i_r}^{(N_m)} \qquad (3.4.48)$$

$$\leq C_r \sum_{k=1}^{r-1} \sum_{l=1}^{k} N_m^k (P(B_m^a))^{k\ell} \sum_{d_1,...,d_l \geq L_a(m)} \prod_{j=1}^{l} ((2d_j)^r e^{-\gamma d_j}) \to 0$$

and

$$\sum_{(i_1,...,i_r) \in \cup_{1 \leq k < r} \cup_{1 \leq \ell \leq k} I_r^{(k,l)}(N_m)} \prod_{j=1}^{r} b_{i_j}^{(N_m)} \leq C_r N_m^{r-1}(P(B_m^a))^{r\ell} \to 0.$$

$$(3.4.49)$$

These complete the proof of (3.4.9) (for i_j's from $\mathcal{N}_N$ which as explained above suffices for our purposes) and Theorem 3.4.3 follows from Lemma 3.4.4. $\qquad \square$

PART 2

Thermodynamic Formalism for Random Complex Operators and applications

Chapter 4

Random complex Ruelle-Perron-Frobenius theorem via cones contractions

In this chapter we will prove a random complex Ruelle-Perron-Frobenius type theorem under rather general assumptions which will be verified in Chapter 5 for random locally expanding covering maps and in Chapter 6 for random complex integral operators.

4.1 Preliminaries

Our setup consists of a complete probability space $(\Omega, \mathcal{F}, P)$, together with an invertible P-preserving transformation $\theta : \Omega \to \Omega$, and of a family $(B_\omega, \mathcal{C}_\omega)$, $\omega \in \Omega$ of complex Banach spaces B_ω and complex, proper and linearly convex cones $\mathcal{C}_\omega \subset B_\omega$ (see Appendix A). Our results will rely on the following notion of measurability. For each $\omega \in \Omega$ denote by B_ω^* the dual space of B_ω and set

$$B^* = \bigcup_{\omega \in \Omega} \{\omega\} \times B_\omega^*. \tag{4.1.1}$$

Let $\mathcal{G}$ be a σ-algebra on B^* and $b = \{b_\omega : \omega \in \Omega\}$ and $\mu = \{\mu_\omega : \omega \in \Omega\}$ be two families so that $b_\omega \in B_\omega$ and $\mu_\omega \in B_\omega^*$ for any ω. The family b defines a function $\hat{b} : B^* \to \mathbb{C}$ by the formula

$$\hat{b}(\omega, \nu) = \nu(b_\omega). \tag{4.1.2}$$

We will say that b is $\mathcal{G}$-measurable if the function $\hat{b}$ is $\mathcal{G}$-measurable and that the family μ is $\mathcal{G}$-measurable if the function $\omega \to \mu_\omega(c_\omega)$ is $\mathcal{F}$-measurable for any $\mathcal{G}$-measurable family $c = \{c_\omega : \omega \in \Omega\}$. Similarly, we will say that a family $\upsilon = \{\upsilon_\omega : \omega \in \Omega\}$ so that $\upsilon_\omega \in (B_\omega^*)^*$ for each ω is $\mathcal{G}$-measurable if the function $\omega \to \upsilon_\omega(\nu_\omega)$ is $\mathcal{F}$-measurable for any $\mathcal{G}$-measurable family $\nu = \{\nu_\omega : \omega \in \Omega\}$. Note that the above definitions naturally extend to families zdefined only P-a.s.

Next, consider a parametric family of complex linear continuous operators $\{A_z^\omega : B_\omega \to B_{\theta\omega} : \omega \in \Omega\}$, $z \in U$ where $U \subset \mathbb{C}$ is a (deterministic) neighborhood of the origin. We assume here that the maps $z \to A_z^\omega$, $\omega \in \Omega$ are analytic with respect to the operator norm of continuous maps from B_ω to $B_{\theta\omega}$ and that for each $z \in U$ the family of operators A_z^ω, $\omega \in \Omega$ is $\mathcal{G}$-measurable in the sense that $A_z b := \{A_z^{\theta^{-1}\omega} b_{\theta^{-1}\omega} : \omega \in \Omega\}$ is $\mathcal{G}$-measurable for any $\mathcal{G}$-measurable family $b = \{b_\omega : \omega \in \Omega\}$. For any $\omega \in \Omega$, $n \in \mathbb{N}$ and $z \in U$ consider the n-th step composition $A_z^{\omega,n} : B_\omega \to B_{\theta^n\omega}$ given by

$$A_z^{\omega,n} = A_z^{\theta^{n-1}\omega} \circ \cdots \circ A_z^{\theta\omega} \circ A_z^\omega. \tag{4.1.3}$$

We will describe now our assumptions concerning the cones $\mathcal{C}_\omega$ and their relation with the operators A_z^ω. For each $\omega \in \Omega$ consider the dual cone $\mathcal{C}_\omega^*$ of $\mathcal{C}_\omega$ given by

$$\mathcal{C}_\omega^* = \left\{ \mu \in B_\omega^* : |\mu(x)| > 0 \text{ for any } x \in \mathcal{C}_\omega' \right\}$$

where $\mathcal{C}_\omega' = \mathcal{C}_\omega \setminus \{0\}$.

Assumption 4.1.1. There exists a $\mathcal{G}$-measurable family b so that $b_\omega \in \mathcal{C}_\omega'$ for any ω. Moreover, there exist random variables $K_\omega, M_\omega \in (0, \infty)$ and $\mathcal{G}$-measurable families $l = \{l_\omega : \omega \in \Omega\}$ and $\kappa = \{\kappa_\omega : \omega \in \Omega\}$ so that $l_\omega \in B_\omega^* \setminus \{0\}$ and $\kappa_\omega \in (B_\omega^*)^* \setminus \{0\}$ for each ω and P-a.s.,

$$\|x\|\|l_\omega\| \le K_\omega |l_\omega(x)| \text{ for any } x \in \mathcal{C}_\omega \tag{4.1.4}$$

and

$$\|\mu\|\|\kappa_\omega\| \le M_\omega |\kappa_\omega(\mu)| \text{ for any } \mu \in \mathcal{C}_\omega^*. \tag{4.1.5}$$

Furthermore, $\|l_\omega\|$ and $\|\kappa_\omega\|$ are $\mathcal{F}$-measurable functions of ω, where $\|l_\omega\|$ and $\|\kappa_\omega\|$ are the operator norms of l_ω and κ_ω, respectively.

The second part of Assumption 4.1.1 means that P-a.s. the complex cones $\mathcal{C}_\omega$ and $\mathcal{C}_\omega^*$ have bounded aperture (see Appendix A). Note that when such l_ω and κ_ω exist then $l_\omega \in \mathcal{C}_\omega^*$ and $\kappa_\omega \in (\mathcal{C}_\omega^*)^*$. We refer the readers to Lemma A.2.7 for conditions on the cone $\mathcal{C}_\omega$ which guarantee that its dual cone $\mathcal{C}_\omega^*$ has bounded aperture.

Next, let $\mathcal{C}$ be a complex cone in some complex Banach space and consider the complex Hilbert projective metric $\delta_\mathcal{C}$ associated with $\mathcal{C}$ defined in Appendix A. For any subset $\mathcal{Q}$ of $\mathcal{C}' = \mathcal{C} \setminus \{0\}$ denote by $\Delta_\mathcal{C}(\mathcal{Q})$ its (Hilbert) diameter with respect to $\delta_\mathcal{C}$ given by

$$\Delta_\mathcal{C}(\mathcal{Q}) = \sup_{q_1, q_2 \in \mathcal{Q}} \delta_\mathcal{C}(q_1, q_2).$$

Assumption 4.1.2. There exist constants $j_0 \in \mathbb{N}$ and $d_0 > 0$ such that P-a.s.,

$$A_z^{\omega,j} \mathcal{C}_\omega' \subset \mathcal{C}_{\theta^j \omega}' \text{ and } \Delta_{\mathcal{C}_{\theta^j \omega}} \left(A_z^{\omega,j} \mathcal{C}_\omega' \right) < d_0 \tag{4.1.6}$$

for any $z \in U$ and $j_0 \leq j \leq 2j_0$.

Some of our results can be obtained only under the following assumption.

Assumption 4.1.3. The complex cone $\mathcal{C}_\omega$ is P-a.s. reproducing of order $k = k_\omega \in \mathbb{N}$, namely there exists a random variable $r_\omega > 0$ such that P-a.s. for any $f \in B_\omega$ there exist $f_1, ..., f_k \in \mathcal{C}_\omega'$ so that $f = f_1 + f_2 + ... + f_k$ and

$$\|f_1\| + \|f_2\| + ... + \|f_k\| \leq r_\omega \|f\|. \tag{4.1.7}$$

Example 4.1.4. Let $\mathcal{X}$ be a topological space together with the Borel σ-algebra $\mathcal{B}$ and a set $\mathcal{E} \subset \Omega \times \mathcal{X}$ measurable with respect to the product σ-algebra $\mathcal{F} \times \mathcal{B}$. Denote by $\mathcal{E}_\omega = \{x \in \mathcal{X} : (\omega, x) \in \mathcal{E}\}$, $\omega \in \Omega$ the fibers of $\mathcal{E}$ and consider the situation when each B_ω is a (Banach) space of functions on $\mathcal{E}_\omega$. Let $\mathcal{P}$ be the σ-algebra induced on $\mathcal{E}$ from the product σ-algebra $\mathcal{F} \times \mathcal{B}$. In this situation a family $b = \{b_\omega : \omega \in \Omega\}$ is naturally identifies with a function $\bar{b} : \mathcal{E} \to \mathbb{C}$ given by $\bar{b}(\omega, x) = b_\omega(x)$ and we say that b is measurable if the function $\bar{b}$ is $\mathcal{P}$-measurable. Suppose now that the (evaluation) functionals given by $\hat{x}(b_\omega) = b_\omega(x)$, $x \in \mathcal{E}_\omega$ are continuous and set

$$\mathcal{Y} = \bigcup_{\omega \in \Omega} \{\omega\} \times \mathcal{Y}_\omega \subset B^*$$

where $\mathcal{Y}_\omega = \{\hat{x} : x \in \mathcal{E}_\omega\}$ for each ω. Identifying $\mathcal{Y}_\omega$ with $\mathcal{E}_\omega$ we endow $\mathcal{Y}$ with the σ-algebra $\mathcal{P}$. Under this identification the function $\hat{b}$ defined in (4.1.2) coincides with the function $\bar{b}$ on the set $\mathcal{Y}$. We can extend $\mathcal{P}$ to B^* by adding all the subsets of $B^* \setminus \mathcal{Y}$ and their unions with elements of $\mathcal{P}$, i.e. say that a set $\Gamma \subset B^*$ is measurable if $\Gamma \cap \mathcal{Y} \in \mathcal{P}$. Denote this extension by $\mathcal{G}$. Then a family b is $\mathcal{G}$-measurable in the sense described at the beginning of this section if and only if the function $\bar{b}$ is $\mathcal{P}$-measurable.

4.2 Main results

Our main result is the following Ruelle-Perron-Frobenius (RPF) theorem.

Theorem 4.2.1. *Suppose that Assumptions 4.1.1 and 4.1.2 hold true. Then P-a.s. for any $z \in U$ there exists a unique triplet $(\lambda_\omega(z), h_\omega(z), \nu_\omega(z))$*

consisting of a complex number $\lambda_\omega(z) \neq 0$ and random elements $h_\omega(z) \in$ $\mathcal{C}_\omega, \nu_\omega(z) \in \mathcal{C}_\omega^*$ such that $\kappa_\omega \nu_\omega(z) = 1$,

$$A_z^\omega h_\omega(z) = \lambda_\omega(z) h_\omega(z), \quad (A_z^\omega)^* \nu_{\theta\omega}(z) = \lambda_\omega(z) \nu_\omega(z) \quad and \quad \nu_\omega(z) h_\omega(z) = 1 \tag{4.2.1}$$

where $\mathcal{C}_\omega$ and κ_ω are specified in Assumption 4.1.1. Moreover, $\lambda_\omega(z)$, $h_\omega(z)$ and $\nu_\omega(z)$ are analytic functions of z, the function $\omega \to \lambda_\omega(z)$ is measurable and the families $\{h_\omega(z) : \omega \in \Omega\}$ and $\{\nu_\omega(z) : \omega \in \Omega\}$ are $\mathcal{G}$-measurable.

The above uniqueness statement is restricted to elements of the cones $\mathcal{C}_\omega$ and $\mathcal{C}_\omega^*$. Nevertheless, when the cones $\mathcal{C}_\omega$ are reproducing and θ is ergodic then we prove that the unique triplet from Theorem 4.2.1 is the only triplet satisfying (4.2.1) such that $\kappa_\omega \nu_\omega(z) = 1$ and $\nu_\omega h_\omega(z) \neq 0$. In particular, it is the unique triplet which satisfies these conditions under the restriction that $\nu_\omega(z) \in \mathcal{C}_\omega^*$.

Next, consider the random variables

$$\lambda_{\omega,n}(z) = \prod_{k=0}^{n-1} \lambda_{\theta^k \omega}(z)$$

where $n \in \mathbb{N}$ and $z \in U$. Let V be a neighborhood of 0 whose closure $\bar{V}$ is contained in U. The second result we prove here is the following exponential convergence theorem.

Theorem 4.2.2. *Suppose that Assumptions 4.1.1, 4.1.2 and 4.1.3 hold true.*

(i) There exists a random variable $k_1(\omega)$ such that P-a.s. for any $n \geq k_1(\omega)$, $z \in \bar{V}$, $q \in B_{\theta^{-n}\omega}$ and $\mu \in B_\omega^*$,

$$\left\| \frac{A_z^{\theta^{-n}\omega, n} q}{\lambda_{\theta^{-n}\omega, n}(z)} - \left(\nu_{\theta^{-n}\omega}(z) q\right) h_\omega(z) \right\| \leq C_{\theta^{-n}\omega} \mathcal{D}_\omega \|q\| \cdot c^{\frac{n}{j_0}} \tag{4.2.2}$$

and

$$\left\| \frac{(A_z^{\theta^{-n}\omega, n})^* \mu}{\lambda_{\theta^{-n}\omega, n}(z)} - \left(\mu h_\omega(z)\right) \nu_{\theta^{-n}\omega}(z) \right\| \leq C_{\theta^{-n}\omega} \mathcal{D}_\omega \|\mu\| \cdot c^{\frac{n}{j_0}} \tag{4.2.3}$$

where $C_\omega = r_\omega M_\omega \|\kappa_\omega\|^{-1}$,

$$\mathcal{D}_\omega = c^{-3} dK_\omega \left(c_\omega + K_\omega M_\omega \|l_\omega\|^{-1} \|\kappa_\omega\|^{-1}\right),$$

$c = \tanh(\frac{1}{4} d_0)$ and $c_\omega = \sup_{z \in \bar{V}} |l_\omega h_\omega(z)|$ which is measurable in ω and P-a.s. finite.

(ii) Let $\delta > 0$ be so that $B(0, 2\delta) \subset V$. Assume that the random variables $|l_\omega h_\omega(0)|$, $K_\omega \|l_\omega\|^{-1}$ and $M_\omega \|\kappa_\omega\|^{-1}$ are bounded. Then there exist

constants $\delta_1 > 0$ and $Q > 0$ *which depend only on δ and the essential supremum of the latter random variables (and this dependence can be recovered explicitly from the proof) such that (4.2.2) and (4.2.3) hold true P-a.s. with $r_{\theta^{-n}\omega} Q$ in place of $C_{\theta^{-n}\omega} D_\omega$ for any $n \geq j_0$ and $z \in B(0, \delta_1)$.*

In the course of the proof of Theorem 4.2.2 we prove additional types of exponential convergences, some of which without assuming that the cones $\mathcal{C}_\omega$ are reproducing while the others require ergodicity of θ. We refer the readers to Section 4.4. Note also that in certain situations $|l_\omega h_\omega(0)|$ can be controlled by means of the random variables $K_\omega, \|l_\omega\|$ etc. (see Chapters 5 and 6).

4.3 Block partitions and RPF triplets

The first step towards proving Theorems 4.2.1 and 4.2.2 is the following.

4.3.1 *Reverse block partitions*

For any $\omega \in \Omega$, $z \in U$ and a natural $n \geq j_0$ consider the following "block partition",

$$A_z^{\theta^{-n}\omega, n} = A_z^{\theta^{-j_0}\omega, j_0} \cdot A_z^{\theta^{-2j_0}\omega, j_0} \circ \cdots \circ A_z^{\theta^{-(k-1)j_0}\omega, j_0} \circ A_z^{\theta^{-n}\omega, j_0+r} \quad (4.3.1)$$

where $n = kj_0 + r$ for some integers $k \geq 1$ and $0 \leq r < j_0$. By Assumption 4.1.6,

$$A_z^{\theta^{-n}\omega, j_0+r} \mathcal{C}'_{\theta^{-n}\omega} \subset \mathcal{C}'_{\theta^{-(k-1)j_0}\omega} \quad \text{and} \quad \Delta_{\mathcal{C}_{\theta^{-(k-1)j_0}\omega}} (A_z^{\theta^{-n}\omega, j_0+r} \mathcal{C}'_{\theta^{-n}\omega}) < d_0 \quad (4.3.2)$$

and for any $1 \leq m \leq k - 1$,

$$A_z^{\theta^{-mj_0}\omega, j_0} \mathcal{C}'_{\theta^{-mj_0}\omega} \subset \mathcal{C}'_{\theta^{-(m-1)j_0}\omega} \quad \text{and} \quad (4.3.3)$$

$$\Delta_{\mathcal{C}_{\theta^{-(m-1)j_0}\omega}} (A_z^{\theta^{-mj_0}\omega, j_0} \mathcal{C}'_{\theta^{-mj_0}\omega}) < d_0.$$

In particular, it follows that

$$A_z^{\theta^{-n}\omega, n} \mathcal{C}_{\theta^{-n}\omega} \subset \mathcal{C}_\omega. \quad (4.3.4)$$

Next, set

$$c = \tanh \left(\frac{1}{4} d_0 \right) \in (0, 1).$$

We claim that P-a.s.,

$$\delta_{\mathcal{C}_\omega} \left(A_z^{\theta^{-n}\omega, n} f_n, A_z^{\theta^{-s}\omega, s} g_s \right) \leq d_0 c^{\left[\frac{n}{j_0} \right] - 2} \quad (4.3.5)$$

for any $s \geq n \geq j_0$, $z \in U$ and sequences $\{f_m : m \geq 1\}$ and $\{g_m : m \geq 1\}$ with $f_m, g_m \in \mathcal{C}'_{\theta^{-m}\omega}$ for any $m \geq 1$. Before proving (4.3.5) we need the following notations. For any finite sequence $G_1, ..., G_m$ of linear operators we write

$$\prod_{i=1}^{m} G_i = G_1 \circ G_2 \circ \cdots \circ G_m$$

and for the sake of convenience set

$$\prod_{G \in \emptyset} G = Identity,$$

i.e. the empty product will always be equal to the identity map, no matter which spaces are under consideration. In order to prove (4.3.5), we first write $n = kj_0 + r$ and $s = qj_0 + p$, for some $1 \leq k \leq q$ and $0 \leq r, p < j_0$. Then,

$$A_z^{\theta^{-n}\omega, n} = \Big(\prod_{m=1}^{k-2} A_z^{\theta^{-mj_0}\omega, j_0} \Big) \circ A_z^{\theta^{-(k-1)j_0}\omega, j_0} \circ B_1 \quad \text{and} \quad (4.3.6)$$

$$A_z^{\theta^{-s}\omega, s} = \Big(\prod_{m=1}^{k-2} A_z^{\theta^{-mj_0}\omega, j_0} \Big) \circ A_z^{\theta^{-(k-1)j_0}\omega, j_0} \circ B_2$$

where

$$B_1 = B_1(n, \omega, z) = A_z^{\theta^{-n}\omega, j_0 + r} \quad \text{and}$$

$$B_2 = B_2(n, s, \omega, z) = \Big(\prod_{m=k}^{q-1} A_z^{\theta^{-mj_0}\omega, j_0} \Big) \circ A_z^{\theta^{-s}\omega, j_0 + r}.$$

Applying repeatedly Theorem A.2.3 *(iii)* and taking into account (4.3.2) and (4.3.3) we derive that

$$\delta_{\mathcal{C}_\omega} \Big(\Big(\prod_{m=1}^{k-2} A_z^{\theta^{-mj_0}\omega, j_0} \Big) h_1, \Big(\prod_{m=1}^{k-2} A_z^{\theta^{-mj_0}\omega, j_0} \Big) h_2 \Big) \leq c^{k-2} \delta_{\mathcal{C}_{\theta^{-(k-2)j_0}\omega}}(h_1, h_2)$$

$$(4.3.7)$$

for any $h_1, h_2 \in \mathcal{C}'_{\theta^{-(k-2)j_0}\omega}$. Let $f_n \in \mathcal{C}'_{\theta^{-n}\omega}$ and $g_s \in \mathcal{C}'_{\theta^{-s}\omega}$. It follows from the inclusions in (4.3.2) and (4.3.3) that $B_1 f_n, B_2 g_s \in \mathcal{C}'_{\theta^{-(k-1)j_0}\omega}$ and that the functions h_1 and h_2 given by

$$h_1 = A_z^{\theta^{-(k-1)j_0}\omega, j_0}(B_1 f_n) \quad \text{and} \quad h_2 = A_z^{\theta^{-(k-1)j_0}\omega, j_0}(B_2 g_s)$$

are members of $\mathcal{C}'_{\theta^{-(k-2)j_0}\omega}$. Applying (4.3.3) we derive that

$$\delta_{\mathcal{C}_{\theta^{-(k-2)j_0}\omega}}(h_1, h_2) < d_0$$

and (4.3.5) follows from (4.3.6), (4.3.7) and the above estimates.

Since $l_\omega \in \mathcal{C}_\omega^*$, we conclude from (4.3.5), Theorem A.2.3 *(ii)* and (4.1.4) that

$$\|\bar{A}_z^{\theta^{-s}\omega,s}g_s - \bar{A}_z^{\theta^{-n}\omega,n}f_n\| \leq \frac{1}{2}K_\omega d_0 c^{[\frac{n}{j_0}]-2} \leq R(\omega)c^{\frac{n}{j_0}} \qquad (4.3.8)$$

for any $s \geq n \geq j_0$, $z \in U$ and sequences $\{f_m : m \geq 1\}$ and $\{g_m : m \geq 1\}$ with $f_m, g_m \in \mathcal{C}_{\theta^{-m}\omega}'$ for any $m \geq 1$. Here $R(\omega) = \frac{1}{2}c^{-3}d_0 K_\omega$ and

$$\bar{A}_z^{\theta^{-j}\omega}h = \frac{A_z^{\theta^{-j}\omega,j}h}{l_\omega(A_z^{\theta^{-j}\omega,j}h)} \qquad (4.3.9)$$

for any $j \geq j_0$, $h \in \mathcal{C}_{\theta^{-j}\omega}'$ and $z \in U$. By considering the case when $f_m = g_m$ for any $m \in \mathbb{N}$, we deduce that the sequence

$$\{\bar{A}_z^{\theta^{-n}\omega}f_n : n \in \mathbb{N}\} \subset B_\omega$$

is a Cauchy sequence, and therefore P-a.s. the limits

$$\hat{h}_\omega(z) := \lim_{n \to \infty} \bar{A}_z^{\theta^{-n}\omega,n}f_n, \quad z \in U \qquad (4.3.10)$$

exist in the Banach space B_ω. Considering now the situation when $n = s$ and then letting $n \to \infty$ in (4.3.8), we deduce that the limits $\hat{h}_\omega(z), z \in U$ do not depend on the choice of the sequence $\{f_m : m \geq 1\}$. Moreover, fixing n and letting $s \to \infty$ it follows from (4.3.8) that for any choice of $f_n \in \mathcal{C}_{\theta^{-n}\omega}'$ and $n \geq j_0$,

$$\|\bar{A}_z^{\theta^{-n}\omega,n}f_n - \hat{h}_\omega(z)\| \leq R(\omega)c^{[\frac{n}{j_0}]}. \qquad (4.3.11)$$

Furthermore, by (4.3.5) and Theorem A.2.3 *(i)* we have $\hat{h}_\omega(z) \in \mathcal{C}_\omega' = \mathcal{C}_\omega \setminus \{0\}$ and the limits $\hat{h}_\omega(z)$ exist also in the projective metric $\delta_{\mathcal{C}_\omega}$. Since l_ω is continuous, for any $z \in U$ we have that

$$l_\omega\big(\hat{h}_\omega(z)\big) = 1.$$

Therefore by (4.1.4) from Assumption 4.1.1 P-a.s. for any $z \in U$,

$$\frac{1}{\|l_\omega\|} \leq \|\hat{h}_\omega(z)\| \leq \frac{K_\omega}{\|l_\omega\|} \qquad (4.3.12)$$

using that $\hat{h}_\omega(z) \in \mathcal{C}_\omega$.

Finally, we show that $\hat{h}_\omega(z)$ is $\mathcal{G}$-measurable and analytic in z. Indeed, by Assumption 4.1.1 there exists a $\mathcal{G}$-measurable family $b = \{b_\omega : \omega \in \Omega\}$ so that $b_\omega \in \mathcal{C}_\omega'$ for each ω. Choosing $f_n = b_{\theta^{-n}\omega}$ the desired measurability of $\hat{h}_\omega(z)$ follows. Since the operators A_z^ω are analytic in z the limit $\hat{h}_\omega(z)$ is analytic in $z \in U$ as it is a uniform limit of analytic functions.

Remark 4.3.1. Applying the continuous operator A_z^ω to both sides of (4.3.10) we derive that the limit

$$\hat{\lambda}_\omega(z) = \lim_{n\to\infty} \frac{l_{\theta\omega}\left(A_z^{\theta^{-n}\omega,n+1}f_n\right)}{l_\omega\left(A_z^{\theta^{-n}\omega,n}f_n\right)}$$

exists and satisfies

$$A_z^\omega \hat{h}_\omega(z) = \hat{\lambda}_\omega(z)\hat{h}_{\theta\omega}(z). \tag{4.3.13}$$

Since $l_{\theta\omega}(\hat{h}_{\theta\omega}(z)) = 1$, $\hat{\lambda}_\omega(z)$ can also be written as

$$\hat{\lambda}_\omega(z) = l_{\theta\omega}\left(A_z^\omega \hat{h}_\omega(z)\right) \tag{4.3.14}$$

which, in particular, implies that the limit $\hat{\lambda}_\omega(z)$ does not depend on the choice of the sequence $\{f_k : k \geq 1\}$ and that $\hat{\lambda}_\omega(z)$ is both measurable in ω and analytic in z since the operators A_z^ω are. The pair $(\hat{\lambda}_\omega(z), \hat{h}_\omega(z))$ is not the one from the resulting RPF triplet $(\lambda_\omega(z), \nu_\omega(z), \nu_\omega(z))$. Still, it will follow (after the construction of these triplets) that $\hat{h}_\omega(z)$ and $h_\omega(z)$ are linearly dependent, namely that $h_\omega(z)$ is (possibly) a different representative from the projective class of the function $\hat{h}_\omega(z)$, and so in the sense of convergence in the projective metric $\delta_{\mathcal{C}_\omega}$ we have arrived to the right limit. As a result, $\hat{\lambda}_\omega(z)$ and $\lambda_\omega(z)$ differ by a (multiplicative) coboundary term, namely P-a.s. for any $z \in U$ we can write

$$\hat{\lambda}_\omega(z) = a_\omega(z)(a_{\theta\omega}(z))^{-1}\lambda_\omega(z) \tag{4.3.15}$$

where $a_\omega(z) = \nu_\omega(z)(\hat{h}_\omega(z))$ which, in fact, satisfies $\hat{h}_\omega(z) = a_\omega(z)h_\omega(z)$.

4.3.2   *Forward block partition and dual operators*

Let $n \geq j_0$ and write $n = kj_0 + r$ for some natural $k \geq 1$ and $0 \leq r < j_0$. For any $z \in U$ consider the following "block partition",

$$A_z^{\omega,n} = A_z^{\theta^{(k-1)j_0}\omega,j_0+r} \circ A_z^{\theta^{(k-2)j_0}\omega,j_0} \circ \cdots \circ A_z^{\theta^{2j_0}\omega,j_0} \circ A_z^{\theta^{j_0}\omega,j_0} \circ A_z^{\omega,j_0}. \tag{4.3.16}$$

By Assumption 4.1.6 for any $0 \leq m \leq k - 2$,

$$A_z^{\theta^{mj_0}\omega,j_0}\mathcal{C}'_{\theta^{mj_0}\omega} \subset \mathcal{C}'_{\theta^{(m+1)j_0}\omega}, \quad \Delta_{\mathcal{C}_{\theta^{(m+1)j_0}\omega}}(A_z^{\theta^{mj_0}\omega,j_0}\mathcal{C}'_{\theta^{mj_0}\omega}) < d_0, \tag{4.3.17}$$

$$A_z^{\theta^{(k-1)j_0}\omega,j_0+r}\mathcal{C}'_{\theta^{(k-1)j_0}\omega} \subset \mathcal{C}'_{\theta^n\omega} \quad \text{and} \tag{4.3.18}$$

$$\Delta_{\mathcal{C}_{\theta^n\omega}}(A_z^{\theta^{(k-1)j_0}\omega,j_0+r}\mathcal{C}'_{\theta^{(k-1)j_0}\omega}) < d_0.$$

In particular, it follows that

$$A_z^{\omega,n}\mathcal{C}'_\omega \subset \mathcal{C}'_{\theta^n\omega}, \quad P\text{-a.s.} \tag{4.3.19}$$

for any $z \in U$ and $n \geq j_0$.

Next, for the sake of convenience, we denote sometimes the dual operator $\left(A_z^{\omega,n}\right)^*$ also by $A_z^{\omega,n,*}$. Consider P-a.s. the dual block partition of (4.3.16),

$$\left(A_z^{\omega,n}\right)^* = \left(A_z^{\omega,j_0}\right)^* \circ \left(A_z^{\theta^{j_0}\omega,j_0}\right)^* \circ \cdots \circ \left(A_z^{\theta^{(k-2)j_0}\omega,j_0}\right)^* \circ \left(A_z^{\theta^{(k-1)j_0}\omega,j_0+r}\right)^* \tag{4.3.20}$$

where $n = kj_0 + r$, $k \in \mathbb{N}$ and $0 \leq r < j_0$. Now we begin with construction of $\nu_\omega(z)$ and $\lambda_\omega(z)$. Note that this construction will not depend on the results from Section 4.3.1.

First, by (A.2.6) from Section A.2.5, taking into account (4.3.17) and (4.3.18), it follows that P-a.s. for any $z \in U$ and $n \geq j_0$,

$$A_z^{\theta^{(k-1)j_0}\omega,j_0+r,*}\mathcal{C}_{\theta^n\omega}^* \subset \mathcal{C}_{\theta^{(k-1)j_0}\omega}^* \quad \text{and} \tag{4.3.21}$$

$$A_z^{\theta^{mj_0}\omega,j_0,*}\mathcal{C}_{\theta^{(m+1)j_0}\omega}^* \subset \mathcal{C}_{\theta^{mj_0}\omega}^*, \quad m = 0,1,2,\ldots$$

and the corresponding δ-diameters of the appropriate images (with respect to the appropriate dual cones) do not exceed d_0. Note that in this application of (A.2.6) we have used that the cones $\mathcal{C}_\omega$ are proper and linearly convex. In particular, it follows that

$$A_z^{\omega,n,*}\mathcal{C}_{\theta^n\omega}^* \subset \mathcal{C}_\omega^*, \quad P\text{-a.s.} \tag{4.3.22}$$

for any $n \geq j_0$ and $z \in U$.

Next, let $s \geq n \geq j_0$, $\mu_1 \in \mathcal{C}_{\theta^s\omega}^*$ and $\mu_2 \in \mathcal{C}_{\theta^n\omega}^*$ for $\omega \in \Omega$ satisfying (4.3.21). We claim that

$$\delta_{\mathcal{C}_\omega^*}\left(A_z^{\omega,s,*}\mu_1, A_z^{\omega,n,*}\mu_2\right) \leq d_0 c^{\lceil \frac{n}{j_0} \rceil - 2}. \tag{4.3.23}$$

Indeed, write $n = kj_0 + r$ and $s = qj_0 + p$ where $1 \leq k \leq q$ and $0 \leq r,p < j_0$. Consider the following block partition

$$\left(A_z^{\omega,s}\right)^* = \left(A_z^{\omega,j_0}\right)^* \circ \left(A_z^{\theta^{j_0}\omega,j_0}\right)^* \circ \cdots \circ \left(A_z^{\theta^{(k-2)j_0}\omega,j_0}\right)^* \circ B_s$$

where

$$B_s = B_s(n,\omega,z) = \left(A_z^{\theta^{(k-1)j_0}\omega,s-(k-1)j_0}\right)^*.$$

Using (4.3.21) with n and s we see that $B_n\mu_1, B_s\mu_2 \in \mathcal{C}_{\theta^{(k-1)j_0}\omega}^*$ for any $z \in U$. Therefore, by (4.3.21) and the diameter estimates following it,

$$\delta_{\mathcal{C}_{\theta^{(k-2)j_0}\omega}^*}\left(A_z^{\theta^{(k-2)j_0}\omega,j_0,*}(B_s\mu_1), A_z^{\theta^{(k-1)j_0}\omega,j_0,*}(B_n\mu_2)\right) < d_0.$$

Similarly to (4.3.5), (4.3.23) follows by a repetitive use of Theorem A.2.3 *(iii)*, taking into account (4.3.21) and the diameter estimates following it. Next, we derive from Theorem A.2.3 *(ii)* and (4.3.23) that

$$\|\bar{A}_z^{\omega,s,*}\mu_1 - \bar{A}_z^{\omega,n,*}\mu_2\| \leq \frac{1}{2} M_\omega d_0 c^{\lceil \frac{n}{j_0} \rceil - 2} \tag{4.3.24}$$

where M_ω is specified in Assumption 4.1.1 and for any $\nu \in \mathcal{C}_\omega^*$ and $m \geq j_0$,

$$\bar{A}_z^{\omega,m,*}\nu = \frac{A_z^{\omega,m,*}\nu}{\kappa_\omega\left(A_z^{\omega,m,*}\nu\right)}$$

with κ_ω coming from (4.1.5) in Assumption 4.1.1. Similarly to (4.3.10), we conclude that for any choice of $\{\mu_n : n \geq 1\}$ with $\mu_n \in \mathcal{C}_{\theta^n\omega}^*$ the limits

$$\nu_\omega(z) := \lim_{n \to \infty} \bar{A}_z^{\omega,n,*}\mu_n, \quad z \in U \qquad (4.3.25)$$

exist and are independent of the choice of $\{\mu_n : n \geq 1\}$. The operators A_z^ω are analytic and so $\nu_\omega(z)$ is a uniform limit of analytic functions which makes it analytic in z. Note that we have used here that $(A_z^\omega)^*$ is analytic in z which holds true since the duality map $A \to A^*$ is an isometry. In order to show that $\nu_\omega(z)$ is $\mathcal{G}$-measurable, let $\nu = \{\nu_\omega : \omega \in \Omega\}$ be a $\mathcal{G}$-measurable family so that $\nu_\omega \in \mathcal{C}_\omega^*$ for any ω (for instance we can take $\nu_\omega = l_\omega$). Taking $\mu_n = \nu_{\theta^n\omega}$, the required measurability follows by (4.3.25). Next, applying Theorem A.2.3 *(i)* and using that any dual cone is linearly convex (see Appendix A) it follows that $\nu_\omega(z) \in \mathcal{C}_\omega^*$. Moreover, by (4.3.24) and (4.3.25),

$$\|\bar{A}_z^{\omega,n,*}\mu_n - \nu_\omega(z)\| \leq \frac{1}{2}M_\omega d_0 c^{\lceil \frac{n}{j_0}\rceil - 2} \qquad (4.3.26)$$

for any $n \geq j_0$. Furthermore, since $\nu_\omega(z) \in \mathcal{C}_\omega^*$ and $\kappa_\omega\left(\nu_\omega(z)\right) = 1$ we deduce from (4.1.5) that

$$\frac{1}{\|\kappa_\omega\|} \leq \|\nu_\omega(z)\| \leq \frac{M_\omega}{\|\kappa_\omega\|}. \qquad (4.3.27)$$

Replacing ω with $\theta\omega$, making the choice $\mu_n = l_{\theta^n(\theta\omega)}$ and then plugging in both sides of (4.3.25) into the continuous operator $\left(A_z^\omega\right)^*$, we derive that the limits

$$\lambda_\omega(z) := \lim_{n \to \infty} \frac{\kappa_\omega\left(A_z^{\omega,n+1,*}l_{\theta^{n+1}\omega}\right)}{\kappa_{\theta\omega}\left(A_z^{\theta\omega,n,*}l_{\theta^{n+1}\omega}\right)}, \quad z \in U \qquad (4.3.28)$$

P-a.s. exist and satisfy

$$(A_z^\omega)^*\nu_{\theta\omega}(z) = \lambda_\omega(z)\nu_\omega(z). \qquad (4.3.29)$$

The operators A_z^ω are analytic in z and therefore the map $z \to \lambda_\omega(z)$ is analytic since it is a uniform limit of analytic functions. Since the operators A_z^ω are $\mathcal{G}$-measurable the limit $\lambda_\omega(z)$ is measurable in ω. Substituting both sides of (4.3.29) in κ_ω we deduce that

$$\lambda_\omega(z) = \lambda_\omega(z)\kappa_\omega\left(\nu_\omega(z)\right) = \kappa_\omega\left((A_z^\omega)^*\nu_{\theta\omega}(z)\right) \qquad (4.3.30)$$

which, in particular, provides a different proof that $\lambda_\omega(z)$ is analytic in z and measurable in ω. Next, applying (4.3.29) repeatedly with $\bar\omega = \theta^{n-1}\omega, \ldots.\theta\omega, \omega$ it follows that for any $n \in \mathbb{N}$,

$$A_z^{\omega,n,*}\nu_{\theta^n\omega}(z) = \lambda_{\omega,n}(z)\nu_\omega(z), \tag{4.3.31}$$

where

$$\lambda_{\omega,n}(z) := \prod_{i=0}^{n-1} \lambda_{\theta^i\omega}(z) = \kappa_\omega\big(A_z^{\omega,n,*}\nu_{\theta^n\omega(z)}\big).$$

Since $\nu_{\theta^n\omega}(z) \in \mathcal{C}_{\theta^n\omega}^*$ we deduce from (4.3.22) that $A_z^{\omega,n,*}\nu_{\theta^n\omega}(z) \in \mathcal{C}_\omega^*$ for any $n \geq j_0$. This together with the fact that $\kappa_\omega \in (\mathcal{C}_\omega^*)^*$ implies that $\lambda_{\omega,n}(z) \neq 0$ for any sufficiently large n, and as a consequence P-a.s. we have $\lambda_\omega(z) \neq 0$ for any $z \in U$.

4.3.3   RPF triplets

For any $z \in U$ and P-a.a. ω set

$$h_\omega(z) = \frac{\hat h_\omega(z)}{\nu_\omega(z)\big(\hat h_\omega(z)\big)}. \tag{4.3.32}$$

The denominator does not vanish since $\hat h_\omega(z) \in \mathcal{C}_\omega'$ and $\nu_\omega(z) \in \mathcal{C}_\omega^*$, and so $h_\omega(z)$ is well defined,

$$\nu_\omega(z)\big(h_\omega(z)\big) = 1 \quad\text{and}\quad h_\omega(z) \in \mathcal{C}_\omega' \tag{4.3.33}$$

where the inclusion follows since $\mathcal{C}_\omega$ is invariant under multiplication of nonzero complex numbers (i.e. it is a complex cone). Since $\nu_\omega(z)$ and $\hat h_\omega(z)$ are $\mathcal{G}$-measurable and analytic in z so is $h_\omega(z)$. We claim that the triplet $\big(\lambda_\omega(z), h_\omega(z), \nu_\omega(z)\big)$ is the RPF triplet from Theorem 4.2.1. The missing ingredient is to show that P-a.s. we have $A_z^\omega h_\omega(z) = \lambda_\omega(z)h_{\theta\omega}(z)$ for any $z \in U$. This will follow from the following claim.

Claim 4.3.2. For a.a. $\omega \in \Omega$, for each $z \in U$ and any choice of $\{q_n : n \geq 1\}$ with $q_n \in \mathcal{C}_{\theta^{-n}\omega}'$,

$$h_\omega(z) = \lim_{n\to\infty} \frac{A_z^{\theta^{-n}\omega,n}q_n}{\lambda_{\theta^{-n}\omega,n}(z)\nu_{\theta^{-n}\omega}(z)q_n}. \tag{4.3.34}$$

Before proving Claim 4.3.2 we assume it to hold true and complete the proof that

$$\big(\lambda_\omega(z), h_\omega(z), \nu_\omega(z)\big)$$

satisfy the conditions specified in Theorem 4.2.1. Indeed, let $q_n = h_{\theta^{-n}\omega}(z)$ and recall that $\nu_{\theta^{-n}\omega}(z)h_{\theta^{-n}\omega}(z) = 1$ and that $h_{\theta^{-n}\omega}(z) \in C'_{\theta^{-n}\omega}$ (see (4.3.33)). Plugging in both sides of (4.3.34) into the continuous operator A_z^ω we deduce that

$$A_z^\omega h_\omega(z) = \lambda_\omega(z) \lim_{n \to \infty} \frac{A_z^{\theta^{-(n+1)}\theta\omega,n+1} h_{\theta^{-(n+1)}\theta\omega}(z)}{\lambda_{\theta^{-(n+1)}\theta\omega,n}(z)} = \lambda_\omega(z) h_{\theta\omega}(z)$$

(4.3.35)

where the second equality follows by applying (4.3.34) with $\theta\omega$ in place of ω and with the sequence $\{\tilde{q}_n : n \geq 1\}$ given by $\tilde{q}_n = q_{n-1} = h_{\theta^{-n}\theta\omega}(z) = h_{\theta^{-(n-1)}\omega}(z)$ in place of $\{q_n : n \geq 1\}$ and using again (4.3.33). We remark that replacing ω with $\theta\omega$ is indeed possible P-a.s. since θ is P-invariant.

Proof of Claim 4.3.2. First, by (4.3.31) P-a.s.,

$$\lambda_{\theta^{-n}\omega,n}(z)\nu_{\theta^{-n}\omega}(z)q_n = \nu_\omega(z)A_z^{\theta^{-n}\omega,n}q_n \qquad (4.3.36)$$

for any $n \in \mathbb{N}$, $q_n \in B_{\theta^{-n}\omega}$ and $z \in U$. Therefore, when $n \geq j_0$ and $q_n \in C'_{\theta^{-n}\omega}$ we can write

$$\frac{A_z^{\theta^{-n}\omega,n}q_n}{\lambda_{\theta^{-n}\omega,n}(z)\nu_{\theta^{-n}\omega}(z)q_n} = \frac{A_z^{\theta^{-n}\omega,n}q_n}{\nu_\omega(z)A_z^{\theta^{-n}\omega,n}q_n} = \frac{\bar{A}_z^{\theta^{-n}\omega,n}q_n}{\nu_\omega(z)\bar{A}_z^{\theta^{-n}\omega,n}q_n}$$

where $\bar{A}_z^{\theta^{-n}\omega,n}$ is given by (4.3.9). By (4.3.10) the numerator converges to $\hat{h}_\omega(z)$ while the denominator converges to $\nu_\omega(z)\hat{h}_\omega(z)$ and therefore by the definition (4.3.32) of $h_\omega(z)$,

$$\lim_{n \to \infty} \frac{\bar{A}_z^{\theta^{-n}\omega,n}q_n}{\nu_\omega(z)\bar{A}_z^{\theta^{-n}\omega,n}q_n} = h_\omega(z)$$

which completes the proof of the claim. $\qquad\qquad\square$

The proof of the statement about uniqueness of this triplet is postponed to Section 4.5 and meanwhile we will prove Theorem 4.2.2.

4.4   Exponential convergences

We will assume in this section that all the conditions of Theorem 4.2.2 hold true. Then $\lambda_\omega(z), h_\omega(z), \nu_\omega(z)$ are both analytic in z and measurable in ω. We need first the following estimates. Let V be a neighborhood of 0 whose closure $\bar{V}$ is contained in U.

4.4.1   Taylor reminders and important bounds

For P-a.a. ω consider the analytic function $\alpha_\omega(\cdot) : U \to \mathbb{C}$ given by

$$\alpha_\omega(z) = \nu_\omega(z)\hat{h}_\omega(z) \tag{4.4.1}$$

which can also be written as $\alpha_\omega(z) = \frac{1}{l_\omega(h_\omega(z))}$. Then by (4.3.12) and (4.3.27),

$$|\alpha_\omega(z)| \le \frac{K_\omega M_\omega}{\|l_\omega\|\|\kappa_\omega\|} := \upsilon_\omega. \tag{4.4.2}$$

On the other hand, since this function does not vanish on $\bar{V}$ it follows that $\beta_\omega := \inf_{z \in \bar{V}} |\nu_\omega(z)(\hat{h}_\omega(z))| > 0$. Since the latter function is continuous then this infimum can be taken over a countable dense set which makes β_ω measurable since $\nu_\omega(z)$ and $\hat{h}_\omega(z)$ are $\mathcal{G}$-measurable. It follows from (4.3.32) and (4.3.12) that

$$\frac{\|\kappa_\omega\|}{K_\omega M_\omega} \le \|h_\omega(z)\| \le \frac{K_\omega}{\|l_\omega\|\beta_\omega} \tag{4.4.3}$$

for any $z \in \bar{V}$. The disadvantage of bounds with β_ω is that the latter is not defined by means of the random variables $K_\omega, \|l_\omega\|, M_\omega$ and $\|\kappa_\omega\|$ appearing in our assumptions which makes it hard to estimate. In what follows we will bound $\|h_\omega(z)\|$ from above around $z_0 = 0$ by means of the latter random variables. Applying Lemma 2.8.2 with $k = 0$ and the analytic function $\alpha_\omega(\cdot)$ it follows that

$$|\alpha_\omega(z) - \alpha_\omega(0)| \le \frac{4|z|\upsilon_\omega}{\delta}$$

for any $z \in B(0,\delta) = \{\zeta \in \mathbb{C} : |\zeta| < \delta\}$ where $\delta > 0$ satisfies that $\bar{B}(0,2\delta) \subset V$. Consider the set $V_\omega = B(0,\delta_\omega) \subset \mathbb{C}$ where

$$\delta_\omega = \frac{1}{8}\delta \min\left(1, |\alpha_\omega(0)|(\upsilon_\omega)^{-1}\right).$$

Then for any $z \in \bar{V}_\omega$,

$$\frac{1}{2}|\alpha_\omega(0)| \le |\alpha_\omega(z)| \le \frac{3}{2}|\alpha_\omega(0)| \tag{4.4.4}$$

and therefore by (4.3.12),

$$\frac{2}{3\|l_\omega\|\,|\alpha_\omega(0)|} \le \|h_\omega(z)\| \le \frac{2K_\omega}{\|l_\omega\|\,|\alpha_\omega(0)|} := \zeta_\omega. \tag{4.4.5}$$

In many situations (e.g. our applications in the next chapters) we will have a positive lower bound on $|\alpha_\omega(0)|$, which will make the above upper bound effective.

4.4.2 *Exponential convergences*

For the sake of convenience set $V^{(1)} = V$ and $V^{(2)} = V_\omega^{(2)} = V_\omega$. We first show that there exist random variables $k_1(\omega)$ and $k_2(\omega)$, such that P-a.s. for any $n \geq k_i(\omega)$, $z \in \bar{V}^{(i)}$, $i = 1, 2$ and $q \in \mathcal{C}_{\theta^{-n}\omega}$,

$$D(\omega, q, z, n) := \left\| \frac{A_z^{\theta^{-n}\omega, n} q}{\lambda_{\theta^{-n}\omega, n}(z) \nu_{\theta^{-n}\omega}(z)q} - h_\omega(z) \right\| \leq R_\omega^{(i)} R(\omega) c^{\frac{n}{j_0}} \quad (4.4.6)$$

where $R(\omega)$ is defined below (4.3.8),

$$R_\omega^{(1)} = 2(\beta_\omega^{-1} + W_\omega), \quad R_\omega^{(2)} = 4(|\alpha_\omega(0)|^{-1} + 2W_\omega)$$

and $W_\omega = K_\omega M_\omega \|l_\omega\|^{-1} \|\kappa_\omega\|^{-1}$. The random variable $k_2(\omega)$ will depend only on $\zeta_\omega, M_\omega \|\kappa_\omega\|^{-1}$ and $R(\omega)$, while the random variable $k_1(\omega)$ will depend also on β_ω.

Before proving (4.4.6), we assume that it holds true and complete the proof of Theorem 4.2.2. Set $C_\omega = r_\omega M_\omega \|\kappa_\omega\|^{-1}$. We claim that P-a.s. for any $n \geq k_i(\omega)$, $z \in \bar{V}^{(i)}$, $i = 1, 2$ and $q \in B_{\theta^{-n}\omega}$,

$$\left\| \frac{A_z^{\theta^{-n}\omega, n} q}{\lambda_{\theta^{-n}\omega, n}(z)} - (\nu_{\theta^{-n}\omega}(z)q) h_\omega(z) \right\| \leq C_{\theta^{-n}\omega} R_\omega^{(i)} R(\omega) \|q\| c^{\frac{n}{j_0}}. \quad (4.4.7)$$

Indeed, in the circumstances of Theorem 4.2.2, the cones $\mathcal{C}_{\theta^{-n}\omega}$ are reproducing of order $k = k_{\theta^{-n}\omega}$ with a constant $r_{\theta^{-n}\omega}$. Therefore, for any $q \in B_{\theta^{-n}\omega}$ we can write $q = q_1 + ... + q_k$ where each q_i is a member of the cone $\mathcal{C}_{\theta^{-n}\omega}$ and

$$\|q_1\| + \|q_2\| + ... + \|q_k\| \leq r_{\theta^{-n}\omega} \|q\|.$$

Inequality (4.4.7) follows by first applying (4.4.6) with the functions $q_i, i = 1, ..., k$, then multiplying the resulting inequality by $\nu_{\theta^{-n}\omega}(z)q_i$ and then using the equality $q = q_1 + q_2 + ... + q_k$ and the inequality

$$\sum_{i=1}^{k} |\nu_{\theta^{-n}\omega}(z)q_i| \leq \sum_{i=1}^{k} \|\nu_{\theta^{-n}\omega}(z)\| \|q_i\| \leq M_{\theta^{-n}\omega} \|\kappa_{\theta^{-n}\omega}\|^{-1} r_{\theta^{-n}\omega} \|q\|$$

where we use (4.3.27).

In order to prove (4.4.6), first we apply (4.3.36) and obtain that for P-a.a. ω and any $n \geq j_0$, $z \in U$ and $q \in \mathcal{C}_{\theta^{-n}\omega}$,

$$\lambda_{\theta^{-n}\omega, n}(z) \nu_{\theta^{-n}\omega}(z)q = \nu_\omega(z) A_z^{\theta^{-n}\omega, n} q.$$

Next, set

$$b_n(q, z) = b_n(\omega, q, z) = \frac{l_\omega(A_z^{\theta^{-n}\omega, n} q)}{\nu_\omega(z) A_z^{\theta^{-n}\omega, n} q} = \frac{1}{\nu_\omega(z) \bar{A}_z^{\theta^{-n}\omega, n} q}.$$

Then,

$$D(\omega, q, z, n) = \left\| b_n(q, z) \cdot \bar{A}_z^{\theta^{-n}\omega, n} q - h_\omega(z) \right\| \quad (4.4.8)$$

$$\leq \left\| b_n(q, z)\big(\bar{A}_z^{\theta^{-n}\omega, n} q - \hat{h}_\omega(z)\big) \right\| + \left\| \big(b_n(q, z) - \frac{1}{\nu_\omega(z)\hat{h}_\omega(z)}\big)\hat{h}_\omega(z) \right\|$$

where we use the definition of $h_\omega(z)$ in (4.3.32). By (4.3.11) and (4.3.27),

$$\left| (b_n(q, z))^{-1} - \nu_\omega(z)\hat{h}_\omega(z) \right| \quad (4.4.9)$$

$$= \left| \nu_\omega(z)\bar{A}_z^{\theta^{-n}\omega, n} q - \nu_\omega(z)\hat{h}_\omega(z) \right|$$

$$\leq \|\nu_\omega(z)\| \left\| \bar{A}_z^{\theta^{-n}\omega, n} q - \hat{h}_\omega(z) \right\| \leq M_\omega \|\kappa_\omega\|^{-1} R(\omega) c^{\frac{n}{j_0}}$$

where $R(\omega)$ was defined below (4.3.8). Using (4.3.11) again, we deduce from (4.4.9) and the definition of $\alpha_\omega(z)$ that for any $z \in \bar{V}$,

$$D(\omega, q, z, n) \leq R(\omega) c^{\frac{n}{j_0}} \mathcal{D}^{-1}\big(1 + M_\omega \|\kappa_\omega\|^{-1}\|\hat{h}_\omega(z)\|.|\alpha_\omega(z)|^{-1}\big)$$

$$= R(\omega) c^{\frac{n}{j_0}} \mathcal{D}^{-1}\big(1 + M_\omega \|\kappa_\omega\|^{-1}\|h_\omega(z)\|\big)$$

assuming that

$$\mathcal{D} := |\alpha_\omega(z)| - M_\omega \|\kappa_\omega\|^{-1} R(\omega) c^{\frac{n}{j_0}} > 0.$$

When $z \in \bar{V}$ then by the definition of β_ω,

$$\mathcal{D} \geq \beta_\omega - M_\omega \|\kappa_\omega\|^{-1} R(\omega) c^{\frac{n}{j_0}}$$

and when in addition $z \in \bar{V}_\omega$ then by (4.4.4),

$$\mathcal{D} \geq \frac{1}{2}|\alpha_\omega(0)| - M_\omega \|\kappa_\omega\|^{-1} R(\omega) c^{\frac{n}{j_0}}.$$

Let $k_1(\omega)$ be the smallest natural number k_1 such that

$$M_\omega \|\kappa_\omega\|^{-1} R(\omega) c^{\frac{k_1}{j_0}} \leq \frac{1}{2}\beta_\omega \quad (4.4.10)$$

and $k_1 \geq j_0$. Similarly, let $k_2(\omega)$ be the smallest natural number k_2 such that

$$M_\omega \|\kappa_\omega\|^{-1} R(\omega) c^{\frac{k_2}{j_0}} \leq \frac{1}{4}|\alpha_\omega(0)| \quad (4.4.11)$$

and $k_2 \geq j_0$. It is clear that $k_1(\omega)$ and $k_2(\omega)$ are measurable, and the proof of (4.4.6) is complete, taking into account (4.4.3) and (4.4.5). $\qquad \square$

4.4.3 Additional types of exponential convergences

We begin with proving (4.2.3) in Theorem 4.2.2. First, for any $\omega \in \Omega$, $\mu \in B_\omega^*$, $n \in \mathbb{N}$ and $q \in B_{\theta^{-n}\omega}$ the duality relation

$$\mu(A_z^{\theta^{-n}\omega,n}q) = (A_z^{\theta^{-n}\omega,n,*}\mu)q$$

holds true. Applying μ to the expression inside the norm in (4.4.7) we deduce that P-a.s. for any $n \geq k_i(\omega)$ and $z \in \bar{V}^{(i)}$, $i = 1, 2$,

$$\left\| \frac{(A_z^{\theta^{-n}\omega,n,*}\mu)q}{\lambda_{\theta^{-n}\omega,n}(z)} - (\mu h_\omega(z))\nu_{\theta^{-n}\omega}(z)q \right\| \leq C_{\theta^{-n}\omega}R_\omega^{(i)}R(\omega)\|q\|\|\mu\|c^{\frac{n}{j_0}}$$

$$\tag{4.4.12}$$

for any μ and q as above. We conclude by taking supremum over all q's with $\|q\| = 1$ that

$$\left\| \frac{A_z^{\theta^{-n}\omega,n,*}\mu}{\lambda_{\theta^{-n}\omega,n}(z)} - (\mu h_\omega(z))\nu_{\theta^{-n}\omega}(z) \right\| \leq C_{\theta^{-n}\omega}R_\omega^{(i)}R(\omega)\|\mu\|c^{\frac{n}{j_0}}. \tag{4.4.13}$$

$\square$

Suppose next that θ is ergodic. Consider the random variable $\delta_\omega = \frac{1}{8}\delta\min\left(1, |\alpha_\omega(0)|(v_\omega)^{-1}\right)$ from the definition of the nieghborhood V_ω, and recall that $|\alpha_\omega(0)|^{-1} = |l_\omega h_\omega(0)|$. Let $a \in (0, 1)$ be sufficiently small so that $p_a = P(G_a) > 0$, where

$$G_a = \{\omega : a \leq \|\kappa_\omega\|, \|l_\omega\| \leq a^{-1}\} \cap \{\omega : |l_\omega h_\omega(0)|, M_\omega, K_\omega \leq a^{-1}\}. \tag{4.4.14}$$

Applying the mean ergodic theorem with the indicator function of the set $G = G_a$ we see that P-a.s. there exists a strictly increasing sequence $\{l_i : i \geq 1\}$, $l_i = l_i(\omega)$ of natural numbers such that $\delta_{\theta^{l_i}\omega} \geq \frac{a^5\delta}{8}$ for any $i \geq 1$ and $\lim_{n\to\infty} \frac{l_n}{n} = \frac{1}{p_a}$.

We observe that the random variable $k_2(\omega)$ is P-a.s. bounded from above by some constant $\bar{k}_2$ (which depends only on a) when $\omega \in G_a$, since $k_2(\omega)$ is the minimal natural number k_2 satisfying (4.4.11) and $k_2 \geq j_0$. Plugging in $\theta^n\omega$ in place of ω in (4.4.7) and then considering n's of the form $n = l_i$ it follows that P-a.s. for any $z \in B(0, \frac{a^5\delta}{8}) := U_2$, $g \in B_\omega$ and $s \geq \bar{k}_2$,

$$\left\| \frac{A_z^{\omega,l_s}g}{\lambda_{\omega,l_s}(z)} - h_{\theta^{l_s}\omega}(z)\nu_\omega(z)g \right\| \leq 8C_1c^{-3}r_\omega M_\omega\|\kappa_\omega\|^{-1}\|g\|a^{-5}c^{\frac{l_s}{j_0}}. \tag{4.4.15}$$

When the random variables K_ω, M_ω are bounded from above, $|\alpha_\omega(0)|$ is bounded away from 0 and $\|\kappa_\omega\|$ and $\|l_\omega\|$ are bounded and bounded away from 0 then $p_a = 1$, assuming that a is sufficiently small. In this case $l_i = i$ for any i and we obtain appropriate estimates on the differences

$$\left\| \frac{A_z^{\omega,n} g}{\lambda_{\omega,n}(z)} - h_{\theta^n \omega}(z) \nu_\omega(z) g \right\|.$$

In fact, these estimates hold in this situation without assuming that θ is ergodic since in this case the random neighborhood V_ω contains a deterministic neighborhood of 0. Corresponding results for the dual operators follow similarly, relying on (4.4.13).

4.5 Uniqueness of RPF triplets

Let $z \in U$ and let $(\tilde{\lambda}_\omega(z), \tilde{h}_\omega(z), \tilde{\nu}_\omega(z))$ be a triplet consisting of a nonzero complex number $\tilde{\lambda}_\omega(z)$ an element $\tilde{h}_\omega(z) \in B_\omega$ and a functional $\tilde{\nu}_\omega(z) \in B_\omega^*$ satisfying P-a.s.,

$$A_z^\omega \tilde{h}_\omega(z) = \tilde{\lambda}_\omega(z) \tilde{h}_\omega(z), \quad (A_z^\omega)^* \tilde{\nu}_{\theta\omega}(z) = \tilde{\lambda}_\omega(z) \tilde{\nu}_\omega(z)$$
$$\text{and} \quad \tilde{\nu}_\omega(z) \tilde{h}_\omega(z) = \kappa_\omega \tilde{\nu}_\omega(z) = 1.$$

Iterating these equalities we obtain that for any natural n,

$$A_z^{\omega,n,*} \tilde{\nu}_{\theta^n \omega}(z) = \tilde{\lambda}_{\omega,n}(z) \tilde{\nu}_\omega(z), \quad P\text{-a.s.} \tag{4.5.1}$$

and

$$A_z^{\omega,n} \tilde{h}_\omega(z) = \tilde{\lambda}_{\omega,n}(z) \tilde{h}_{\theta^n \omega}(z), \quad P\text{-a.s.} \tag{4.5.2}$$

where

$$\tilde{\lambda}_{\omega,n}(z) = \prod_{k=0}^{n-1} \tilde{\lambda}_\omega(z).$$

We begin with proving the uniqueness statement from Theorem 4.2.1. Suppose that

$$\tilde{h}_\omega(z) \in \mathcal{C}'_\omega \quad \text{and} \quad \tilde{\nu}_\omega(z) \in \mathcal{C}^*_\omega, \quad P\text{-a.s.}$$

Substituting $\mu_n = \tilde{\nu}_{\theta^n \omega}(z)$ in (4.3.25) and using the equalities (4.5.1), $\bar{A}^{\omega,n,*} \tilde{\nu}_{\theta^n \omega}(z) = \tilde{\nu}_\omega(z)$ and $\kappa_\omega \tilde{\nu}_\omega(z) = 1$ it follows that

$$\tilde{\nu}_\omega(z) = \nu_\omega(z), \quad P\text{-a.s.} \tag{4.5.3}$$

Using the equality $\kappa_\omega \nu_\omega(z) = \kappa_\omega \tilde{\nu}_\omega(z) = 1$ we obtain that $\lambda_\omega(z) = \kappa_\omega (A_z^\omega)^* \nu_{\theta\omega}(z)$ and $\tilde{\lambda}_\omega(z) = \kappa_\omega (A_z^\omega)^* \tilde{\nu}_{\theta\omega}(z)$ and therefore

$$\lambda_\omega(z) = \tilde{\lambda}_\omega(z), \quad P\text{-a.s.} \tag{4.5.4}$$

Finally, substituting $q_n = \tilde{h}_{\theta^{-n}\omega}(z)$ in (4.3.34) and taking into account (4.5.2), (4.5.3) and (4.5.4) we derive that

$$\tilde{h}_\omega(z) = h_\omega(z), \quad P\text{-a.s.} \tag{4.5.5}$$

Next, we prove a stronger type of uniqueness under the assumption that the cones $\mathcal{C}_\omega$ are reproducing and that θ is ergodic. We claim that (4.5.3)-(4.5.5) hold true assuming that $\tilde{\nu}_\omega(z)h_\omega(z) \neq 0$ and that $\|\tilde{\nu}_\omega(z)\| \leq J_\omega$ for some random variable J_ω (e.g. when $\tilde{\nu}_\omega(z) \in \mathcal{C}_\omega^*$). Indeed, let $L > 0$ be such that $P(\mathcal{T}_L) > 0$, where

$$\mathcal{T}_L = \{\omega : r_\omega, M_\omega\|\kappa_\omega\|^{-1}, \|\kappa_\omega\|, J_\omega \leq L\}$$

and r_ω, M_ω and κ_ω were introduced in Assumptions 4.1.1 and 4.1.3. By ergodicity of θ^{-1} we have

$$P\left(\bigcup_{i=1}^\infty \theta^i \mathcal{T}_L\right) = 1.$$

Let $\iota_L(\omega)$ be the first backward visiting time to the set $\mathcal{T}_L$, namely the smallest index $i \in \mathbb{N}$ such that $\theta^{-i}\omega \in \mathcal{T}_L$. Consider the conditional probability measure P_L on $\mathcal{T}_L$ which is given by

$$P_L(\mathcal{Q}) = \frac{P(\mathcal{Q} \cap \mathcal{T}_L)}{P(\mathcal{T}_L)}, \quad \mathcal{Q} \in \mathcal{F}$$

and the map $\vartheta = \vartheta_L = \theta^{-\iota_L}$. Then ϑ preserves P_L and the measure preserving system $(\Omega, \mathcal{F}, P_L, \vartheta)$ is ergodic (see Section 5 in Chapter 1 of [14]). Plugging in $\mu = \tilde{\nu}_\omega(z)$ in (4.4.13) and then using (4.5.1) yields

$$\lim_{n\to\infty} \left\| \frac{\tilde{\lambda}_{\vartheta^n\omega,n}(z)}{\lambda_{\vartheta^n\omega,n}(z)} \tilde{\nu}_{\vartheta^n\omega}(z) - \left(\tilde{\nu}_\omega(z)h_\omega(z)\right)\nu_{\vartheta^n\omega}(z) \right\| = 0, \quad P_L\text{-a.s.} \tag{4.5.6}$$

Next, denote by $l_{n,\omega}(z)$ the linear functional inside the norm in the above left hand-side. Since $\kappa_{\vartheta^n\omega}\nu_{\vartheta^n\omega}(z) = \kappa_{\vartheta^n\omega}\tilde{\nu}_{\vartheta^n\omega}(z) = 1$ we have

$$\kappa_{\vartheta^n\omega}l_{n,\omega}(z) = \frac{\tilde{\lambda}_{\vartheta^n\omega,n}(z)}{\lambda_{\vartheta^n\omega,n}(z)} - \tilde{\nu}_\omega(z)h_\omega(z)$$

and, taking into account that $\lim_{n\to\infty}\|l_{n,\omega}(z)\| = 0$ and $\|\kappa_{\vartheta^n\omega}\| \leq L$, it follows that

$$|\kappa_{\vartheta^n\omega}l_{n,\omega}(z)| \leq L\|l_{n,\omega}(z)\| \to 0 \quad \text{as } n \to \infty$$

and therefore

$$\lim_{n\to\infty} \frac{\tilde{\lambda}_{\vartheta^n\omega,n}(z)}{\lambda_{\vartheta^n\omega,n}(z)} = \tilde{\nu}_\omega(z)h_\omega(z). \tag{4.5.7}$$

Since $\tilde{\nu}_\omega(z)h_\omega(z) \neq 0$, we conclude from (4.5.6) and (4.5.7) that

$$\lim_{n\to\infty} \|\tilde{\nu}_{\vartheta^n\omega}(z) - \nu_{\vartheta^n\omega}(z)\| = 0$$

where we also used that $\max\left(\|\tilde{\nu}_{\vartheta^n\omega}(z)\|, \|\nu_{\vartheta^n\omega}(z)\|\right) \leq L$ (recall (4.3.27)). Applying the mean ergodic theorem with the function $g(\omega) = \|\tilde{\nu}_\omega(z) - \nu_\omega(z)\|$ and the ergodic map ϑ we conclude that

$$\tilde{\nu}_\omega(z) = \nu_\omega(z), \quad P_L\text{-a.s.}$$

Letting $L \to \infty$ it follows that $\tilde{\nu}_\omega(z) = \nu_\omega(z)$, P-a.s. Similarly to the beginning of this section, we deduce that $\tilde{\lambda}_\omega(z) = \lambda_\omega(z)$, P-a.s. In order to show that $\tilde{h}_\omega(z) = h_\omega(z)$, P-a.s. we consider again the systems $(\Omega, \mathcal{F}, P_L, \vartheta_L)$, use (4.4.7) and that $\tilde{\lambda}_\omega(z) = \lambda_\omega(z)$ and $\tilde{\nu}_\omega(z) = \nu_\omega(z)$ in order to deduce that $\tilde{h}_\omega(z) = h_\omega(z)$, P_L-a.s. We complete the proof by letting $L \to \infty$. $\quad\square$

4.6   The largest characteristic exponents

We recall that when θ is ergodic then by Kigman's subadditive ergodic theorem, for any $z \in \mathbb{C}$ the so-called largest characteristic exponent of the cocycle $\{A_z^\omega : \omega \in \Omega\}$,

$$\Lambda(z) = \lim_{n\to\infty} \frac{1}{n} \ln \|A_z^{\omega,n}\|$$

exists P-a.s. and does not depend on ω. In this section we will show that there exists a ball $B(0, \delta_0)$, $\delta_0 > 0$ around 0 such that with probability one all the limits $\Lambda(z)$, $z \in B(0, \delta_0)$ exist and we will describe the behavior of the function $\Lambda(\cdot)$. These results will be obtained under the following assumptions.

Assumption 4.6.1. The map θ is ergodic and the random variables

$$\|\kappa_\omega\|, \|l_\omega\|, \|\kappa_\omega\|^{-1}, \|l_\omega\|^{-1}, M_\omega, K_\omega \text{ and } |l_\omega h_\omega(0)|$$

are bounded from above by some $a^{-1} > 1$, i.e. $P(G_a) = 1$ where G_a is the set defined in (4.4.14).

Assumption 4.6.2. There exists a constant $C > 1$ such that

$$\sup_{z\in\bar{V}} \|A_z^\omega\| \leq C|\lambda_\omega(0)|, \quad P\text{-a.s.}$$

where V is the neighborhood of 0 specified before Theorem 4.2.2.

We will show that (under reasonable conditions) these assumptions hold true in our application to random transfer operators in Chapter 5.

4.6.1 *Analyticity of the largest characteristic exponent around* 0

Our main result here is the following.

Theorem 4.6.3. *Suppose that Assumptions 4.6.1 and 4.6.2 and the assumptions of Theorem 4.2.2 hold true.*

(i) There exists a constant $\delta_0 > 0$ such that P-a.s. for any $z \in B(0, \delta_0)$ we have $\frac{1}{2} \leq \ln |\lambda_\omega(z)/\lambda_\omega(0)| \leq \frac{3}{2}$ and

$$\lim_{k \to \infty} \frac{1}{k} \ln \|A_z^{\omega,k}\|/|\lambda_{\omega,k}(0)| = \int \ln |\lambda_\omega(z)/\lambda_\omega(0)| dP(\omega) := L_0(z).$$

As a consequence,

$$\lim_{k \to \infty} \frac{1}{k} \ln |\lambda_{\omega,k}(0)| = \Lambda(0), \quad P\text{-a.s.} \tag{4.6.1}$$

and

$$\Lambda(z) = \Lambda(0) + L_0(z) \tag{4.6.2}$$

where we use the convention $-\infty + c = -\infty$ for any $c < \infty$. The function $L_0(\cdot)$ is harmonic function of the variable $(\Re(z), \Im(z))$.

(ii) Suppose, in addition, that $\ln |\lambda_\omega(0)|$ is integrable. Then the random variables $\ln |\lambda_\omega(z)|$, $z \in B(0, \delta_0)$ are integrable and

$$\lim_{n \to \infty} \frac{1}{n} \ln \|A_z^{\omega,n}\| = \int \ln |\lambda_\omega(z)| dP(\omega) := L(z), \quad P\text{-a.s.}$$

for any $z \in B(0, \delta_0)$. Moreover, for P-a.a. ω and any $q \in B_\omega$ and $z \in B(0, \delta_0)$ so that $\nu_\omega(z)q \neq 0$,

$$\lim_{n \to \infty} \frac{1}{n} \ln \|A_z^{\omega,n} q\| = L(z). \tag{4.6.3}$$

On the other hand, when $\nu_\omega(z)q = 0$ then

$$\limsup_{n \to \infty} \frac{1}{n} \ln \|A_z^{\omega,n} q\| \leq L(z) + \frac{\ln c}{j_0} < L(z) \tag{4.6.4}$$

where $c = \tanh(\frac{1}{4}d_0)$ and d_0 is specified in Assumption 4.1.6.

Note that Theorem 4.6.3 *(i)* states, in particular, that P-a.s. the limit $\Lambda(z)$ exist for any $z \in B(0, \delta_0)$.

Remark 4.6.4. Suppose that the conditions from Theorem 4.6.3 *(ii)* hold true. Consider the following decomposition $B_\omega = Ker(\nu_\omega(z)) + Span\{h_\omega(z)\}$. Then

$$Ker(\nu_\omega(z)) = \{q : \limsup_{n \to \infty} \|A_z^{\omega,n} q\|^{\frac{1}{n}} \leq (1 - c^{\frac{1}{j_0}})r(z)\} \tag{4.6.5}$$

while $\lim_{n\to\infty} \|A_z^{\omega,n} h_\omega(z)\|^{\frac{1}{n}} = r(z)$, where $r(z) = e^{L(z)} > 0$. When the spaces $B_\omega = B$ and the operators $A_z^\omega = A_z$ are not random, the above decomposition of B means that A_z has an isolated eigenvector $h_\omega(z) = h(z)$ with eigenvalue $r(z)$ and that the rest of its spectrum is contained in the closed disk around 0 with radius $(1 - c^{\frac{1}{j_0}})r(z)$ (see [28]), namely A_z has a spectral gap. In the random case we have decompositions of more than one operator and so the results obtained in this section can be viewed as a generalization of spectral gap to the case of decomposition of random operators.

Before starting to prove Theorem 4.6.3, we need the following.

4.6.2   *The pressure function*

Proposition 4.6.5. *Suppose that the Assumptions of Theorem 4.2.1 hold true. Then, P-a.s. there exist $s = s_\omega > 0$ and an analytic function $\bar{\Pi}_\omega(\cdot)$: $B(0, s) \to \mathbb{C}$ satisfying*

$$e^{\bar{\Pi}_\omega(z)} = \frac{\lambda_\omega(z)}{\lambda_\omega(0)}, \quad \bar{\Pi}_\omega(0) = 0 \qquad (4.6.6)$$

and

$$|\bar{\Pi}_\omega(z)| \le \ln 2 + \pi. \qquad (4.6.7)$$

Moreover, when also Assumptions 4.6.1 and 4.6.2 hold true then there exists a constant $\delta_1 > 0$ such that $\delta_1 \le s_\omega$, P-a.s.

Proof. For P-a.a. ω set

$$\bar{\lambda}_\omega(z) = \frac{\lambda_\omega(z)}{\lambda_\omega(0)}, \quad z \in U.$$

Let $V_1 = B(0, 2\varepsilon)$ be an open ball around $0 \in \mathbb{C}$ whose closure $\bar{V}_1$ is contained in U. The map $z \to \|A_z^\omega\|$ is continuous and therefore achieves a maximum on $\bar{V}_1$, which we denote by $S = S(\omega)$. By (4.2.1),

$$\lambda_\omega(z) = \kappa_\omega (A_z^\omega)^* \nu_{\theta\omega}(z)$$

which together with (4.3.27) yields

$$|\lambda_\omega(z)| \le \|\kappa_\omega\| \cdot \|(A_z^\omega)^*\| \cdot \|\nu_{\theta\omega}(z)\| \le \|\kappa_\omega\| \cdot S \cdot M_{\theta\omega} \cdot \|\kappa_{\theta\omega}\|^{-1} := \epsilon_\omega$$

and it follows that $|\bar{\lambda}_\omega(z)| \le \bar{\epsilon}_\omega := \frac{\epsilon_\omega}{|\lambda_\omega(0)|}$. Applying Lemma 2.8.2 with $k = 0$ we deduce that

$$|\bar{\lambda}_\omega(z) - 1| \le \frac{4\bar{\epsilon}_\omega |z|}{\delta}$$

for any $z \in B(0, \varepsilon)$. Set $s = \frac{\varepsilon}{8\bar{\varepsilon}_\omega}$. Then for any $z \in B(0, s)$,

$$\frac{1}{2} \leq |\bar{\lambda}_\omega(z)| \leq \frac{3}{2} \tag{4.6.8}$$

and therefore we can define on $B(0, s)$ a branch of the logarithm of the function $\bar{\lambda}_\omega(\cdot)$, i.e. an analytic function $\bar{\Pi}_\omega : B(0, s) \to \mathbb{C}$ satisfying (4.6.6) and (4.6.7).

Finally, suppose that Assumptions 4.6.1 and 4.6.2 hold true. Let a be as specified in Assumption 4.6.1 and C as specified in Assumption 4.6.2. Then $s_\omega \geq \frac{a^3 \varepsilon}{8C}$, P-a.s.   □

4.6.3   *Proof of Theorem 4.6.3*

Suppose that all the conditions of Theorem 4.6.3 *(i)* hold true. Consider the neighborhood $U_2 = B(0, \frac{a^5 \delta}{8}) := B(0, \delta_2)$ of 0 specified before (4.4.15). By taking a sufficiently small ε in the proof of Proposition 4.6.5 we can always assume that $\delta_1 < \delta_2$. Next, by (4.6.8),

$$\frac{1}{2} \leq |\bar{\lambda}_\omega(z)| \leq \frac{3}{2} \tag{4.6.9}$$

for any $z \in B(0, \delta_1)$. In particular, $\ln |\bar{\lambda}_\omega(z)|$ is integrable. Set $\delta_0 = \frac{1}{2}\delta_1$. We claim that P-a.s.,

$$\lim_{k \to \infty} \frac{1}{k} \sum_{j=0}^{k-1} \ln |\bar{\lambda}_{\theta^j \omega}(z)| = L_0(z) \tag{4.6.10}$$

for any $z \in B(0, \delta_0)$. Indeed, applying the mean ergodic theorem with the random variables $\lambda_\omega(z)$ we deduce that (4.6.10) holds true P-a.s. for z's belonging to a countable dense subset of $B(0, \delta_0)$. Applying Lemma 2.8.2 with $k = 0$ and then using (4.6.9) we infer that there exists a constant $d > 0$ such that P-a.s. for any $z_0 \in B(0, \delta_0)$ and $h \in B(0, \frac{1}{2}\delta_0)$,

$$\left| \frac{\bar{\lambda}_\omega(z_0 + h)}{\bar{\lambda}_\omega(z_0)} - 1 \right| \leq d|h|.$$

We conclude that the function $L_0(\cdot)$ is continuous and that the family of functions

$$\frac{1}{k} \sum_{j=0}^{k-1} \ln |\bar{\lambda}_{\theta^j \omega}(\cdot)| : B(0, \delta_0) \to \mathbb{C}$$

is equicontinuous, where ω belongs to a full P-probability set and $k \in \mathbb{N}$, and our claim follows.

Next, in the case when $l_i = i$, taking supremum in (4.4.15) over all g's such that $\|g\| = 1$ and taking into account (4.3.27), the lower bound from (4.4.3) and the upper bound from (4.4.5), we deduce that for any sufficiently large k,

$$C_1(\omega)|\lambda_{\omega,k}(z)| \leq \|A_z^{\omega,k}\| \leq C_2(\omega)|\lambda_{\omega,k}(z)| \qquad (4.6.11)$$

for any $z \in U_2$, where $C_1(\omega), C_2(\omega) > 0$. Taking the logarithms of the expressions in (4.6.11) and observing that

$$\ln|\bar{\lambda}_{\omega,k}(z)| = \sum_{j=0}^{k-1} \ln|\bar{\lambda}_{\theta^j\omega}(z)|$$

it follows that

$$\lim_{k\to\infty} \frac{1}{k} \ln\|A_z^{\omega,k}\||\lambda_{\omega,k}(0)|^{-1} = \int \ln|\bar{\lambda}_{\omega_0}(z)|dP(\omega_0) = L_0(z), \quad P\text{-a.s.}$$
$$(4.6.12)$$

for any $z \in B(0,\delta_0)$, where we also used (4.6.10). It is clear now from (4.6.12) that equalities (4.6.1) and (4.6.2) hold true. In order to complete the proof of Theorem 4.6.3 *(i)*, we show now that the function $L_0(\cdot)$ is harmonic. Indeed, since

$$L_0(z) = \Re\Big(\int \bar{\Pi}_\omega(z)dP(\omega)\Big)$$

it is sufficient to show that the function $z \to \int \bar{\Pi}_\omega(z)dP(\omega)$ is analytic. Indeed, applying Lemma 2.8.2 with $Q = \bar{\Pi}_\omega$, $U = B(0,2\delta_0)$, $k = 1$, $z_0 = 0$ and $\delta = \delta_0$ and taking into account that $\sup_{z\in B(0,2\delta_0)} |\bar{\Pi}_\omega(z)| \leq \ln 2 + \pi$, we obtain uniform in ω estimates on Taylor reminders (around 0) of $\bar{\Pi}\omega$ which guarantees that L_0 is analytic in $B(0,\delta_0)$.

In order to prove Theorem 4.6.3 *(ii)*, suppose that $\ln|\lambda_\omega(0)|$ is integrable. Then by the mean ergodic theorem,

$$\lim_{k\to\infty} \frac{1}{k} \ln|\lambda_{\omega,k}(0)| = \lim_{k\to\infty} \frac{1}{k} \sum_{j=0}^{k-1} \ln|\lambda_{\theta^j\omega}(0)|$$

$$= \int \ln|\lambda_\omega(0)|dP(\omega), \quad P\text{-a.s.}$$

and we conclude from (4.6.12) that P-a.s.,

$$\lim_{k\to\infty} \frac{1}{k} \ln\|A_z^{\omega,k}\| = \int \ln|\lambda_\omega(z)|dP(\omega) := L(z) \qquad (4.6.13)$$

for any $z \in B(0,\delta_0)$.

Next, we prove the rest of Theorem 4.6.3 *(ii)*. Let $z \in B(0, \delta_0)$ and $q \in B_\omega$ be such that $|\nu_\omega(z)(q)| > 0$. Taking into account (4.3.27), the lower bound in (4.4.3) and the upper bound in (4.4.5), we deduce from (4.4.15) in the case when $P(G_a) = 1$ that

$$C_1(q, \omega, z)|\lambda_{\omega,k}(z)| \leq \|A_z^{\omega,k} q\| \leq C_2(q, \omega, z)|\lambda_{\omega,k}(z)| \qquad (4.6.14)$$

for any sufficiently large k, where $C_1(q, \omega, z), C_2(q, \omega, z) > 0$. This together with (4.6.11) implies that

$$D_1(q, \omega, z)\|A_z^{\omega,k}\| \leq \|A_z^{\omega,k} q\| \leq D_2(q, \omega, z)\|A_z^{\omega,k}\|$$

for any sufficiently large k, for some $D_1(q, \omega, z), D_2(q, \omega, z) > 0$, and (4.6.3) follows from (4.6.13).

On the other hand, if $\nu_\omega(z)q = 0$ then by (4.4.15),

$$\|A_z^{\omega,k} q / \lambda_{\omega,k}(z)\| \leq C(\omega)\|q\| c^{\frac{k}{j_0}}$$

for some $C(\omega) > 0$. This together with the previous arguments shows that

$$\limsup_{k \to \infty} \frac{1}{k} \ln \|A_z^{\omega,k} q / \lambda_{\omega,k}(0)\| \leq L_0(z) + j_0^{-1} \ln c < L_0(z)$$

and since $\ln |\lambda_\omega(0)|$ is integrable we derive that

$$\limsup_{n \to \infty} \frac{1}{n} \ln \|A_z^{\omega,n} q\| \leq L(z) + j_0^{-1} \ln c < L(z).$$

$\square$

Chapter 5

Application to random locally distance expanding covering maps

5.1 Random locally expanding covering maps

In this chapter we apply the results from Chapter 4 in the setup which consists of a complete probability space $(\Omega, \mathcal{F}, P)$ together with an invertible P-preserving transformation $\theta : \Omega \to \Omega$, of a compact metric space $(\mathcal{X}, \rho)$ normalized in size so that $\mathrm{diam}\mathcal{X} \leq 1$ together with the Borel σ-algebra $\mathcal{B}$, and of a set $\mathcal{E} \subset \Omega \times \mathcal{X}$ measurable with respect to the product σ-algebra $\mathcal{F} \times \mathcal{B}$ such that the fibers $\mathcal{E}_\omega = \{x \in \mathcal{X} : (\omega, x) \in \mathcal{E}\}$, $\omega \in \Omega$ are compact. The latter yields (see [16] Chapter III) that the mapping $\omega \to \mathcal{E}_\omega$ is measurable with respect to the Borel σ-algebra induced by the Hausdorff topology on the space $\mathcal{K}(\mathcal{X})$ of compact subspaces of $\mathcal{X}$ and the distance function $d(x, \mathcal{E}_\omega)$ is measurable in ω for each $x \in \mathcal{X}$. Furthermore, the projection map $\pi_\Omega(\omega, x) = \omega$ is measurable and it maps any $\mathcal{F} \times \mathcal{B}$-measurable set to a $\mathcal{F}$-measurable set (see "measurable projection" Theorem III.23 in [16]). Let

$$\{T_\omega : \mathcal{E}_\omega \to \mathcal{E}_{\theta\omega}, \ \omega \in \Omega\}$$

be a collection of continuous bijective maps between the metric spaces $\mathcal{E}_\omega$ and $\mathcal{E}_{\theta\omega}$ so that the map $(\omega, x) \to T_\omega x$ is measurable with respect to the σ-algebra $\mathcal{P}$ which is the restriction of $\mathcal{F} \times \mathcal{B}$ on $\mathcal{E}$. Consider the skew product transformation $T : \mathcal{E} \to \mathcal{E}$ given by

$$T(\omega, x) = (\theta\omega, T_\omega x). \tag{5.1.1}$$

For any $\omega \in \Omega$ and $n \in \mathbb{N}$ consider the n-th step iterates T_ω^n given by

$$T_\omega^n = T_{\theta^{n-1}\omega} \circ \cdots \circ T_{\theta\omega} \circ T_\omega : \mathcal{E}_\omega \to \mathcal{E}_{\theta^n\omega}. \tag{5.1.2}$$

Our additional requirements concerning the family of maps $\{T_\omega : \omega \in \Omega\}$ are collected in the following assumptions.

Assumption 5.1.1 (Topological exactness). There exist a constant $\xi > 0$ and a random variable $n_\omega \in \mathbb{N}$ such that P-a.s.,

$$T_\omega^{n_\omega}(B_\omega(x, \xi)) = \mathcal{E}_{\theta^{n_\omega}\omega} \text{ for any } x \in \mathcal{E}_\omega \qquad (5.1.3)$$

where for any $\omega \in \Omega$, $x \in \mathcal{E}_\omega$ and $r > 0$, $B_\omega(x, r)$ denotes a ball in $\mathcal{E}_\omega$ around x with radius r.

Assumption 5.1.2 (The pairing property). There exist random variables $\gamma_\omega > 1$ and $D_\omega \in \mathbb{N}$ such that P-a.s. for any $x, x' \in \mathcal{E}_{\theta\omega}$ with $\rho(x, x') < \xi$ we can write

$$T_\omega^{-1}\{x\} = \{y_1, ..., y_k\} \text{ and } T_\omega^{-1}\{x'\} = \{y_1', ..., y_k'\} \qquad (5.1.4)$$

where ξ is specified in Assumption 5.1.1,

$$k = k_{\omega, x} = |T_\omega^{-1}\{x\}| \leq D_\omega$$

and

$$\rho(y_i, y_i') \leq (\gamma_\omega)^{-1}\rho(x, x') \qquad (5.1.5)$$

for any $1 \leq i \leq k$.

Next, for any $n \in \mathbb{N}$ set

$$\gamma_{\omega, n} = \prod_{i=0}^{n-1} \gamma_{\theta^i\omega} \text{ and } D_{\omega, n} = \prod_{i=0}^{n-1} D_{\theta^i\omega}. \qquad (5.1.6)$$

When the pairing property holds true then it follows by induction on n that for P-a.a. ω and for any $x, x' \in \mathcal{E}_{\theta^n\omega}$ with $\rho(x, x') < \xi$ we can write

$$(T_\omega^n)^{-1}\{x\} = \{y_1, ..., y_k\} \text{ and } (T_\omega^n)^{-1}\{x'\} = \{y_1', ..., y_k'\} \qquad (5.1.7)$$

where

$$k = k_{\omega, x, n} = |(T_\omega^n)^{-1}\{x\}| \leq D_{\omega, n},$$

$|\Gamma|$ denotes the cardinality of a finite set Γ and

$$\rho(T_\omega^j y_i, T_\omega^j y_i') \leq (\gamma_{\theta^j\omega, n-j})^{-1}\rho(x, x') \qquad (5.1.8)$$

for any $1 \leq i \leq k$ and $0 \leq j < n$.

We will also use the following. By Lemma 4.11 in [53] (applied with $r = \xi$), there exists an integer valued random variable $L_\omega \geq 1$ and $\mathcal{F}$-measurable functions $\omega \to x_{\omega, i} \in \mathcal{X}$, $i = 1, 2, 3, ...$ so that $x_{\omega, i} \in \mathcal{E}_\omega$ for each i and

$$\bigcup_{k=1}^{L_\omega} B_\omega(x_{\omega, k}, \xi) = \mathcal{E}_\omega, \quad P\text{-a.s.} \qquad (5.1.9)$$

Next, let $g : \mathcal{E} \to \mathbb{C}$ be a measurable function. For any $\omega \in \Omega$ consider the function $g_\omega : \mathcal{E}_\omega \to \mathbb{C}$ given by $g_\omega(x) = g(\omega, x)$. For any $0 < \alpha \leq 1$ set

$$v_{\alpha,\xi}(g_\omega) = \inf\{R : |g_\omega(x) - g_\omega(x')| \leq R\rho^\alpha(x, x') \text{ if } \rho(x, x') < \xi\}$$

$$\text{and} \quad \|g_\omega\|_{\alpha,\xi} = \|g_\omega\|_\infty + v_{\alpha,\xi}(g_\omega)$$

where $\| \cdot \|_\infty$ is the supremum norm and $\rho^\alpha(x, x') = (\rho(x, x'))^\alpha$. These norms are $\mathcal{F}$-measurable as a consequence of the following result (cf. Lemma 4.1 in [53]) together with Theorem III.23 in [16].

Lemma 5.1.3. *Let $(\Omega, \mathcal{F})$ be a measurable space, $\mathcal{X}$ be another space and $\mathcal{Q}$ be a σ-algebra on the product space $\Omega \times \mathcal{X}$ so that for any $\Gamma \in \mathcal{Q}$,*

$$\pi_\Omega \Gamma = \{\omega \in \Omega : (\omega, x) \in \Gamma \text{ for some } x \in \mathcal{X}\} \in \mathcal{F}. \qquad (5.1.10)$$

Then for any $\mathcal{Q}$-measurable function $g = g_\omega(x) = g(\omega, x)$ on $\Omega \times \mathcal{X}$ the supremum norm

$$\|g_\omega\|_\infty = \sup_{x \in \mathcal{X}} |g(\omega, x)|$$

is $\mathcal{F}$-measurable as a function of $\omega \in \Omega$.

Proof. Since $|g|$ is also $\mathcal{Q}$-measurable, each set $\Gamma_k^{(n)} = \{(\omega, x) : \frac{k}{n} \leq |g(\omega, x)| < \frac{k+1}{n}\}$ belongs to $\mathcal{Q}$. Then the function $g^{(n)} = \sum_{k=0}^\infty \frac{k}{n} \mathbb{I}_{\Gamma_k^{(n)}}$, where $\mathbb{I}_\Gamma$ is the indicator of Γ, is $\mathcal{Q}$-measurable. Consider the supremum norms $\|g_\omega^{(n)}\|_\infty = \sup_{x \in \mathcal{X}} g^{(n)}(\omega, x)$. Then

$$\{\omega : \|g_\omega^{(n)}\|_\infty \leq \frac{k}{n}\} = \bigcup_{j=0}^k \pi_\Omega \Gamma_k^{(n)} \in \mathcal{F}$$

and so $\|g_\omega^{(n)}\|_\infty$ is $\mathcal{F}$-measurable. Since

$$\sup_{(\omega,x) \in \Omega \times \mathcal{X}} \left||g(\omega, x)| - g^{(n)}(\omega, x)\right| \leq \frac{1}{n}$$

we have also for $\omega \in \Omega$ that

$$\left|\|g_\omega\|_\infty - \|g_\omega^{(n)}\|_\infty\right| \leq \frac{1}{n}.$$

Hence $\|g_\omega\|_\infty = \lim_{n\to\infty} \|g_\omega^{(n)}\|_\infty$ implying the assertion of the lemma. $\qquad \square$

Next, consider the Banach spaces

$$(B_\omega, \| \cdot \|) = (\mathcal{H}_\omega^{\alpha,\xi}, \| \cdot \|_{\alpha,\xi})$$

of all functions $h : \mathcal{E}_\omega \to \mathbb{C}$ such that $\|h\|_{\alpha,\xi} < \infty$ and denote by $\mathcal{H}_{\omega,\mathbb{R}}^{\alpha,\xi}$ the space of all real valued functions in $\mathcal{H}_\omega^{\alpha,\xi}$. Let $H = H_\omega \geq 1$ be a random variable so that $\int \ln H_\omega dP(\omega) < \infty$ and let $\mathcal{H}^{\alpha,\xi}(H)$ be the set of all measurable functions $g : \mathcal{E} \to \mathbb{C}$ such that $v_{\alpha,\xi}(g_\omega) \leq H_\omega$ for any ω. For any $\omega \in \Omega$ consider the set $\mathcal{Y}_\omega$ of all complex continuous linear functionals $\hat{x} \in B_\omega^*$ which have the form $\hat{x}(g) = g(x)$ for some $x \in \mathcal{E}_\omega$. Consider the space

$$\mathcal{Y} = \bigcup_{\omega \in \Omega} \{\omega\} \times \mathcal{Y}_\omega.$$

Identifying $\hat{x}$ and x, we endow $\mathcal{Y}$ with the σ-algebra $\mathcal{P}$ on $\mathcal{E}$ induced by $\mathcal{F} \times \mathcal{B}$. Then we are in the situation of Example 4.1.4 in Chapter 4 and so we can extend $\mathcal{P}$ to a σ-algebra $\mathcal{G}$ defined on B^* and with this extension a measurable family in the sense of Chapter 4 is just a measurable function $g : \mathcal{E} \to \mathbb{C}$ so that $g(\omega, \cdot) \in B_\omega$ for each ω and a measurable family of elements of $(B_\omega)^*$ is just a family of continuous linear functionals $\mu_\omega \in (B_\omega)^*$ so that the map $\omega \to \mu_\omega(g(\omega, \cdot))$ is measureable for any measurable function $g : \mathcal{E} \to \mathbb{C}$ with $g(\omega, \cdot) \in B_\omega$ for any ω.

Next, for any function $\psi : \mathcal{E} \to \mathbb{C}$ consider the random functions $S_n^\omega \psi : \mathcal{E}_\omega \to \mathbb{C}$, $\omega \in \Omega$, $n \in \mathbb{N}$ given by

$$S_n^\omega \psi = \sum_{i=0}^{n-1} \psi_{\theta^i \omega} \circ T_\omega^i$$

where $\psi_\omega = \psi(\omega, \cdot)$. Let the random variable $Q_\omega(H)$ be defined by

$$Q_\omega(H) = \sum_{j=1}^{\infty} H_{\theta^{-j}\omega}(\gamma_{\theta^{-j}\omega,j})^{-\alpha}. \tag{5.1.11}$$

Note that Q_ω is bounded when H_ω is a bounded random variable and $\gamma_\omega \geq 1 + \delta$, P-a.s. for some $\delta > 0$. The following simple distortion property is a direct consequence of (5.1.8).

Lemma 5.1.4. *Let* $\in \mathcal{H}^{\alpha,\xi}(H)$. *For any ω and $n \geq 1$, let $x, x' \in \mathcal{E}_{\theta^n \omega}$ with $\rho(x, x') < \xi$ and $y_1, ..., y_k$ and $y_1', ..., y_k'$ satisfy (5.1.7) and (5.1.8). Then for any $1 \leq i \leq k$,*

$$|S_n^\omega \psi(y_i) - S_n^\omega \psi(y_i')| \tag{5.1.12}$$

$$\leq \rho^\alpha(x, x') \sum_{j=0}^{n-1} H_{\theta^j \omega}(\gamma_{\theta^j \omega, n-j})^{-\alpha} \leq \rho^\alpha(x, x') Q_{\theta^n \omega}(H).$$

5.2 Transfer operators

Let $\phi, u : \mathcal{E} \to \mathbb{R}$ be measurable functions.

Assumption 5.2.1. For any $\omega \in \Omega$, the functions $\phi_\omega, u_\omega : \mathcal{E}_\omega \to \mathbb{R}$ given by $\phi_\omega(\cdot) = \phi(\omega, \cdot)$ and $u_\omega(\cdot) = u(\omega, \cdot)$ are continuous and satisfy (5.1.12).

By Lemma 5.1.4 this assumption holds true when $\phi_\omega, u_\omega \in \mathcal{H}_\omega^{\alpha, \xi}(H)$. Still, Assumption 5.2.1 holds true in more general situations which will make the results from this section applicable for the random transfer operators introduced in the proof of the nonconventional LLT from Chapter 2 in Part 1.

Let $z \in \mathbb{C}$ and consider the transfer operators $\mathcal{L}_z^\omega$, $\omega \in \Omega$ which map functions from $\mathcal{E}_\omega$ to functions from $\mathcal{E}_{\theta\omega}$ by the formula

$$\mathcal{L}_z^\omega g(x) = \sum_{y \in T_\omega^{-1}\{x\}} e^{\phi_\omega(y) + z u_\omega(y)} g(y). \tag{5.2.1}$$

Note that under Assumption 5.1.2 the operators $\mathcal{L}_z^\omega$, $z \in \mathbb{C}$ are well defined and when also Assumption 5.2.1 holds true they map a continuous function on $\mathcal{E}_\omega$ to a continuous function on $\mathcal{E}_{\theta\omega}$. For any $n \in \mathbb{N}$ and $z \in \mathbb{C}$ consider the n-th step iterates $\mathcal{L}_z^{\omega, n}$ of the transfer operator given by

$$\mathcal{L}_z^{\omega, n} = \mathcal{L}_z^{\theta^{n-1}\omega} \circ \cdots \circ \mathcal{L}_z^{\theta\omega} \circ \mathcal{L}_z^\omega. \tag{5.2.2}$$

Then

$$\mathcal{L}_z^{\omega, n} g(x) = \sum_{y \in (T_\omega^n)^{-1}\{x\}} e^{S_n^\omega \phi(y) + z S_n^\omega u(y)} g(y) \tag{5.2.3}$$

namely $\mathcal{L}_z^{\omega, n}$ is the transfer operator generated by the map T_ω^n and the function $S_n^\omega \phi + z S_n^\omega u$.

Next, consider the (global) transfer operator $\mathcal{L}_z$ acting on functions $g : \mathcal{E} \to \mathbb{C}$ by the formula

$$\mathcal{L}_z g(s) = \sum_{s' \in T^{-1}\{s\}} e^{\phi_{\theta^{-1}\omega}(y) + z u_{\theta^{-1}\omega}(y)} g(s') = \mathcal{L}_z^{\theta^{-1}\omega} g_{\theta^{-1}\omega}(x), \quad s = (\omega, x),$$

$$\tag{5.2.4}$$

namely $\mathcal{L}_z$ is generated by the skew product map T and the function $\phi + zu$. For any $\psi : \mathcal{E} \to \mathbb{C}$ and $n \in \mathbb{N}$ set

$$S_n \psi = \sum_{i=0}^{n-1} \circ T^i.$$

Then, $S_n \psi(\omega, y) = S_n^\omega \psi(y)$ and the iterates of the operator $\mathcal{L}_z$ satisfy

$$\mathcal{L}_z^n g(s) = \sum_{s' \in T^{-n}\{s\}} e^{S_n \phi(s') + z S_n u(s')} g(s') = \mathcal{L}_z^{\theta^{-n}\omega, n} g_{\theta^{-n}\omega}(x), \quad s = (\omega, x).$$

Assumption 5.2.2. The transfer operators $\mathcal{L}_z$, $z \in \mathbb{C}$ map measurable functions on $\mathcal{E}$ to measurable functions on $\mathcal{E}$.

5.3   Real and complex cones

In this section we will describe the cones for which we apply the result from Chapter 4. Set

$$Q_\omega = Q_\omega(2H) = 2Q_\omega(H) \tag{5.3.1}$$

and let ω be so that $Q_\omega < \infty$. Following [53], let $s > 1$ and consider the real proper convex cones (see Appendix A),

$$\mathcal{C}_{\omega,\mathbb{R}} = \mathcal{C}_{\omega,\mathbb{R},s} \tag{5.3.2}$$

$$= \{g : \mathcal{E}_\omega \to [0,\infty) : g(x) \le e^{sQ_\omega \rho^\alpha(x,x')} g(x') \text{ if } \rho(x,x') < \xi\}.$$

Next, let the functional $l_\omega : \mathbb{C}^{\mathcal{E}_\omega} \to \mathbb{C}$ be given by the formula

$$l_\omega(g) = \sum_{j=1}^{L_\omega} g(x_{\omega,j}) \tag{5.3.3}$$

where the $x_{\omega,j}$'s are the points in $\mathcal{E}_\omega$ satisfying (5.1.9). It is clear that l_ω is measurable in the sense described in Section 5.1 with the σ-algebra $\mathcal{G}$ on B^* defined there. Note that $\|l_\omega\|_{\alpha,\xi} \le \|l_\omega\|_\infty = l_\omega(\mathbf{1}) = L_\omega$ and therefore

$$\|l_\omega\|_{\alpha,\xi} = L_\omega$$

where the (operator) norm $\|l_\omega\|_{\alpha,\xi}$ of l_ω is either with respect to $\mathcal{H}_{\omega,\mathbb{R}}^{\alpha,\xi}$ or with respect to $\mathcal{H}_\omega^{\alpha,\xi}$. In particular, $\|l_\omega\|_{\alpha,\xi}$ is measurable in ω.

Theorem 5.3.1. *Let $\omega \in \Omega$ be such that $Q_\omega < \infty$.*

(i) The cone $\mathcal{C}_{\omega,\mathbb{R}}$ is a closed subsets of the real Banach space $\mathcal{H}_{\omega,\mathbb{R}}^{\alpha,\xi}$. Let $\mathcal{C}_\omega$ be the canonical complexification of $\mathcal{C}_{\omega,\mathbb{R}}$ (see Appendix A). Then the complex cone $\mathcal{C}_\omega$ is linearly convex. Moreover, when Q_ω is P-a.s. finite then the first part of Assumption 4.1.1 holds true with the family $b_\omega \equiv \mathbf{1}$, where $\mathbf{1}$ is the function which takes the constant value 1.

(ii) Set $K_\omega = 2\sqrt{2}L_\omega(sQ_\omega e^{sQ_\omega \xi^\alpha} + 1)e^{sQ_\omega \xi^\alpha}$ and $M_\omega = 8(1 - e^{-sQ_\omega \xi^\alpha})^{-2}$. Then

$$\|x\|_{\alpha,\xi}\|l_\omega\|_{\alpha,\xi} \le K_\omega |l_\omega(x)| \quad \text{for any } x \in \mathcal{C}_\omega \tag{5.3.4}$$

and

$$\|\mu\|_{\alpha,\xi} \le M_\omega |\mu(\mathbf{1})| \quad \text{for any } \mu \in \mathcal{C}_\omega^*. \tag{5.3.5}$$

Therefore when Q_ω is P-a.s. finite the cones $\mathcal{C}_\omega$ and $\mathcal{C}_\omega^*$ satisfy Assumption 4.1.1 with the above K_ω, M_ω, the functional l_ω defined in (5.5.11) and the functional $\kappa_\omega \in (B_\omega^*)^*$ given by $\kappa_\omega \mu = \mu \mathbf{1}$ which is $\mathcal{G}$-measurable, where $\mathcal{G}$ was defined in Section 5.1.

(iii) For any $f \in B_\omega$ there exist $f_1, ..., f_8 \in C_\omega$ so that $f = f_1 + f_2 + ... + f_8$ and

$$\|f_1\|_{\alpha,\xi} + \|f_2\|_{\alpha,\xi} + ... + \|f_8\|_{\alpha,\xi} \le r_\omega \|f\|_{\alpha,\xi} \qquad (5.3.6)$$

where $r_\omega = 4(1 + \frac{2}{sQ_\omega})$. Hence Assumption 4.1.3 holds true with $k_\omega = 8$ and the above r_ω when Q_ω is P-a.s. finite.

5.4 RPF triplets

Theorem 5.4.1. *Suppose that Assumption 5.2.1 holds true and that Q_ω is P-a.s. finite.*

(i) The random operators $\mathcal{L}_z^\omega$, $z \in \mathbb{C}$, map B_ω to $B_{\theta\omega}$, they are continuous with respect to the $\| \cdot \|_{\alpha,\xi}$-norms and are analytic in z. As a consequence, under Assumption 5.2.2 the family A_z^ω, $\omega \in \Omega$ is $\mathcal{G}$-measurable in the sense of Section 4.1 where $\mathcal{G}$ is defined in Section 5.1.

(ii) Suppose, in addition, that there exists $\delta > 0$ so that $\gamma_\omega \ge 1 + \delta$, P-a.s. and that the random variables

$$H_\omega, n_\omega, \|\phi_\omega\|_\infty, \|u_\omega\|_\infty \text{ and } D_{\omega, n_\omega} \qquad (5.4.1)$$

are bounded. Then Q_ω is a bounded random variable and Assumption 4.1.2 holds true with $A_z^\omega = \mathcal{L}_z^\omega$ in some deterministic neighborhood U of 0.

Corollary 5.4.2. *Suppose that all the assumptions of Theorem 5.4.1 hold true. Let U be the neighborhood of 0 from Theorem 5.4.1 (ii). Then P-a.s. for any $z \in U$ there exists a unique triplet $\lambda_\omega(z)$, $h_\omega(z)$ and $\nu_\omega(z)$ consisting of a nonzero complex number $\lambda_\omega(z)$, a complex function $h_\omega(z) \in C_\omega$ and a complex continuous linear functional $\nu_\omega(z) \in C_\omega^*$ such that*

$$A^\omega h_\omega(z) = \lambda_\omega(z) h_{\theta\omega}(z), \ (A_z^\omega)^* \nu_{\theta\omega}(z) = \lambda_\omega(z) \nu_\omega(z) \text{ and} \qquad (5.4.2)$$
$$\nu_\omega(z) h_\omega(z) = \nu_\omega(z) \mathbf{1} = 1.$$

For any $z \in U$ the maps $\omega \to \lambda_\omega(z)$ and $(\omega, x) \to h_\omega(z)(x)$, $(\omega, x) \in \mathcal{E}$ are measurable and the family $\nu_\omega(z)$ is measurable in ω. When $z = t \in \mathbb{R}$ then $\lambda_\omega(t) > 0$, the function $h_\omega(t)$ is strictly positive, $\nu_\omega(t)$ is a probability measure and the equality $\nu_{\theta\omega}(t)(\mathcal{L}_t^\omega g) = \lambda_\omega(t) \nu_\omega(t)(g)$ holds true for any bounded Borel function $g : \mathcal{E}_\omega \to \mathbb{C}$ and when the maps T_ω satisfy the assumptions from [53] then the triplet $(\lambda_\omega(t), h_\omega(t), \nu_\omega(t))$ coincides with the one constructed there with the function $\phi + tu$.

Moreover, when L_ω is bounded then this triplet is analytic and uniformly bounded around 0. Namely, the maps

$$\lambda_\omega(\cdot) : U \to \mathbb{C}, h_\omega(\cdot) : U \to \mathcal{H}_\omega^{\alpha,\xi} \text{ and } \nu_\omega(\cdot) : U \to \left(\mathcal{H}_\omega^{\alpha,\xi}\right)^*$$

are analytic, where $(\mathcal{H}_\omega^{\alpha,\xi})^*$ is the dual space of $\mathcal{H}_\omega^{\alpha,\xi}$, and there exists a deterministic neighborhood $U_0 \subset U$ of 0 such that for any $k \geq 0$ there is a constant $C_k > 0$ so that

$$\max\left(\sup_{z \in U_0} |\lambda_\omega^{(k)}(z)|, \sup_{z \in U_0} \|h_\omega^{(k)}(z)\|_{\alpha,\xi}, \sup_{z \in U_0} \|\nu_\omega^{(k)}(z)\|_{\alpha,\xi} \right) \leq C_k, \ P\text{-}a.s.$$

$$(5.4.3)$$

where $g^{(k)}$ stands for the k-th derivative of a function on the complex plane which takes values in some Banach space and $\|\nu\|_{\alpha,\xi}$ is the operator norm of a linear functional $\nu : \mathcal{H}_\omega^{\alpha,\xi} \to \mathbb{C}$.

Furthermore, the exponential convergences (4.2.2) and (4.2.3) hold true with $A_z^\omega = \mathcal{L}_z^\omega$ and all the random variables $k_1(\omega), C_\omega, \mathcal{D}_\omega$ and c_ω described in Theorem 4.2.2 can be replaced by constants when $z \in U_0$.

Theorems 5.3.1 and 5.4.1 will be proved in the remaining part of this chapter. Before proving them we will describe their application to the situation of random transfer operators arising in the proof of the nonconventional LLT in Chapter 2. Let $(\mathcal{X}, \rho)$ be the metric space and $T : \mathcal{X} \to \mathcal{X}$ be the locally distance expanding map described in Section 2.5. Consider the measure preserving system $(\Omega_\Theta, \mathcal{F}_\Theta, P_\Theta, \theta)$, $\theta = \vartheta^{-1}$ and the random family of transfer operators $\mathbf{L}_z^\omega$ defined in Section 2.7.1 using the map T^ℓ and the Hölder continuous functions $f : \mathcal{X} \to \mathbb{R}$ and $F : \mathcal{X}^\ell \to \mathbb{R}$. Let $\mathcal{E} = \Omega \times \mathcal{X}$, namely we consider the case when $\mathcal{E}_\omega = \mathcal{X}$ for any ω. Consider the random transfer operators $\mathcal{L}_z^\omega$ given by $\mathcal{L}_z^\omega = \mathbf{L}_z^{\vartheta^{-1}\omega} = \mathbf{L}_z^{\theta\omega}$. Then for any $n \in \mathbb{N}$,

$$\mathcal{L}_z^{\theta^{-n}\omega, n} = \mathbf{L}_z^\omega \circ \mathbf{L}_z^{\vartheta\omega} \circ \cdots \circ \mathbf{L}_z^{\vartheta^{n-1}\omega}$$

and therefore Theorems 2.7.1 and 2.7.2 will follow if Corollary 5.4.2 holds true for the random transfer operators $\mathcal{L}_z^\omega$ presented above. For any $\omega \in \Omega_\Theta$, set

$$T_\omega = T^\ell, \ \phi(\omega, x) = \sum_{k=0}^{\ell-1} f(T^k x) \text{ and } u(\omega, x) = \mathbf{F}_\omega(x) = F(p_0(\theta^{-1}\omega), x),$$

where p_0 is defined in Section 2.7.1. We claim that the assumptions of Corollary 5.4.2 hold true for this choice of T_ω, ϕ and u. Indeed, ϕ and u are measurable since f and F are continuous and the random variables $\|\phi_\omega\|_\infty$ and $\|u_\omega\|_\infty$ are bounded since f and F are bounded. Moreover, the transfer operators $\mathcal{L}_z^\omega$ are measurable in the sense required in Assumption 5.2.2 since T_ω does not depend on ω and ϕ and u are measurable. Furthermore, it follows by the definition of the map T introduced in Section 2.5.1 together with the pairing property discussed in Section 2.5.2 that the map $T_\omega = T^\ell$ satisfies the assumptions from Section 5.1 with constant n_ω, γ_ω and D_ω. Finally, Assumption 5.2.1 holds true with a constant H_ω since $f \in \mathcal{H}^{\alpha,\xi}(\mathcal{X})$ and F satisfies (2.5.18) and we can guarantee that L_ω would not depend on ω since $\mathcal{E}_\omega = \mathcal{X}$. We conclude that Corollary 5.4.2 holds true with the random transfer operators $\mathcal{L}_z^\omega = \mathbf{L}_z^{\theta\omega}$ and the system $(\Omega_\Theta, \mathcal{F}_\Theta, P_\Theta, \theta)$.

5.5 Properties of cones: proof of Theorem 5.3.1

Let $\omega \in \Omega$ be such that $Q_\omega < \infty$. We begin with the following lemma.

Lemma 5.5.1. *For any $g \in \mathcal{C}_{\omega,\mathbb{R}}$,*

$$L_\omega \|g\|_{\alpha,\xi} \le K_{\omega,\mathbb{R}} l_\omega(g) \tag{5.5.1}$$

where

$$K_{\omega,\mathbb{R}} = L_\omega (sQ_\omega e^{sQ_\omega \xi^\alpha} + 1)e^{sQ_\omega \xi^\alpha}. \tag{5.5.2}$$

In particular $\mathcal{C}_{\omega,\mathbb{R}} \subset \mathcal{H}_{\omega,\mathbb{R}}^{\alpha,\xi}$.

This lemma means that the aperture of the real cone $\mathcal{C}_{\omega,\mathbb{R}}$ (when considered as a subset of $\mathcal{H}_{\omega,\mathbb{R}}^{\alpha,\xi}$) does not exceed $K_{\omega,\mathbb{R}}$. The proof of the lemma goes exactly as the proof of Lemma 9.7 from [53] and is given here for readers' convenience.

Proof. Let $g \in \mathcal{C}_{\omega,\mathbb{R}}$. We first show that

$$\|g\|_\infty \le l_\omega(g)e^{sQ_\omega \xi^\alpha}. \tag{5.5.3}$$

Indeed, for any $x \in \mathcal{E}_\omega$ there exists an index $1 \le j \le L_\omega$ so that $x \in B(x_{\omega,j}, \xi)$. Therefore

$$g(x) \le g(x_{\omega,j})e^{sQ_\omega \xi^\alpha} \le l_\omega(g)e^{sQ_\omega \xi^\alpha}$$

and (5.5.3) follows by taking supremum over all $x \in \mathcal{E}_\omega$.

Next, let $x, x' \in \mathcal{E}_\omega$ be such that $\rho(x, x') < \xi$ and $g(x) \ge g(x')$. Then

$$|g(x) - g(x')| = g(x) - g(x') \le \left(e^{sQ_\omega \rho^\alpha(x,x')} - 1\right)g(x')$$

and so by the mean value theorem,

$$|g(x) - g(x')| \le sQ_\omega e^{sQ_\omega \rho^\alpha(x,x')}\rho^\alpha(x,x')\|g\|_\infty.$$

Reversing the roles of x and x', we conclude that this inequality holds true for any x and x' such that $\rho(x, x') < \xi$. Dividing by $\rho^\alpha(x, x')$ and taking supremum over all possible choices of x and x' we conclude that

$$v_{\alpha,\xi}(g) \le sQ_\omega e^{sQ_\omega \xi^\alpha} \|g\|_\infty$$

which implies that

$$\|g\|_{\alpha,\xi} \le (sQ_\omega e^{sQ_\omega \xi^\alpha} + 1)\|g\|_\infty$$

and (5.5.1) follows now from (5.5.3). □

Next, by Lemma 5.5.1 the cone $\mathcal{C}_{\omega,\mathbb{R}}$ is a closed subset of the real Banach space $\mathcal{H}_{\omega,\mathbb{R}}^{\alpha,\xi}$, namely it is closed with respect to the norm $\|\cdot\|_{\alpha,\xi}$. We conclude that the cone $\mathcal{C}_{\omega,\mathbb{R}}$ is a real proper closed convex cone (see Appendix A) when considered as a subset of $\mathcal{H}_{\omega,\mathbb{R}}^{\alpha,\xi}$. Let $\mathcal{C}_\omega \subset \mathcal{H}_\omega^{\alpha,\xi}$ be the canonical complexification of $\mathcal{C}_{\omega,\mathbb{R}}$ (see Appendix A) which according to (A.2.2) can also be written in the form

$$\mathcal{C}_\omega = \mathbb{C}'(\mathcal{C}_{\omega,\mathbb{R}} + i\mathcal{C}_{\omega,\mathbb{R}}) = \mathbb{C}'\{x + iy : x \pm y \in \mathcal{C}_{\omega,\mathbb{R}}\}, \quad \mathbb{C}' = \mathbb{C} \setminus \{0\}.$$

The following result is a corollary of Lemma 5.5.1 above and Lemma 5.3 in [58].

Corollary 5.5.2. *For any $g \in \mathcal{C}_\omega$,*

$$\|l_\omega\|_{\alpha,\xi}\|g\|_{\alpha,\xi} \le K_\omega |l_\omega(g)| \tag{5.5.4}$$

where

$$K_\omega = 2\sqrt{2}K_{\omega,\mathbb{R}} \tag{5.5.5}$$

and therefore the aperture of the cone $\mathcal{C}_\omega$ (as a subset of $\mathcal{H}_\omega^{\alpha,\xi}$) does not exceed K_ω.

Another corollary of Lemma 5.5.1 goes as follows.

Corollary 5.5.3. *The cone $\mathcal{C}_\omega$ is linearly convex, namely for any $g \notin \mathcal{C}_\omega$ there exist $\mu \in \mathcal{C}_\omega^*$ such that $\mu(g) = 0$.*

Proof. Since Q_ω is finite it follows from (5.5.1) that l_ω does not vanish on $\mathcal{C}'_{\omega,\mathbb{R}}$ and so the canonical complexification $\mathcal{C}_\omega$ of $\mathcal{C}_{\omega,\mathbb{R}}$ is linearly convex by Lemma 4.1 from [19]. □

We prove next the following.

Lemma 5.5.4. *The statement of Theorem 5.3.1 (iii) holds true.*

This lemma is a consequence of Lemma 3.13 from [53], and for readers' convenience we provide the proof here.

Proof. Let $g \in \mathcal{H}_\omega^{\alpha,\xi}$ be a nonnegative function. Consider the function

$$q = g + \frac{v_{\alpha,\xi}(g)}{sQ_\omega} \qquad (5.5.6)$$

which satisfies

$$v_{\alpha,\xi}(g) = v_{\alpha,\xi}(q) \text{ and } \|q\|_{\alpha,\xi} \leq \big(1 + \frac{1}{sQ_\omega}\big)\|g\|_{\alpha,\xi}.$$

As in Lemma 3.13 from [53], this function is a member of $\mathcal{C}_{\omega,\mathbb{R},s}$. Indeed, let $x, x' \in \mathcal{E}_\omega$ be such that $\rho(x, x') < \xi$. We have to show that

$$q(x) \leq q(x')e^{sQ_\omega \rho^\alpha(x,x')}. \qquad (5.5.7)$$

If $q(x) \leq q(x')$ then this inequality trivially holds true, while when $q(x) > q(x')$ we have

$$0 \leq \frac{q(x)}{q(x')} - 1 = \frac{|q(x) - q(x')|}{q(x')} \leq \frac{v_{\alpha,\xi}(g)\rho^\alpha(x,x')}{q(x')} \leq sQ_\omega\rho^\alpha(x,x')$$

where we have used that $q \geq \frac{v_{\alpha,\xi}(g)}{sQ_\omega}$. Inequality (5.5.7) follows now from the inequality $1 + t \leq e^t$ which holds true for any $t \geq 0$. The constant function $g_2 \equiv -\frac{v_{\alpha,\xi}(g)}{sQ_\omega}$ is a member of the complex cone $\mathcal{C}_\omega$, and we conclude that any nonnegative member of $\mathcal{H}_\omega^{\alpha,\xi}$ can be represented as a sum of two members $q = g_1$ and g_2 of $\mathcal{C}_\omega$ which satisfy

$$\|g_1\|_{\alpha,\xi} + \|g_2\|_{\alpha,\xi} \leq \big(1 + \frac{2}{sQ_\omega}\big)\|g\|_{\alpha,\xi}.$$

Finally, any complex valued function $g \in \mathcal{H}_\omega^{\alpha,\xi}$ can be written as a linear combination of four nonnegative members of $\mathcal{H}_\omega^{\alpha,\xi}$ whose $\|\cdot\|_{\alpha,\xi}$ norms do not exceed $\|g\|_{\alpha,\xi}$, and thus can be written as a sum of eight members $g_1, ..., g_8$ of the cone $\mathcal{C}_{\omega,s}$ which satisfy

$$\sum_{i=1}^{8} \|g_i\|_{\alpha,\xi} \leq r_\omega \|g\|_{\alpha,\xi}$$

where

$$r_\omega = 4(1 + \tfrac{2}{sQ_\omega}). \qquad (5.5.8)$$

$\square$

Next, we will show that dual cone $\mathcal{C}_\omega^*$ has bounded aperture. More precisely, we prove the following.

Lemma 5.5.5. *For any $\mu \in \mathcal{C}_\omega^*$,*

$$\|\mu\|_{\alpha,\xi} \leq M_\omega |\mu(\mathbf{1})| = M_\omega |\kappa_\omega(\mu)| \tag{5.5.9}$$

where

$$M_\omega = 8(1 - e^{-sQ_\omega \xi^\alpha})^{-2} \tag{5.5.10}$$

and κ_ω is the linear functional given by $\kappa_\omega(\mu) = \mu(\mathbf{1})$ which satisfies $\|\kappa_\omega\|_{\alpha,\xi} = 1$.

When Q_ω is P-a.s. finite this lemma means that (4.1.5) in Assumption 4.1.1 holds true with the above κ_ω and M_ω. We remark that in this case κ_ω is clearly measurable in ω in the sense of the definition at the beginning of Section 4.1 with the σ-algebra $\mathcal{G}$ introduced in Section 5.1.

Before proving this lemma we need the following notations. Let $x, x' \in \mathcal{E}_\omega$ be such that $0 < \rho(x, x') < \xi$ and consider the linear functional $l_{\omega,x,x'} : \mathbb{C}^{\mathcal{E}_\omega} \to \mathbb{C}$ given by the formula

$$l_{\omega,x,x'}(g) = g(x) - e^{-sQ_\omega \rho^\alpha(x,x')} g(x') \tag{5.5.11}$$

where $\mathbb{C}^{\mathcal{E}_\omega}$ is the space of all functions $g : \mathcal{E}_\omega \to \mathbb{C}$. Denote by $\mathbf{\Lambda}_\omega$ the set of linear functionals $\lambda \in (\mathcal{H}_\omega^{\alpha,\xi})^*$ which either have the form $\lambda(g) = g(x)$ for some $x \in \mathcal{E}_\omega$ or the form $\lambda = l_{\omega,x,x'}$ for some distinct $x, x' \in \mathcal{E}_\omega$ such that $\rho(x, x') < \xi$. Then we can write

$$\mathcal{C}_{\omega,\mathbb{R}} = \{g \in \mathcal{H}_{\omega,\mathbb{R}}^{\alpha,\xi} : \lambda(g) \geq 0 \;\; \forall \lambda \in \mathbf{\Lambda}_\omega\}, \tag{5.5.12}$$

namely, the family $\mathbf{\Lambda}_\omega$ generates cone $\mathcal{C}_{\omega,\mathbb{R}}$. Therefore, by (A.2.3) we can write

$$\mathcal{C}_\omega = \{g \in \mathcal{H}_\omega^{\alpha,\xi} : \Re(\bar{\lambda}_1(g)\lambda_2(g)) \geq 0 \;\; \forall \lambda_1, \lambda_2 \in \mathbf{\Lambda}_\omega\} \tag{5.5.13}$$

where $\bar{\lambda}_1(g)$ stands for the complex conjugate of $\lambda_1(g)$.

Proof of Lemma 5.5.5. Relying on Lemma A.2.7, in order to prove (5.5.9) it is sufficient to show that

$$\{g \in \mathcal{H}_\omega^{\alpha,\xi} : \|g - \mathbf{1}\|_{\alpha,\xi} < (M_\omega)^{-1}\} \subset \mathcal{C}_\omega. \tag{5.5.14}$$

For the proof of (5.5.14), let $g \in \mathcal{H}_\omega^{\alpha,\xi}$ be such that $\|g - \mathbf{1}\|_{\alpha,\xi} < (M_\omega)^{-1}$ and write $g = \mathbf{1} + h$. Then by (5.5.13), the function $\mathbf{1} + h$ is a member of the cone $\mathcal{C}_\omega$ if and only if

$$\Re(\bar{\lambda}_1(\mathbf{1} + h), \lambda_2(\mathbf{1} + h)) \geq 0 \tag{5.5.15}$$

for any $\lambda_1, \lambda_2 \in \Lambda_\omega$. We distinguish between four cases. When λ_1 and λ_2 are evaluation functionals at some points $a_1, a_2 \in \mathcal{E}_\omega$, respectively, then

$$\Re(\bar{\lambda}_1(1+h), \lambda_2(1+h)) = \Re((1+\bar{h}(a_1))(1+h(a_2))) \geq 1 - 2\|h\|_\infty - (\|h\|_\infty)^2$$

and so (5.5.15) holds true when $\|h\|_\infty \leq \frac{1}{3}$, and in particular when $\|h\|_{\alpha,\xi} \leq (M_\omega)^{-1}$.

Suppose next that $\lambda_1 = l_{\omega,a_1,b_1}$ and $\lambda_2 = l_{\omega,a_2,b_2}$ for some $a_1, b_1, a_2, b_2 \in \mathcal{E}_\omega$ such that $0 < \rho(a_i, b_i) < \xi$, $i = 1, 2$. We have to show that

$$\Re(\bar{U}_{a_1,b_1} U_{a_2,b_2}) \geq 0$$

where

$$U_{a,b} = 1 + h(a) - e^{-sQ_\omega \rho^\alpha(a,b)}(1 + h(b)) = 1 - e^{-sQ_\omega \rho^\alpha(a,b)} + l_{\omega,a,b}(h)$$

for any $a, b \in \mathcal{E}_\omega$ such that $0 < \rho(a,b) < \xi$, and $\bar{U}_{a,b}$ is the complex conjugate of $U_{a,b}$. For such pairs (a,b) we have

$$|l_{\omega,a,b}(h)| \leq 2\|h\|_\infty \quad \text{and} \quad 1 \geq 1 - e^{-sQ_\omega \rho^\alpha(a,b)} \geq 1 - e^{-sQ_\omega \xi^\alpha} := e_\omega$$

and observe also that $\|h\|_\infty \leq \|h\|_{\alpha,\xi} < (M_\omega)^{-1} \leq 1$. Therefore,

$$\Re(\bar{U}_{a_1,b_1} U_{a_2,b_2}) \geq (e_\omega)^2 - 4\|h\|_\infty - 4(\|h\|_\infty)^2 \geq (e_\omega)^2 - 8\|h\|_\infty,$$

and so (5.5.15) holds true when $8\|h\|_{\alpha,\xi} \leq (e_\omega)^2$, i.e. when $\|h\|_{\alpha,\xi} \leq (M_\omega)^{-1}$.

Finally, consider the case when one of the λ_i's is an evaluation functional of the form $e(g) = g(a_1)$, $a_1 \in \mathcal{E}_\omega$ and the other has the form $l_{\omega,a,b}$ for some $a, b \in \mathcal{E}_\omega$ such that $0 < \rho(a,b) < \xi$. We assume without loss of generality that λ_1 is the evaluation functional and $\lambda_2 = l_{\omega,a,b}$. We have to show that

$$\Re((1 + \bar{h}(a_1))U_{a,b}) \geq 0.$$

Similarly to the previous case, the expression from the above left-hand side is not less than $e_\omega - 5\|h\|_\infty$ when $\|h\|_\infty \leq 1$, and in particular (5.5.15) holds true when $\|h\|_{\alpha,\xi} \leq (M_\omega)^{-1}$.   □

Remark that, in fact

$$M_\omega \leq 8s^{-2}Q_\omega^{-2}e^{2sQ_\omega \xi^\alpha} \tag{5.5.16}$$

since $1 - e^{-t} \geq te^{-t}$ for any $t \geq 0$ and observe also that $Q_\omega \geq H_{\theta^{-1}\omega}(\gamma_{\theta^{-1}\omega})^{-\alpha}$.

5.6   Properties of transfer operators: proof of Theorem 5.4.1 *(i)*

5.6.1   *Continuity and analyticity*

We begin with the following random Lasota-Yorke type inequality.

Lemma 5.6.1. *P-a.s. for any $n \in \mathbb{N}$, $z \in \mathbb{C}$ and $g \in \mathcal{H}_\omega^{\alpha,\xi}$,*

$$v_{\alpha,\xi}(\mathcal{L}_z^{\omega,n} g) \leq \|\mathcal{L}_0^{\omega,n} \mathbf{1}\|_\infty e^{|\Re(z)|\|S_n^\omega u\|_\infty}$$
$$\times \big(v_{\alpha,\xi}(g)(\gamma_{\omega,n})^{-\alpha} + 2Q_{\theta^n\omega}(H)(1 + \|z\|_1)\|g\|_\infty\big)$$

where $\|z\|_1 = |\Re(z)| + |\Im(z)|$ and $\Re(z)$ ($\Im(z)$) is the real (imaginary) part of z. As a consequence,

$$\|\mathcal{L}_z^{\omega,n} g\|_{\alpha,\xi} \leq \|\mathcal{L}_0^{\omega,n} \mathbf{1}\|_\infty e^{|\Re(z)|\|S_n^\omega u\|_\infty} \qquad (5.6.1)$$
$$\times \big(v_{\alpha,\xi}(g)(\gamma_{\omega,n})^{-\alpha} + (1 + 2Q_{\theta^n\omega}(H))(1 + \|z\|_1)\|g\|_\infty\big).$$

In particular when $Q_{\theta^n\omega} < \infty$ then $\mathcal{L}_z^{\omega,n}|_{B_\omega} : B_\omega \to B_{\theta^n\omega}$ is a continuous linear map, where $B_\omega = \mathcal{H}_\omega^{\alpha,\xi}$.

Proof. Let n, z and g be as in the statement of the lemma. We begin with approximating $v_{\alpha,\xi}(\mathcal{L}_z^{\omega,n}g)$. Let $x, x' \in \mathcal{E}_{\theta^n\omega}$ be such that $\rho(x, x') < \xi$ and let $y_1, ..., y_k$ and $y_1', ..., y_k'$ be the points in $\mathcal{E}_\omega$ satisfying (5.1.4) and (5.1.5). We can write

$$\big|\mathcal{L}_z^{\omega,n} g(x) - \mathcal{L}_z^{\omega,n} g(x')\big|$$

$$= \big|\sum_{t=1}^k \big(e^{S_n^\omega \phi(y_t) + zS_n^\omega u(y_t)} g(y_t) - e^{S_n^\omega \phi(y_t') + zS_n^\omega u(y_t')} g(y_t')\big)\big|$$

$$\leq \sum_{t=1}^k e^{S_n^\omega \phi(y_t) + \Re(z)S_n^\omega u(y_t)} \big|e^{i\Im(z)S_n^\omega u(y_t)} g(y_t) - e^{i\Im(z)S_n^\omega u(y_t')} g(y_t')\big|$$

$$+ \sum_{t=1}^k \big|e^{i\Im(z)S_n^\omega u(y_t')} g(y_t')\big| \big|e^{S_n^\omega \phi(y_t) + \Re(z)S_n^\omega u(y_t)} - e^{S_n^\omega \phi(y_t') + \Re(z)S_n^\omega u(y_t')}\big|$$

$$:= I_1 + I_2.$$

In order to estimate I_1, observe that for any $1 \leq t \leq k$,

$$\big|e^{i\Im(z)S_n^\omega u(y_t)} g(y_t) - e^{i\Im(z)S_n^\omega u(y_t')} g(y_t')\big|$$

$$\leq |g(y_t)| \cdot |e^{i\Im(z)S_n^\omega u(y_t)} - e^{i\Im(z)S_n^\omega u(y_t')}| + |g(y_t) - g(y_t')| := J_1 + J_2.$$

By the mean value theorem and then by (5.1.12),

$$J_1 \leq 2\|g\|_\infty |\Im(z)|Q_{\theta^n\omega}(H)\rho^\alpha(x, x'),$$

while by (5.1.5),

$$J_2 \leq v_{\alpha,\xi}(g)\rho^\alpha(y_t, y_t') \leq v_{\alpha,\xi}(g)(\gamma_{\omega,n})^{-\alpha}\rho^\alpha(x, x)$$

and it follows that

$$I_1 \leq L_{\Re(z)}^{\omega,n}\mathbf{1}(x)\big(2\|g\|_\infty|\Im(z)|Q_{\theta^n\omega}(H) + v_{\alpha,\xi}(g)(\gamma_{\omega,n})^{-\alpha}\big)\rho^\alpha(x, x').$$

Finally, observe that

$$\mathcal{L}_{\Re(z)}^{\omega,n}\mathbf{1}(x) \leq \mathcal{L}_0^{\omega,n}\mathbf{1}(x)e^{|\Re(z)|\|S_n^\omega u\|_\infty}, \qquad (5.6.2)$$

and so

$$I_1 \leq \|\mathcal{L}_0^{\omega,n}\mathbf{1}\|_\infty e^{|\Re(z)|\|S_n^\omega u\|_\infty}$$
$$\times \big(2\|g\|_\infty|\Im(z)|Q_{\theta^n\omega}(H) + v_{\alpha,\xi}(g)(\gamma_{\omega,n})^{-\alpha}\big)\rho^\alpha(x, x').$$

Next, we estimate I_2. First, by the mean value theorem and (5.1.12),

$$\big|e^{S_n^\omega\phi(y_t)+\Re(z)S_n^\omega u(y_t)} - e^{S_n^\omega\phi(y_t')+\Re(z)S_n^\omega u(y_t')}\big| \leq (1 + |\Re(z)|)Q_{\theta^n\omega}(H)$$
$$\times \max\{e^{S_n^\omega\phi(y_t)+\Re(z)S_n^\omega u(y_t)}, e^{S_n^\omega\phi(y_t')+\Re(z)S_n^\omega u(y_t')}\}\rho^\alpha(x, x')$$

and therefore

$$I_2 \leq (1 + |\Re(z)|)\|g\|_\infty(\mathbf{L}_{\Re(z)}^{\omega,n}\mathbf{1}(x) + \mathcal{L}_{\Re(z)}^{\omega,n}\mathbf{1}(x'))Q_{\theta^n\omega}(H)\rho^\alpha(x, x')$$
$$\leq 2(1 + |\Re(z)|)\|g\|_\infty\|\mathcal{L}_0^{\omega,n}\mathbf{1}\|_\infty e^{|\Re(z)|\|S_n^\omega u\|_\infty}Q_{\theta^n\omega}(H)\rho^\alpha(x, x')$$

where in the last inequality we used (5.6.2), yielding the first statement of the lemma.

Finally, by (5.6.2) we have

$$\|\mathcal{L}_z^{\omega,n}g\|_\infty \leq \|g\|_\infty\|\mathcal{L}_0^{\omega,n}\mathbf{1}\|_\infty \cdot e^{|\Re(z)|\|S_n^\omega u\|_\infty}$$

and the lemma follows from the above estimates, taking into account that

$$\|\mathcal{L}_0^{\omega,n}\mathbf{1}\|_\infty \leq \deg T_\omega^n e^{\|S_n^\omega\phi\|_\infty} \leq D_{\omega,n}e^{\|S_n^\omega\phi\|_\infty} < \infty.$$

$\square$

5.6.2 *Analyticity in z*

We show that the transfer operators $\mathcal{L}_z^\omega$ are analytic in z and that for any $g \in \mathcal{H}_\omega^{\alpha,\xi}$ and $z \in \mathbb{C}$,

$$\frac{d\mathcal{L}_z^\omega}{dz}(g) = \mathcal{L}_z^\omega(u_\omega g). \qquad (5.6.3)$$

Indeed, let $h \in \mathbb{C}$ be such that $0 < |h| \leq 1$ and consider the function $\psi_h : \mathbb{R} \to \mathbb{C}$ given by

$$\psi_h(t) = \frac{e^{ht} - 1}{h} - t.$$

Let $z_0 \in \mathbb{C}$. For any $g \in \mathcal{H}_\omega^{\alpha,\xi}$ set

$$\delta_{z_0,h}^\omega g = h^{-1}(\mathcal{L}_{z_0+h}^\omega g - \mathcal{L}_{z_0}^\omega g) - \mathcal{L}_{z_0}^\omega(u_\omega g) = \mathcal{L}_{z_0}^\omega(g \cdot (\psi_h \circ u_\omega)).$$

In order to obtain (5.6.3) with $z = z_0$ we need to show that

$$\lim_{h \to 0} \sup_{g:\|g\|_{\alpha,\xi} \leq 1} \|\delta_{z_0,h}^\omega g\|_{\alpha,\xi} = 0. \tag{5.6.4}$$

Let $g \in \mathcal{H}_\omega^{\alpha,\xi}$ be such that $\|g\|_{\alpha,\xi} \leq 1$. Since $|h| \leq 1$, there exists $C_{\omega,1} > 0$ which depends only on $\|u_\omega\|_\infty$ so that

$$|h\psi_h(t)| = |e^{th} - 1 - th| \leq C_{\omega,1}|th|^2 \tag{5.6.5}$$

for any $t \in [-\|u_\omega\|_\infty, \|u_\omega\|_\infty]$. Therefore,

$$\|\delta_{z_0,h}^\omega g\|_\infty \leq e^{|z_0|\|u_\omega\|_\infty} \|\mathcal{L}_0^\omega 1\|_\infty \|u_\omega\|_\infty^2 C_{\omega,1}|h|. \tag{5.6.6}$$

Next, let $x, x' \in \mathcal{E}_{\theta\omega}$ be such that $0 < \rho(x, x') < \xi$. We claim that there exists a constant a_{ω,z_0} which depends only on ω and z_0 such that

$$|\delta_{z_0,h}^\omega g(x) - \delta_{z_0,h}^\omega g(x')| \leq a_{\omega,z_0}|h|\rho^\alpha(x, x'). \tag{5.6.7}$$

It is clear that (5.6.4) follows from (5.6.6) and (5.6.7). In order to prove (5.6.7), let $\{y_1, ..., y_k\}$ and $\{y_1', ..., y_k'\}$ be the points in $\mathcal{E}_\omega$ satisfying (5.1.4) and (5.1.5), where $k = |T_\omega^{-1}\{x\}| \leq D_\omega$. Then

$$\delta_{z_0,h}^\omega g(x) - \delta_{z_0,h}^\omega g(x')$$

$$= \sum_{i=1}^k e^{\phi_\omega(y_i)+z_0 u_\omega(y_i)} g(y_i)\psi_h(u_\omega(y_i)) - \sum_{i=1}^k e^{\phi_\omega(y_i')+z_0 u_\omega(y_i')} g(y_i')\psi_h(u_\omega(y_i')).$$

Since $0 < |h| \leq 1$ and u_ω satisfies (5.1.12), it follows by the mean value theorem that for any $1 \leq i \leq k$,

$$|\psi_h(u_\omega(y_i)) - \psi_h(u_\omega(y_i'))| \leq 2|u_\omega(y_i) - u_\omega(y_i')| \sup_{t \in [-\|u_\omega\|_\infty, \|u_\omega\|_\infty]} |e^{ht} - 1|$$

$$\leq (2H_\omega \gamma_\omega^{-\alpha}\rho^\alpha(x, x')) \cdot (4|h|\|u_\omega\|_\infty e^{\|u_\omega\|_\infty}).$$

Next, by (5.6.5),

$$|\psi_h(u_\omega(y_i))| \leq C_{\omega,1}|h|\|u_\omega\|_\infty^2$$

for any $1 \leq i \leq k$, and therefore, with $q_\omega = e^{\phi_\omega + z_0 u_\omega}g$,

$$|\delta_{z_0,h}^\omega g(x) - \delta_{z_0,h}^\omega g(x')| \leq C_{\omega,1}|h|\|u_\omega\|_\infty^2 \sum_{i=1}^k |q_\omega(y_i) - q_\omega(y_i')|$$

$$+ (8D_\omega H_\omega e^{\|\phi_\omega\|_\infty + |z_0|\|u_\omega\|_\infty}\|u_\omega\|_\infty|h|e^{\|u_\omega\|_\infty})\rho^\alpha(x, x').$$

Since u_ω and ϕ_ω satisfy (5.1.12) and $\|g\|_{\alpha,\xi} \leq 1$, applying the mean value theorem we obtain that for $1 \leq i \leq k$,

$$|q_\omega(y_i) - q_\omega(y_i')| \leq (1 + |z_0|)C_{\omega,2}\rho^\alpha(x, x')$$

for some $C_{\omega,2}$ depending only on H_ω, $\|\phi_\omega\|_\infty$ and $\|u_\omega\|_\infty$. Combining the above estimates we obtain (5.6.7). $\qquad\square$

5.7 Real Hilbert metric estimates

The first part of the proof of Theorem 5.4.1 goes as follows.

5.7.1 *General estimates*

We recall the definition of the Hilbert projective metric associated with a real cone (see also Appendix A). Let $\mathcal{C}_\mathbb{R}$ be a real proper closed convex cone in some real Banach space. For any nonzero members f, g of $\mathcal{C}_\mathbb{R}$ set

$$\beta_{\mathcal{C}_\mathbb{R}}(f, g) = \inf\{t > 0 : tf - g \in \mathcal{C}_\mathbb{R}\} \tag{5.7.1}$$

which is strictly positive (see Appendix A) and we use here the convention $\inf \emptyset = \infty$. Then the real Hilbert (projective) metric $d_{\mathcal{C}_\mathbb{R}} : \mathcal{C}_\mathbb{R} \times \mathcal{C}_\mathbb{R} \to [0, \infty]$ associated with the cone is given by

$$d_{\mathcal{C}_\mathbb{R}}(f, g) = \ln\big(\beta_{\mathcal{C}_\mathbb{R}}(f, g)\beta_{\mathcal{C}_\mathbb{R}}(g, f)\big). \tag{5.7.2}$$

We refer to Appendix A for references about contraction properties (with respect to the Hilbert metric) of linear maps between two real proper closed convex cones.

We will constantly use the following general diameter estimates which can be obtained along the lines of [53] and resemble the estimates from [40], [58] and [19] (see also references therein). Let $0 < s' < s$, $C > 0$ and consider the real cone

$$\mathcal{K}_{\omega, \mathbb{R}, C} = \{g : \mathcal{E}_\omega \to [0, \infty) : g(x) \le g(y) \text{ for all } x, y \in \mathcal{E}_\omega\}. \tag{5.7.3}$$

Lemma 5.7.1. *Let* $\omega \in \Omega$ *be such that* $Q_\omega < \infty$. *Then for any strictly positive* $f, g \in \mathcal{C}_{\omega, \mathbb{R}, s}$ *we have*

$$d_{\mathcal{C}_{\omega, \mathbb{R}, s}}(f, g) = \ln\big(R(g, f)R(f, g)\big) \tag{5.7.4}$$

where

$$R(g, f) = \max\left(\left\|\frac{g}{f}\right\|_\infty, \sup_{x, x' : 0 < \rho(x, x') < \xi} W(g, f, x, x')\right),$$

$\|\cdot\|_\infty$ *stands for the supremum norm and*

$$W(g, f, x, x') = \frac{e^{sQ_\omega \rho^\alpha(x, x')}g(x) - g(x')}{e^{sQ_\omega \rho^\alpha(x, x')}f(x) - f(x')} \ge 0.$$

As a consequence, for any nonzero members f *and* g *of*

$$\mathcal{K}_{\omega, \mathbb{R}, C} \cap \mathcal{C}_{\omega, s', \mathbb{R}}$$

we have

$$d_{\mathcal{C}_{\omega, \mathbb{R}, s}}(f, g) \le 2\ln\left(\frac{C(s + s)}{s - s'}\right). \tag{5.7.5}$$

We remark that the formula (5.7.4), in fact, follows from (5.5.12) and the general formula given in (2.2) from [18], and for the sake of completeness we will prove it in our specific situation.

Proof. Let ω be such that $Q_\omega < \infty$ and let $f, g \in \mathcal{C}_{\omega,\mathbb{R},s}$ be strictly positive. For any $t > 0$, the function $tf - g$ is a member of $\mathcal{C}_{\omega,\mathbb{R},s}$ if and only if $tf(x) \geq g(x)$ for any $x \in \mathcal{E}_\omega$ and

$$t \geq \frac{e^{sQ_\omega \rho^\alpha(x,x')}g(x) - g(x')}{e^{sQ_\omega \rho^\alpha(x,x')}f(x) - f(x')} = W(g, f, x, x')$$

for any $x, x' \in \mathcal{E}_\omega$ such that $0 < \rho(x,x') < \xi$ (i.e. if and only if $\lambda(tf - g) \geq 0$ for any $\lambda \in \Lambda_\omega$). As a consequence,

$$\beta_{\mathcal{C}_{\omega,\mathbb{R},s}}(f, g) = \max \left(\left\| \frac{g}{f} \right\|_\infty, W(g, f) \right), \tag{5.7.6}$$

where

$$W(g, f) = \sup_{x,x':0<\rho(x,x')<\xi} W(g, f, x, x')$$

and (5.7.4) follows.

Next, in order to prove (5.7.5), let

$$f, g \in \mathcal{C}_{\omega,s',\mathbb{R}} \cap \mathcal{K}_{\omega,\mathbb{R},C} \setminus \{0\}$$

be two nonzero functions. Since $f, g \in \mathcal{K}_{\omega,\mathbb{R},C} \setminus \{0\}$ then both f and g are strictly positive and so (5.7.6) holds true. On the other hand, since $f, g \in \mathcal{C}_{\omega,s',\mathbb{R}}$ we have

$$W(g, f, x, x') \leq \frac{(e^{sQ_\omega \rho^\alpha(x,x')} - e^{-s'Q_\omega \rho^\alpha(x,x')})g(x)}{(e^{sQ_\omega \rho^\alpha(x,x')} - e^{s'Q_\omega \rho^\alpha(x,x')})f(x)} \leq \frac{g(x)}{f(x)}\tau \tag{5.7.7}$$

for any $x, x' \in \mathcal{E}_\omega$ such that $0 < \rho(x,x') < \xi$, where

$$\tau = \tau_{\omega,s,s'} = \sup_{0<r<\xi} \frac{e^{sQ_\omega r^\alpha} - e^{-s'Q_\omega r^\alpha}}{e^{sQ_\omega r^\alpha} - e^{s'Q_\omega r^\alpha}}.$$

It follows that

$$W(g, f) \leq \left\| \frac{g}{f} \right\|_\infty \tau$$

which together with (5.7.6) and the inequality $\tau > 1$ implies that

$$\beta_{\mathcal{C}_{\omega,\mathbb{R},s}}(f, g) \leq \left\| \frac{g}{f} \right\|_\infty \tau. \tag{5.7.8}$$

Now, let $x_1, x_2 \in \mathcal{E}_\omega$ be such that

$$\left\| \frac{g}{f} \right\|_\infty = \frac{g(x_1)}{f(x_1)} \quad \text{and} \quad \left\| \frac{f}{g} \right\|_\infty = \frac{f(x_2)}{g(x_2)}.$$

Such x_1 and x_2 exist since f and g are continuous and strictly positive functions on the compact space $\mathcal{E}_\omega$. Reversing the roles of f and g in (5.7.8), we conclude from the definition of $d_{\mathcal{C}_{\omega,\mathbb{R},s}}(f,g)$ that

$$e^{d_{\mathcal{C}_{\omega,\mathbb{R},s}}(f,g)} \leq \tau^2 \frac{g(x_1)f(x_2)}{g(x_2)f(x_1)} \leq \tau^2 C^2 \tag{5.7.9}$$

where the last inequality holds true since $f, g \in \mathcal{K}_{\omega,\mathbb{R},C}$. Finally, we will show that

$$\tau \leq \frac{s+s'}{s-s'}. \tag{5.7.10}$$

The diameter estimate (5.7.5) clearly follows from (5.7.9) and (5.7.10). In order to prove (5.7.10), observe that by the Cauchy mean value theorem for any $t > 0$ there exists $t_0 > 0$ such that

$$\frac{e^{st} - e^{-s't}}{e^{st} - e^{s't}} = \frac{1 - e^{-(s+s')t}}{1 - e^{-(s-s')t}} = \frac{s+s'}{s-s'}e^{-2s't_0} \leq \frac{s+s'}{s-s'}$$

and (5.7.10) follows. $\qquad\square$

Remark 5.7.2. When any open ball of radius ξ in $\mathcal{E}_\omega$ contains at least two points then, in fact,

$$d_{\mathcal{C}_{\omega,\mathbb{R},s}}(f,g) = \ln\left(W(g,f)W(f,g)\right) \tag{5.7.11}$$

for any strictly positive f and g in $\mathcal{C}_{\omega,\mathbb{R},s}$. Indeed, consider the continuous function $h = \frac{g}{f}$, let $x_1 \in \mathcal{E}_\omega$ be such that $h(x_1) = \|h\|_\infty$ and $x' \in \mathcal{E}_\omega$ such that $0 < \rho(x_1, x') < \xi$. Then,

$$\begin{aligned}
W(g,f,x_1,x') &= \frac{e^{sQ_\omega\rho^\alpha(x_1,x')}h(x_1)f(x_1) - h(x')f(x')}{e^{sQ_\omega\rho^\alpha(x_1,x')}f(x_1) - f(x')} \\
&\geq \frac{e^{sQ_\omega\rho^\alpha(x_1,x')}h(x_1)f(x_1) - h(x_1)f(x')}{e^{sQ_\omega\rho^\alpha(x_1,x')}f(x_1) - f(x')} = h(x_1).
\end{aligned}$$

and therefore by (5.7.6),

$$\beta_{\mathcal{C}_{\omega,\mathbb{R},s}}(f,g) = W(g,f)$$

and (5.7.11) follows.

5.7.2   Real cones invariances and diameter of image estimates

We need first the following notations. Let $\omega \in \Omega$ be such that $Q_\omega < \infty$, where we recall that $Q_\omega = Q_\omega(2H) = 2Q_\omega(H)$, which satisfies

$$Q_{\theta\omega} = (Q_\omega + 2H_\omega)\gamma_\omega^{-\alpha}. \tag{5.7.12}$$

Set

$$s'_{\theta\omega} = \frac{s\gamma_\omega^{-\alpha}Q_\omega + 2\gamma_\omega^{-\alpha}H_\omega}{Q_{\theta\omega}} = \frac{s\gamma_\omega^{-\alpha}Q_\omega + 2\gamma_\omega^{-\alpha}H_\omega}{\gamma_\omega^{-\alpha}Q_\omega + 2\gamma_\omega^{-\alpha}H_\omega} < s \tag{5.7.13}$$

and $s''_\omega = \frac{s+s'_\omega}{s-s'_\omega}$, which can also be written in the form

$$s''_\omega = \frac{2sQ_{\theta^{-1}\omega} + 2(s+1)H_{\theta^{-1}\omega}}{2(s-1)H_{\theta^{-1}\omega}} = \frac{2s}{s-1} \cdot \frac{Q_{\theta^{-1}\omega}}{2H_{\theta^{-1}\omega}} + 1 + \frac{2}{s-1}. \tag{5.7.14}$$

Next, for P-a.a. ω and any $m \in \mathbb{N}$ set

$$C_\omega(m) = D_{\omega,m}e^{sQ_\omega + 2\|S_m^\omega\phi\|_\infty + 2\|S_m^\omega u\|_\infty} \tag{5.7.15}$$

where $D_{\omega,n}$ is given by (5.1.6).

Lemma 5.7.3. *P-a.s. for any $t \in [-1,1]$,*

$$\mathcal{L}_t^\omega \mathcal{C}'_{\omega,\mathbb{R},s} \subset \mathcal{C}'_{\theta\omega,\mathbb{R},s'_{\theta\omega}} \subset \mathcal{C}'_{\theta\omega,\mathbb{R},s} \tag{5.7.16}$$

and for any $n \geq n_\omega$,

$$\mathcal{L}_t^{\omega,n} \mathcal{C}'_{\omega,\mathbb{R},s} \subset \mathcal{K}'_{\theta^n\omega,\mathbb{R},C_\omega(n)} \tag{5.7.17}$$

and

$$D(\omega,n,t) := \sup_{f,g \in \mathcal{C}'_{\omega,\mathbb{R},s}} d_{\mathcal{C}_{\theta^n\omega,\mathbb{R},s}}(\mathcal{L}_t^{\omega,n}f, \mathcal{L}_t^{\omega,n}g) \leq 2\ln\left(s''_{\theta^{n_\omega}\omega}C_\omega(n_\omega)\right). \tag{5.7.18}$$

The proof of (5.7.16), (5.7.17) and (5.7.18) proceeds similarly to the proof of Lemmas 9.8 and 9.9 in [53]. For readers' convenience we provide it here.

Proof. In order to prove (5.7.16) let $t \in [-1,1]$, $g \in \mathcal{C}'_{\omega,\mathbb{R},s}$ and $x,x' \in \mathcal{E}_{\theta\omega}$ be such that $0 < \rho(x,x') < \xi$. Consider the points $y_1, ..., y_k$ and $y'_1, ..., y'_k$ in $\mathcal{E}_\omega$ satisfying (5.1.4) and (5.1.5). Then we can write

$$\mathcal{L}_t^\omega g(x) = \sum_{i=1}^k e^{\phi_\omega(y_i)+tu_\omega(y_i)}g(y_i) \text{ and } \mathcal{L}_t^\omega g(x') = \sum_{i=1}^k e^{\phi_\omega(y'_i)+tu_\omega(y'_i)}g(y'_i).$$

Applying the first inequality from (5.1.12) with $n = 1$, the function $\phi_t = f + tu$ and $2H$ in place of H and using that $g \in C_{\omega,\mathbb{R},s}$ we deduce that for any $1 \leq i \leq k$,

$$e^{\phi_\omega(y_i)+tu_\omega(y_i)}g(y_i) \leq e^{(2H_\omega+sQ_\omega)\gamma_\omega^{-\alpha}\rho^\alpha(x,x')}e^{\phi_\omega(y_i')+tu_\omega(y_i')}g(y_i')$$

where in the corresponding upper bound of $g(y_i)$ we used that $\rho(y_i, y_i') \leq \gamma_\omega^{-1}\rho(x, x')$. We conclude from the definition of s'_{θ_ω} and the above estimate that

$$e^{\phi_\omega(y_i)+tu_\omega(y_i)}g(y_i) \leq e^{s'_{\theta_\omega}Q_{\theta_\omega}\rho^\alpha(x,x')}e^{\phi_\omega(y_i')+tu_\omega(y_i')}g(y_i')$$

and (5.7.16) follows, taking into account the above formulas for $\mathcal{L}_t^\omega g(x)$ and $\mathcal{L}_t^\omega g(x')$.

In order to prove (5.7.17), we first notice that P-a.s. for any natural m, real t and a nonnegative function $g : \mathcal{E}_\omega \to \mathbb{R}$ the function $\mathcal{L}_t^{\omega,m}g$ is not identically 0, unless $g \equiv 0$. Indeed if $g(y) > 0$ for some $y \in \mathcal{E}_\omega$ then $\mathcal{L}_t^{\omega,m}g(T_\omega^m y) \geq e^{-\|S_m^\omega(\phi+tu)\|_\infty}g(y) > 0$. Next, by Assumption 5.1.3 P-a.s. for any $n \geq n_\omega$,

$$T_\omega^n(B_\omega(a, \xi)) = \mathcal{E}_{\theta^n\omega} \text{ for any } a \in \mathcal{E}_\omega \qquad (5.7.19)$$

where we also used that each T_ω is bijective. We claim that for any $x_1, x_2 \in \mathcal{E}_\omega$, $g \in C'_{\omega,\mathbb{R},s}$ and $t \in [-1, 1]$,

$$\mathcal{L}_t^{\omega,n}g(x_2) \leq C_\omega(n)\mathcal{L}_t^{\omega,n}g(x_1). \qquad (5.7.20)$$

Indeed, let $t \in [-1, 1]$ and consider again the function $\phi_t = \phi + tu$. Let $a \in T_\omega^{-n}\{x_2\}$ be such that

$$e^{S_n^\omega\phi_t(a)}g(a) = \max_{y \in T_\omega^{-n}\{x_2\}} e^{S_n^\omega\phi_t(y)}g(y).$$

By (5.7.19), there exists $b \in T_\omega^{-n}\{x_1\} \cap B_\omega(a, \xi)$. Since $g \in C'_{\omega,\mathbb{R},s}$ we can write now

$$\mathcal{L}_t^{\omega,n}g(x_1) \geq e^{S_n^\omega\phi_t(b)}g(b) \geq e^{S_n^\omega\phi_t(b)-S_n^\omega\phi_t(a)-sQ_\omega}g(a)e^{S_n^\omega\phi_t(a)}$$

$$\geq \frac{e^{-2\|S_n^\omega\phi_t\|_\infty-sQ_\omega}}{\deg T_\omega^n}\mathcal{L}_t^{\omega,n}g(x_2) \geq \left(C_\omega(n)\right)^{-1}\mathcal{L}_t^{\omega,n}g(x_2)$$

where we used that $\deg T_\omega^n \leq D_{\omega,n}$ and that $\|S_n^\omega\phi_t\|_\infty \leq \|S_n^\omega\phi\|_\infty+\|S_n^\omega u\|_\infty$ (recall that $|t| \leq 1$).

The inclusion (5.7.17) follows from (5.7.20) and the diameter estimate (5.7.18) with $n = n_\omega$ follows from Lemma 5.7.1 together with (5.7.17). The estimate (5.7.18) when $n > n_\omega$ follows from the case when $n = n_\omega$ together with (5.7.16) since

$$\mathcal{L}_t^{\omega,n} = \mathcal{L}_t^{\theta^{n_\omega}\omega,n-n_\omega} \circ \mathcal{L}_t^{\omega,n_\omega}$$

and any linear map between two real cones weakly contracts the real Hilbert metrics corresponding to the cones (see Appendix A). $\qquad\qquad\square$

5.8   Comparison of real and complex operators

Suppose that Q_ω is P-a.s. finite. The following result is necessary for our future application of Proposition A.2.4. For any $z = \Re(z) + i\Im(z) \in \mathbb{C}$ and $m \geq 1$ set

$$V = V(m, z, \omega) = |\Im(z)|(\|S_m^\omega u\|_\infty + (s-1)^{-1}) \qquad (5.8.1)$$

and

$$\varepsilon_\omega(m, z) = \frac{|\Im(z)|}{s+1} + 2|V|.$$

Let Λ_ω be the set of linear functionals defined before (5.5.13).

Lemma 5.8.1. *For P-a.a. ω and any $m \in \mathbb{N}$, $g \in \mathcal{C}'_{\omega,\mathbb{R},s}$, $\lambda \in \Lambda_{\theta^m\omega}$ and $z \in \mathbb{C}$ so that $|\Re(z)| \leq 1$,*

$$|\lambda(\mathcal{L}_z^{\omega,m} g) - \lambda(\mathcal{L}_{\Re(z)}^{\omega,m} g)| \leq \lambda(\mathcal{L}_{\Re(z)}^{\omega,m} g)\varepsilon_\omega(m, z). \qquad (5.8.2)$$

The proof of Lemma 5.8.1 for the case when $\lambda = l_{\theta^m\omega,x,x'}$ proceeds essentially along the lines of the proof of Lemma 9.15 from [53], and the proof for the other case is straightforward. For readers' convenience we provide the proof here.

Proof. Let m, z, g and λ be as in the statement of the lemma. We first consider the case when λ has the form $\lambda(g) = g(x)$ for some $x \in \mathcal{E}_{\theta^m\omega}$. Since

$$\mathcal{L}_z^{\omega,m} g = \mathcal{L}_{\Re(z)}^{\omega,m}(e^{i\Im(z)S_m^\omega u} g)$$

and g is nonnegative,

$$|\mathcal{L}_z^{\omega,m} g(x) - \mathcal{L}_{\Re(z)}^{\omega,m} g(x)| \leq \|e^{i\Im(z)S_m^\omega u} - 1\|_\infty \mathcal{L}_{\Re(z)}^{\omega,m} g(x).$$

By the mean value theorem,

$$\|e^{i\Im(z)S_m^\omega u} - 1\|_\infty \leq 2|\Im(z)|\|S_m^\omega u\|_\infty \leq \varepsilon_\omega(m, z)$$

and the proof for this case is complete.

Next, we consider the case when $\lambda = l_{\theta^m\omega,x,x'}$ for some distinct $x, x' \in \mathcal{E}_{\theta^m\omega}$ such that $\rho(x, x') < \xi$. We have to show that

$$\left| \frac{l_{\theta^m\omega,x,x'}(\mathcal{L}_z^{\omega,m} g)}{l_{\theta^m\omega,x,x'}(\mathcal{L}_{\Re(z)}^{\omega,m} g)} - 1 \right| \leq \varepsilon_\omega(m, z)$$

where we note that the denominator does not vanish since $\mathcal{L}_{\Re(z)}^{\omega,m} g \in \mathcal{C}_{\theta^n\omega, s'_{\theta^n\omega}, \mathbb{R}}$. Consider the functions $\phi_z : \mathcal{E} \to \mathbb{C}$, $\Re(z) \in [-1, 1]$ given by

$\phi_z = \phi + zu$. Consider now the points $y_1, ..., y_k$ and $y'_1, ..., y'_k$ in $\mathcal{E}_\omega$ satisfying (5.1.7) and (5.1.8) with $n = m$ and write

$$l_{\theta^m \omega, x, x'}(\mathcal{L}_z^{\omega, m} g) = \sum_{j=1}^{k} n_{m,j}(z, g)$$

where for each $j = 1, 2, ..., k$,

$$n_{m,j}(z, g) = e^{S_m^\omega \phi_z(y_j)} g(y_j) - e^{-sQ_{\theta^m \omega} \rho^\alpha(x, x')} e^{S_m^\omega \phi_z(y'_j)} g(y'_j).$$

Fix some $1 \le j \le k$. By the definition of the cone $\mathcal{C}_{\omega, \mathbb{R}}$, $g(y_j) = 0$ if and only if $g(y_j) = 0$ since $\rho(y_j, y'_j) < \xi$. In the case when $g(y_j) > 0$ define a_j by the formula

$$n_{m,j}(\Re(z), g) e^{i\Im(S_m^\omega \phi_z(y_j))} a_j = n_{m,j}(z, g).$$

We will prove in what follows that $n_{m,j}(\Re(z), g) > 0$ and so a_j is indeed well defined. Set

$$z_1 = S_m^\omega \phi_z(y_j) + \ln g(y_j) \text{ and } z_2 = -sQ_{\theta^m \omega} \rho^\alpha(x, x') + S_m^\omega \phi_z(y'_j) + \ln g(y'_j).$$

Then

$$e^{i\Im(z_1)} a_j = \frac{e^{z_1} - e^{z_2}}{e^{\Re(z_1)} - e^{\Re(z_2)}}.$$

Therefore by Lemma 9.3 from [58], if $\Re(z_1) > \Re(z_2)$ then

$$1 \le |a_j|^2 \le 1 + \left(\frac{\Im(z_1 - z_2)}{\Re(z_1 - z_2)}\right)^2 \text{ and } |\arg(a_j)| \le \left|\frac{\Im(z_1 - z_2)}{\Re(z_1 - z_2)}\right|. \quad (5.8.3)$$

Next, observe that

$$\Im(z_1 - z_2) = \Im(z)\left(S_m^\omega u(y_j) - S_m^\omega u(y'_j)\right) \quad (5.8.4)$$

and, therefore, applying (5.1.12) we obtain

$$|\Im(z_1 - z_2)| \le |\Im(z)| \rho^\alpha(x, x') \sum_{t=0}^{m-1} H_{\theta^t \omega}(\gamma_{\theta^t \omega, m-t})^{-\alpha}. \quad (5.8.5)$$

In order to estimate $\Re(z_1 - z_2)$, observe that

$$Q_{\theta^m \omega} = Q_{\theta^m \omega}(2H) = Q_\omega(\gamma_{\omega, m})^{-\alpha} + \sum_{t=0}^{m-1} 2H_{\theta^t \omega}(\gamma_{\theta^t \omega, m-t})^{-\alpha}. \quad (5.8.6)$$

Next, since $g \in \mathcal{C}_{\omega, \mathbb{R}, s}$,

$$\ln g(y_j) - \ln g(y'_j) \ge -sQ_\omega \rho^\alpha(y_j, y'_j) \ge -sQ_\omega(\gamma_{\omega, m})^{-\alpha} \rho^\alpha(x, x')$$

where in the second inequality we have used (5.1.8). Applying the first inequality from (5.1.12) with $n = m$, the function $\phi_{\Re(z)}$ and with $2H$ in place of H it follows that

$$\Re(S_m^\omega \phi_z(y_j)) - \Re(S_m^\omega \phi_z(y_j')) + \ln g(y_j) - \ln g(y_j')$$

$$\geq -\rho^\alpha(x, x')\Big(\sum_{t=0}^{m-1} 2H_{\theta^t \omega}(\gamma_{\theta^t \omega, m-t})^{-\alpha} \Big) - \rho^\alpha(x, x') s Q_\omega(\gamma_{\omega, m})^{-\alpha}$$

$$= -\rho^\alpha(x, x')\big(Q_{\theta^m \omega} + (s-1) Q_\omega(\gamma_{\omega, m})^{-\alpha} \big)$$

where in the last equality we used (5.8.6). As a consequence,

$$\Re(z_1 - z_2) = \Re(S_m^\omega \phi_z(y_j)) - \Re(S_m^\omega \phi_z(y_j')) + \ln g(y_j) - \ln g(y_j') \quad (5.8.7)$$

$$+ s Q_{\theta^m \omega} \rho^\alpha(x, x') \geq \rho^\alpha(x, x') \cdot (s-1)\big(Q_{\theta^m \omega} - Q_\omega(\gamma_{\omega, m})^{-\alpha} \big)$$

$$= (s-1) \rho^\alpha(x, x') \sum_{t=0}^{m-1} 2H_{\theta^t \omega}(\gamma_{\theta^t \omega, m-t})^{-\alpha} > 0.$$

In particular, we obtained that $\Re(z_1) > \Re(z_2)$. By (5.8.3), (5.8.5) and (5.8.7) it follows that

$$1 \leq |a_j|^2 \leq 1 + s_0^2 \quad \text{and} \quad |\arg(a_j)| \leq s_0$$

where $s_0 = s_0(s, z) = \frac{|\Im(z)|}{s-1}$. Finally, since

$$l_{\theta^m \omega, x, x'}(\mathcal{L}_z^{\omega, m} g) = \sum_{j=1}^k n_{m,j}(z, g)$$

$$= \sum_{1 \leq j \leq k : g(y_j) > 0} e^{i\Im(S_m^\omega \phi_z(y_j))} a_j n_{m,j}(\Re(z), g)$$

we can write

$$\frac{l_{\theta^m \omega, x, x'}(\mathcal{L}_z^{\omega, m} g)}{l_{\theta^m \omega, x, x'}(\mathcal{L}_{\Re(z)}^{\omega, m} g)} = Z$$

where

$$Z \in \mathbf{A}_{m, z, \omega} := \text{Conv}\{ r e^{iv} : 1 \leq r \leq 1 + s_0 \text{ and } |v| \leq V \}$$

and $V = V(m, z, \omega) = |\Im(z)| \cdot \|S_m^\omega u\|_\infty + s_0$. Let $1 \leq r \leq 1 + s_0$ and $|v| \leq V$. Then by the mean value theorem,

$$|r e^{iv} - 1| \leq |r - 1| + |e^{iv} - 1| \leq s_0 + 2|v| \leq s_0 + 2V.$$

A convex combination of numbers of the form $r e^{iv}$ satisfying the inequality $|r e^{iv} - 1| \leq s_0 + 2V$ also satisfies it, and therefore

$$|Z - 1| \leq s_0 + 2V$$

and the lemma follows. $\qquad\square$

5.9 Complex image diameter estimates

In this section we will prove Theorem 5.4.1 *(ii)* and Corollary 5.4.2. Let $z \in \mathbb{C}$ be such that $|\Re(z)| \leq 1$. For P-a.a. ω set

$$d_{1,\omega} = 2\ln\left(s''_{\theta^{n}\omega}C_{\omega}(n_{\omega})\right), \ d_{\omega} = d_{1,\omega} + 6\ln 2$$

and

$$a_{\omega} = \frac{1}{4}\left(1 + \cosh(\frac{1}{2}d_{1,\omega})\right)^{-1}$$

where $C_{\omega}(m), m \in \mathbb{N}$ and s''_{ω} are given by (5.7.14) and (5.7.15), respectively. Let $n \geq n_{\omega}$. Then by (5.7.18), Lemma 5.8.1 and Theorem A.2.4 if

$$\varepsilon_{\omega}(n, z) < a_{\omega}$$

then

$$\mathcal{L}_{z}^{\omega,n}\mathcal{C}'_{\omega} \subset \mathcal{C}'_{\theta^{n}\omega} \ \text{ and } \ \Delta_{\mathcal{C}_{\theta^{n}\omega}}\left(\mathcal{L}_{z}^{\omega,n}\mathcal{C}'_{\omega}\right) < d_{\omega}. \tag{5.9.1}$$

Next, suppose that the conditions of Theorem 5.4.1 hold true. Let j_0 and d_0 be constants so that $n_{\omega} \leq j_0$ and $\max(\|u_{\omega}\|_{\infty}, d_{\omega}) \leq d_0$, P-a.s. Then $a_{\omega} \geq d_1$ for some positive constant d_1 which depends only on d_0 and $\varepsilon_{\omega}(n, z) \leq |\Im(z)|(2nd_0 + 3(s-1)^{-1})$ for any natural n. Since $n_{\omega} \leq j_0$ we can apply (5.9.1) with any $n \geq j_0$ which together with the latter estimates implies that Assumption 4.1.2 holds true for the cones $\mathcal{C}_{\omega}$ with the constants j_0 and d_0 and with $U = [-1, 1] \times [-r, r]$, where r depends only on s, j_0 and d_0.

Now we derive Corollary 5.4.2. By Theorem 5.4.1, Assumption 4.1.2 holds true and by Theorem 5.3.1, Assumptions 4.1.1 and 4.1.3 hold true. The functionals l_{ω} and κ_{ω} are measurable in ω and so using also Theorem 5.4.1 *(i)* we see that all measurability assumptions in Section 4.1 are satisfied. Therefore, all the statements of Corollary 5.4.2 except for the uniform boundedness (5.4.3) of the RPF triplets and the properties of the triplets $(\lambda_{\omega}(t), h_{\omega}(t), \nu_{\omega}(t))$ for real t's follow from Theorems 4.2.1 and 4.2.2. The proof of the latter properties is delayed to the next section (see Corollary 5.10.2). In order to prove (5.4.3), we first observe that the neighborhood V_{ω} defined before (4.4.4) contains a deterministic neighborhood U_0 of 0 when the random variables $|\alpha_{\omega}(0)|^{-1}$ and

$$v_{\omega} = \frac{K_{\omega}M_{\omega}}{\|l_{\omega}\|_{\alpha,\xi}\|\kappa_{\omega}\|_{\alpha,\xi}} = \frac{K_{\omega}M_{\omega}}{L_{\omega}}$$

are bounded. Note that v_{ω} is indeed bounded under the assumptions of Corollary 5.4.2. In the paragraph below we will show that $|\alpha_{\omega}(0)|^{-1}$ is a bounded random variable, and assuming this is true we complete the proof of (5.4.3). Boundedness in ω of $\sup_{z \in U_0} \|h_\omega(z)\|_{\alpha,\xi}$ follows from (4.4.5), while $\sup_{z \in U} \|\nu_\omega(z)\|_{\alpha,\xi}$ is a bounded random variable by (4.3.27), the definitions of M_ω and Q_ω and our assumption that $\gamma_\omega \geq 1 + \delta$ and that H_ω is bounded. Since $\lambda_\omega(z) = \nu_{\theta\omega}(z)(\mathcal{L}_z^\omega 1)$ it follows that $|\lambda_\omega(z)| \leq M_{\theta\omega}\|\mathcal{L}_z^\omega\|_{\alpha,\xi}$. Under the assumptions of Corollary 5.4.2 the random variables Q_ω, $\|u_\omega\|_\infty$ and $\|\mathcal{L}_0^\omega 1\|_\infty$ are bounded and therefore by Lemma 5.6.1 $\sup_{z \in U} \|\mathcal{L}_z^\omega\|_{\alpha,\xi}$ is bounded in ω, yielding that $\sup_{z \in U} |\lambda_\omega(z)|$ is bounded in ω. Applying now Lemma 2.8.2 we obtain (5.4.3).

Now we will show that $|\alpha_\omega(0)|^{-1}$ is bounded assuming that the conditions of Theorem 5.4.1 *(ii)* hold true. Recall first that

$$\alpha_\omega(0) = \nu_\omega(0)\hat{h}_\omega(0) = \frac{1}{l_\omega(0)h_\omega(0)}.$$

Secondly, by (5.7.16) and (4.3.10) P-a.s.,

$$\hat{h}_\omega(t) \in \mathcal{C}'_{\omega,\mathbb{R}} = \mathcal{C}'_{\omega,\mathbb{R},s} \qquad (5.9.2)$$

for any $t \in U \cap \mathbb{R}$ since the real cone $\mathcal{C}_{\omega,\mathbb{R}}$ is closed with respect to the norm $\|\cdot\|_\infty$ and $\hat{h}_\omega(t)$ is not identically 0. Let $C_0 > 1$ and $j_0 \in \mathbb{N}$ be so that $n_\omega \leq j_0$ and $C_\omega(n_\omega) \leq C_0$ P-a.s., where $C_\omega(m)$'s are defined in (5.7.15). Then by Lemma 5.7.3,

$$\mathcal{L}_t^{\omega,j_0}\mathcal{C}'_{\omega,\mathbb{R}} \subset \mathcal{K}'_{\theta^{j_0}\omega,\mathbb{R},C_0} \cap \mathcal{C}'_{\theta^{j_0}\omega,\mathbb{R}}$$

for any $t \in U \cap \mathbb{R}$. Combining this with (5.9.2) and (4.3.13) we arrive at

$$\hat{\lambda}_{\omega,j_0}(t)\hat{h}_{\theta^{j_0}\omega}(t) \in \mathcal{K}'_{\theta^{j_0}\omega,\mathbb{R},C_0} \cap \mathcal{C}'_{\theta^{j_0}\omega,\mathbb{R}} \qquad (5.9.3)$$

where $\hat{\lambda}_{\omega,j_0}(t) = \prod_{i=0}^{j_0-1} \hat{\lambda}_{\theta^i\omega}(t)$. By (4.3.15) (or by (4.3.14)) the term $\hat{\lambda}_{\omega,j_0}(t)$ does not vanish and by (4.3.14) it is nonnegative. We conclude that it is strictly positive and therefore P-a.s. the functions $\hat{h}_{\theta^{j_0}\omega}(t)$, $t \in U \cap \mathbb{R}$ are members of the intersection of the cone in the right-hand side of (5.9.3). Since θ is P-invariant the same holds true with ω in place of $\theta^{j_0}\omega$ and so, taking into account that $l_\omega\hat{h}_\omega(t) = 1$ and $\hat{h}_\omega(t) \in \mathcal{K}'_{\omega,\mathbb{R},C_0}$, we obtain that

$$L_\omega^{-1} = L_\omega^{-1}l_\omega\hat{h}_\omega(t) \geq \min_{x \in \mathcal{E}_\omega} \left(\hat{h}_\omega(t)\right)(x) \geq \|\hat{h}_\omega(t)\|_\infty C_0^{-1} \qquad (5.9.4)$$

$$\geq l_\omega(\hat{h}_\omega(t))L_\omega^{-1}C_0^{-1} = L_\omega^{-1}C_0^{-1}$$

where $L_\omega^{-1} = (L_\omega)^{-1}$, and in particular,

$$\min_{x \in \mathcal{E}_\omega} \left(\hat{h}_\omega(t)\right)(x) \geq \frac{1}{L_\omega C_0} \qquad (5.9.5)$$

for any $t \in U \cap \mathbb{R}$. Finally, observe that $\nu_\omega(t)$ is a positive linear functional (which satisfies $\nu_\omega(t)\mathbf{1} = 1$) when t is real. For instance, we can always take $\mu_n = l_{\theta^n \omega}$ in (4.3.25). Therefore,

$$\alpha_\omega(t) = \nu_\omega(t)\hat{h}_\omega(t) \geq \frac{1}{C_0 L_\omega} \qquad (5.9.6)$$

implying that $\alpha_\omega(t)$ is bounded away from 0 under the assumptions of Corollary 5.4.2 since in these circumstances the random variable L_ω is bounded from above. $\qquad \square$

5.10 Real RPF triplets and Gibbs measures

In Theorem 5.4.2 we claimed that the functionals $\nu_\omega(t)$ for real t's have some special properties and in particular coincide with the ones from [53] (in the setup from there). This will follow from

Lemma 5.10.1. *Under the assumptions Corollary 5.4.2, P-a.s. for any $t \in U \cap \mathbb{R}$ the functional $\nu_\omega(t)$ is a probability measure and*

$$\lambda_\omega(t)\nu_\omega(t)g = \nu_{\theta\omega}(t)(\mathcal{L}_t^\omega g) \qquad (5.10.1)$$

for any bounded Borel function $g : \mathcal{X}_\omega \to \mathbb{C}$.

Moreover, the random variable $\lambda_\omega(t)$ and the function $h_\omega(t)$ are strictly positive.

Proof. Making the choice of $\mu_n = l_{\theta^n \omega}$ in (4.3.25) it follows that $\nu_\omega(t)$, $t \in U \cap \mathbb{R}$ are positive linear functionals on the space $\mathcal{H}_\omega^{\alpha,\xi}$ satisfying $\nu_\omega(t)\mathbf{1} = 1$. By the Stone-Weierstrass theorem, any continuous function $g : \mathcal{X}_\omega \to \mathbb{C}$ is a uniform limit of functions from $\mathcal{H}_\omega^{\alpha,\xi}$. Since $\nu_\omega(t)$ is positive and $\mathbf{1} \in \mathcal{H}_\omega^{\alpha,\xi}$ we can extend it to a positive linear functional on the space of all continuous functions from $\mathcal{X}_\omega$ to $\mathbb{C}$. Since $\nu_\omega(t)\mathbf{1} = 1$ we conclude that it is continuous with respect to the supremum norm and that $\|\nu_\omega(t)\|_\infty = 1$. Therefore, by the Riesz-Markov-Kakutani representation theorem (see [57]) $\nu_\omega(t)$ is a positive measure and since $\nu_\omega(t)\mathbf{1} = 1$ it is a probability measure. Finally, (5.10.1) extends from functions $g \in \mathcal{H}_\omega^{\alpha,\xi}$ to continuous functions since $\lambda_\omega(t)\nu_\omega(t)$ and $(\mathcal{L}_t)^*\nu_{\theta\omega}(t)$ are continuous linear functionals on the space of continuous functions agreeing on a dense set. By the uniqueness part of Riesz-Markov-Kakutani representation theorem we deduce that they agree on the space of all bounded Borel functions, and the proof of the first statement of the lemma is complete.

Next, since $\lambda_\omega(t)\nu_\omega(t) = (\mathcal{L}_t^\omega)^*\nu_{\theta\omega}(t)$ we can integrate the function $\mathbf{1}$ with respect to both sides and obtain that $\lambda_\omega(t) = \nu_{\theta\omega}(t)(\mathcal{L}_t^\omega \mathbf{1})$ which is strictly positive since $\mathcal{L}_t^\omega 1 \geq e^{-\|\phi_\omega\|_\infty - |t| \|u_\omega\|_\infty}$. In order to show that $h_\omega(t)$ is strictly positive recall that (see (4.3.32)),

$$h_\omega(t) = \frac{\hat{h}_\omega(t)}{\nu_\omega(t)\hat{h}_\omega(t)}.$$

The function $\hat{h}_\omega(t)$ is strictly positive (see the paragraph below (5.9.3)) and by (5.9.5) the denominator is positive and so $h_\omega(t)$ is a strictly positive function.   □

The following corollary is a direct consequence of Lemma 5.10.1 and the uniqueness statement from [53].

Corollary 5.10.2. *Suppose that T_ω is the random map presented in [53] and that the assumptions of Corollary 5.4.2 hold true. Then for any $t \in U \cap \mathbb{R}$ the triplet $(\lambda_\omega(t), h_\omega(t), \nu_\omega(t))$ P-a.s. coincides with the one constructed in [53] with the random function $\phi_\omega + t u_\omega$.*

Next, consider the Gibbs probability measures $\mu_\omega(t)$, $t \in U \cap \mathbb{R}$ given by $d\mu_\omega(t) = h_\omega(t) d\nu_\omega(t)$. The following lemma will be used in Chapter 7.

Lemma 5.10.3. *Suppose that the Assumptions of Corollary 5.4.2 hold true. Then P-a.s. for any $t \in U \cap \mathbb{R}$ the measures $\nu_\omega(t)$ and $\mu_\omega(t)$ assign positive mass to open sets.*

Proof. We first show that for any $r > 0$ there exists $n_1 > 0$ such that P-a.s.,

$$T_\omega^{n_1}(B_\omega(y, r)) = \mathcal{E}_{\theta^{n_1}\omega} \quad \text{for any } y \in \mathcal{E}_\omega. \tag{5.10.2}$$

In order to prove (5.10.2), let $n_0 \in \mathbb{N}$ and $\delta > 0$ be such that $n_\omega \leq n_0$ and $\gamma_\omega \geq 1 + \delta$, P-a.s. Let m_1 be the smallest natural number so that $\gamma_0^{m_1}\xi < r$, where $\gamma_0 = (1 + \delta)^{-1}$. We claim that $n_1 = n_0 + m_1$ satisfies (5.10.2). Indeed, let $y \in \mathcal{E}_\omega$ and $m \in \mathbb{N}$. Consider the point $x = T_\omega^m y \in \mathcal{E}_{\theta^m\omega}$ and write $y = y_i$ for some $1 \leq i \leq k = |(T_\omega^m)^{-1}\{x\}|$, where $\{y_1, ..., y_k\} = (T_\omega^m)^{-1}\{x\}$. For any $x' \in B_{\theta^m\omega}(x, \xi)$ let $y_i' \in \mathcal{E}_\omega$ be the corresponding member of $(T_\omega^m)^{-1}\{x'\}$ from (5.1.8) with $n = m$. Using (5.1.8) with $j = 0$ we derive that

$$y_i' \subset B_\omega(y_i, (\gamma_{\omega,m})^{-1}\xi) = B_\omega(y, (\gamma_{\omega,m})^{-1}\xi).$$

Applying the map T_ω^m to both sides of the above equality and then taking the union over all possible choices of x' we conclude that

$$B_{\theta^m\omega}(x, \xi) = B_{\theta^m\omega}(T_\omega^m y, \xi) \subset T_\omega^m\big(B_\omega(y, (\gamma_{\omega,m})^{-1}\xi)\big). \tag{5.10.3}$$

Taking $m = m_1$ and applying the map $T^{n_0}_{\theta^m \omega}$ to both sides we obtain (5.10.2), taking into account that $n_{\theta^m \omega} \leq n_0$ and $(\gamma_{\omega,m})^{-1} \leq \gamma_0^m$.

Next, let $B = B_\omega(y, r)$ be an open ball in $\mathcal{E}_\omega$ with rational radius. Then by (5.10.2) for any $x \in \mathcal{E}_{\theta^{n_1} \omega}$ there exists $y \in B$ such that $T^{n_1}_\omega y = x$ and so $\mathcal{L}^{\omega,n_1}_t \mathbb{I}_B(x) \geq e^{-\|S^\omega_{n_1} \phi\|_\infty}$ where $\mathbb{I}_B$ is the indicator function of B. Therefore by (5.10.1),

$$\nu_\omega(t)(B) = \frac{\int \mathcal{L}^{\omega,n_1}_t \mathbb{I}_B d\nu_{\theta^{n_1}\omega}(t)}{\lambda_{\omega,n_1}(t)} \geq \frac{e^{-\|S^\omega_{n_1}\phi\|_\infty}}{\lambda_{\omega,n_1}(t)} > 0$$

and hence $\nu_\omega(t)$ assigns positive mass to open sets. The measures $\nu_\omega(t)$ and $\mu_\omega(t)$ are equivalent since $h_\omega(t)$ is a continuous and strictly positive function and thus the measure $\mu_\omega(t)$ assigns positive mass to open sets, as well. $\qquad \square$

We will state now several results concerning Gibbs measures which will be used in Chapter 7 and are proved exactly as in [53]. We begin with the following T_ω-invariance and exponential decay of correlations.

Lemma 5.10.4. *Suppose that the assumptions of Corollary 5.4.2 hold true.*

(i) The families of measures $\{\mu_\omega(t) : \omega \in \Omega\}$, $t \in U \cap \mathbb{R}$ are T_ω invariant in the sense that P-a.s.,

$$\int g \circ T_\omega d\mu_\omega(t) = \int g d\mu_{\theta\omega}(t) \tag{5.10.4}$$

for any $t \in U \cap \mathbb{R}$ and a bounded Borel function $g : \mathcal{X}_{\theta\omega} \to \mathbb{C}$.

(ii) P-a.s. for any $n \in \mathbb{N}$ and $t \in U \cap \mathbb{R}$,

$$\left| \mu_{\theta^{-n}\omega}(t)\left((q \circ T^n_{\theta^{-n}\omega})g\right) - \mu_\omega(t)(q) \cdot \mu_{\theta^{-n}\omega}(t)(g) \right| \tag{5.10.5}$$

$$\leq \mu_\omega(t)|q| \cdot \left\| \frac{\mathcal{L}^{\theta^{-n}\omega,n}_t(g \cdot h_{\theta^{-n}\omega}(t))}{\lambda_{\theta^{-n}\omega,n}(t)h_\omega(t)} - \mu_{\theta^{-n}\omega}(t)(g)\mathbf{1} \right\|_\infty$$

for any bounded Borel functions $g : \mathcal{E}_{\theta^{-n}\omega} \to \mathbb{C}$ and $q : \mathcal{E}_\omega \to \mathbb{C}$, where $\mathbf{1}$ is the function which takes the constant value 1.

Note that the second factor on the right-hand side of (5.10.5) can be approximated using the exponential convergence type estimates stated in Corollary 5.4.2 when $g \in \mathcal{H}^{\alpha,\xi}_{\theta^{-n}\omega}$.

Next, since $\mu_\omega(t)$, $t \in U \cap \mathbb{R}$ are measurable in ω we can define probability measures $\mu(t)$, $t \in U \cap \mathbb{R}$ on $\mathcal{E}$ by the formula

$$\int g d\mu(t) = \int \int g_\omega d\mu_\omega(t) dP(\omega) \tag{5.10.6}$$

where $g : \mathcal{E} \to \mathbb{R}$ is a bounded measurable function. It is clear from (5.10.4) that the measures $\mu(t)$ are T-invariant, where T is the skew product map defined in (5.1.1). The following proposition is proved exactly as Proposition 4.7 in [53]. We will not use it in this manuscript and it is formulated here for the sake of completeness.

Proposition 5.10.5. *Suppose that the conditions of Corollary 5.4.2 hold true and that θ is ergodic. Then the measure preserving system $(\mathcal{E}, \mathcal{P}, \mu(t), T)$ is egrodic for any $t \in U \cap \mathbb{R}$ where $\mathcal{P}$ is the σ-algebra on $\mathcal{E}$ induced from $\mathcal{F} \times \mathcal{B}$.*

5.11 Complex Gibbs functionals

For any $z \in U$ consider the the complex (Gibbs) linear functional $\mu_\omega(z) : \mathcal{H}_\omega^{\alpha,\xi} \to \mathbb{C}$ given by

$$\mu_\omega(z)(g) = \nu_\omega(z)(gh_\omega(z)).$$

In this section we will show that some of the properties of the Gibbs measures $\mu_\omega(t)$, $t \in U \cap \mathbb{R}$ hold true for complex z's. We begin with the following.

Proposition 5.11.1. *Suppose that the conditions of Corollary 5.4.2 hold true. Then P-a.s. for any $z \in U$,*

$$\|\mu_\omega(z)\|_{\alpha,\xi} \leq 3M_\omega \|h_\omega(z)\|_{\alpha,\xi}$$

where M_ω is defined in Theorem 5.3.1. Moreover, $\mu_\omega(z)$ is T_ω invariant in the sense that

$$\mu_\omega(z)(g \circ T_\omega) = \mu_{\theta\omega}(z)(g) \tag{5.11.1}$$

for any $g \in \mathcal{H}_{\theta\omega}^{\alpha,\xi}$ such that $g \circ T_\omega \in \mathcal{H}_\omega^{\alpha,\xi}$.

Proof. First, for any two functions g_1, g_2 we have $\|g_1 g_2\|_{\alpha,\xi} \leq 3\|g_1\|_{\alpha,\xi}\|g_2\|_{\alpha,\xi}$ and therefore

$$\|\mu_\omega(z)\|_{\alpha,\xi} = \|h_\omega(z)\nu_\omega(z)\|_{\alpha,\xi} \leq 3\|h_\omega(z)\|_{\alpha,\xi}\|\nu_\omega(z)\|_{\alpha,\xi} \leq 3\|h_\omega(z)\|_{\alpha,\xi}M_\omega$$

where in the last inequality we used (4.3.27). Now we prove the statement about T_ω-invariance of $\mu_\omega(z)$. Let $g \in \mathcal{H}_{\theta\omega}^{\alpha,\xi}$ be such that $g \circ T_\omega \in \mathcal{H}_\omega^{\alpha,\xi}$. Since

$$\mathcal{L}_z^\omega h_\omega(z) = \lambda_\omega(z)h_{\theta\omega}(z) \quad \text{and} \quad (\mathcal{L}_z^\omega)^* \nu_{\theta\omega}(z) = \lambda_\omega(z)\nu_\omega(z)$$

we obtain that

$$\mu_{\theta\omega}(z)(g) = \nu_{\theta\omega}(z)(h_{\theta\omega}(z)g) = (\lambda_\omega(z))^{-1}\nu_{\theta\omega}(z)(g \cdot \mathcal{L}_z^\omega h_\omega(z))(5.11.2)$$
$$= (\lambda_\omega(z))^{-1}\nu_{\theta\omega}(z)\big(\mathcal{L}_z^\omega(h_\omega(z) \cdot g \circ T_\omega)\big)$$
$$= \nu_\omega(z)(h_\omega(z) \cdot g \circ T_\omega)) = \mu_\omega(z)(g \circ T_\omega)$$

and the proposition follows. □

Note that under the conditions of Proposition 5.11.1 the random variable Q_ω is bounded, which makes M_ω a bounded random variable. Therefore we can define (global) linear functionals $\mu(z)$, $z \in U$ by the formula

$$\mu(z)(g) = \int \mu_\omega(z)\big(g(\omega, \cdot)\big)dP(\omega)$$

where g is a member of the space $\mathcal{H} = \mathcal{H}^{\alpha,\xi}$ of all measurable functions so that $\int \|g(\omega, \cdot)\|_{\alpha,\xi}dP(\omega) < \infty$. The functional $\mu(z)$ is T invariant when restricted to the space $\mathcal{H} \cap \hat{T}^{-1}\mathcal{H}$ where $\hat{T}g = g \circ T$.

Example 5.11.2. Consider the situation when the maps T_ω are uniformly Lipschitz continuous, i.e. there exists a constant $K > 1$ such that $\rho(T_\omega x, T_\omega x') \leq K\rho(x, x')$ for any ω and $x, x' \in \mathcal{E}_\omega$. For instance, we can consider the case of a random topologically mixing subshift of finite type (see [40]). Since the spaces $\mathcal{E}_\omega$ are compact and their diameters do not exceed 1, there exists a constant $c_\xi > 0$ which depends only on ξ such that

$$c_\xi\|g\|_{\alpha,1} \leq \|g\|_{\alpha,\xi} \leq \|g\|_{\alpha,1}$$

for any $g \in \mathcal{H}_\omega^{\alpha,\xi}$. Therefore, $\hat{T}$ maps the space $\mathcal{H}^{\alpha,\xi}$ to itself, preserves $\mu(z)$ and it is continuous with respect to the norm $\|g\|_{\alpha,\xi,1} := \int \|g(\omega, \cdot)\|_{\alpha,\xi}dP(\omega)$ where we identify two functions g and h so that $g(\omega, \cdot) = h(\omega, \cdot)$, P-a.s.

The complex Gibbs functionals also satisfy the following exponential decay of correlations.

Proposition 5.11.3. *Under the assumptions of Corollary 5.4.2, P-a.s. for any $n \in \mathbb{N}$ and $z \in U$,*

$$\left|\mu_{\theta^{-n}(\omega)}(z)\big(q \circ T_{\theta^{-n}\omega}^n \cdot g\big) - \mu_\omega(z)(q) \cdot \mu_{\theta^{-n}\omega}(z)(g)\right| \leq 3M_\omega\|q\|_{\alpha,\xi}$$
$$\times \left\|\frac{\mathcal{L}_z^{\theta^{-n}\omega,n}(g \cdot h_{\theta^{-n}\omega}(z))}{\lambda_{\theta^{-n}\omega,n}(z)} - \mu_{\theta^{-n}\omega}(z)(g) \cdot h_\omega(z)\right\|_{\alpha,\xi}$$

for any $g \in \mathcal{H}_{\theta^{-n}\omega}^{\alpha,\xi}$ and $q \in \mathcal{H}_\omega^{\alpha,\xi}$ such that $q \circ T_{\theta^{-n}\omega} \in \mathcal{H}_{\theta^{-n}\omega}^{\alpha,\xi}$.

The last factor on the above right-hand side can be approximated using the exponential convergence type estimates stated in Corollary 5.4.2.

Proof. First, P-a.s. for any n and $z \in U$,

$$\left(\mathcal{L}_z^{\theta^{-n}\omega, n}\right)^* \nu_\omega(z) = \lambda_{\theta^{-n}\omega, n}(z) \nu_{\theta^{-n}\omega}(z)$$

and therefore

$$\mu_{\theta^{-n}\omega}(z)\left((q \circ T_{\theta^{-n}\omega}^n)g\right) = \nu_\omega(z)\left(q \cdot \frac{\mathcal{L}_z^{\theta^{-n}\omega, n}(g \cdot h_{\theta^{-n}\omega}(z))}{\lambda_{\theta^{-n}\omega, n}(z)}\right) \qquad (5.11.3)$$

for any g and q as in the statement of the proposition, where we used here that $\mathcal{L}_z^{\theta^{-n}\omega, n}\left((q \circ T_{\theta^{-n}\omega}^n)g_1\right) = q \cdot \mathcal{L}_z^{\theta^{-n}\omega, n}(g_1)$, for any function g_1. On the other hand, we can write

$$\mu_\omega(z)(q) \cdot \mu_{\theta^{-n}\omega}(z)(g) = \nu_\omega(z)\left(q \cdot \mu_{\theta^{-n}\omega}(z)(g) \cdot h_\omega(z)\right) \qquad (5.11.4)$$

and the proposition follows. $\qquad \square$

5.12   The largest characteristic exponents

Suppose that θ is ergodic. In this section we will show that the assumptions of Theorem 4.6.3 are satisfied under the conditions of Corollary 5.4.2. We begin with the following.

Lemma 5.12.1. *Under the conditions of Corollary 5.4.2 the random variable* $|\ln \lambda_\omega(0)|$ *is bounded and in particular* $\ln \lambda_\omega(0)$ *is integrable.*

Proof. We first recall that P-a.s.,

$$\lambda_\omega(0) = \nu_{\theta\omega}(0)(\mathcal{L}_0^\omega 1) > 0.$$

Since $\nu_{\theta\omega}(0)$ is probability measure, we obtain that

$$\min_{x \in \mathcal{E}_{\theta\omega}} \mathcal{L}_0^\omega 1(x) \leq \lambda_\omega(0) \leq \|\mathcal{L}_0^\omega 1\|_\infty.$$

Since

$$\min_{x \in \mathcal{E}_{\theta\omega}} \mathcal{L}_0^\omega 1(x) \geq e^{-\|\phi_\omega\|_\infty} \qquad (5.12.1)$$

we conclude that

$$-\|\phi_\omega\|_\infty \leq \ln \lambda_\omega(0) \leq \ln \|\mathcal{L}_0^\omega 1\|_\infty. \qquad (5.12.2)$$

Under the assumptions of Corollary 5.4.2 the random variables $\|\phi_\omega\|_\infty$ and D_ω are bounded, and the lemma follows by the inequality

$$\ln \|\mathcal{L}_0^\omega 1\|_\infty \leq \|\phi_\omega\|_\infty + \ln \deg T_\omega \leq \|\phi_\omega\|_\infty + \ln D_\omega.$$

$\qquad \square$

Next we will show that Assumption 4.6.2 holds true. We first need the following. Let $\tilde{\phi}_\omega : \mathcal{E}_\omega \to \mathbb{C}$ be defined by

$$\tilde{\phi}_\omega = \phi_\omega + \ln h_\omega(0) - \ln(h_{\theta\omega}(0) \circ T_\omega) - \ln \lambda_\omega(0) \qquad (5.12.3)$$

and consider the transfer operators $\tilde{\mathcal{L}}_z^\omega$ generated by the random functions $\tilde{\phi}_\omega + z u_\omega$ and the map T_ω which is given by the formula

$$\tilde{\mathcal{L}}_z^\omega g = \mathcal{L}_z^\omega \left(\frac{g h_\omega(0)}{\lambda_\omega(0) \cdot h_{\theta\omega} \circ T_\omega} \right) = \frac{\mathcal{L}_z^\omega \left(g \cdot h_\omega(0) \right)}{\lambda_\omega(0) \cdot h_{\theta\omega}(0)}. \qquad (5.12.4)$$

Then $\tilde{\mathcal{L}}_z^\omega 1 = 1$. The proof of the following lemma goes similarly to the proof of Lemma 5.6.1 using the function $\tilde{\phi}$ in place of ϕ and taking into account that $h_\omega(0) \in \mathcal{C}'_{\omega,\mathbb{R},s}$ which guarantees that its logarithm is a member of $\mathcal{H}_\omega^{\alpha,\xi}$.

Lemma 5.12.2. *P-a.s. for any $z = \Re(z) + i\Im(z) \in \mathbb{C}$ and $n \in \mathbb{N}$,*

$$\|\tilde{\mathcal{L}}_z^{\omega,n}\|_{\alpha,\xi} \leq 2 \max \left(|\Im(z)|, 1 + |\Re(z)| \right) e^{|\Re(z)| \|S_n^\omega u\|_\infty} \left(1 + 6 Q_{\theta^n\omega} \right). \quad (5.12.5)$$

Corollary 5.12.3. *Suppose that the conditions of Corollary 5.4.2 hold true and that the random variable L_ω is bounded. Then Assumption 4.6.2 holds true with some constant C which depends only on the essential supremums of the random variables in (5.4.1) and on δ specified in Theorem 5.4.1.*

Proof. For any $g : \mathcal{E}_\omega \to \mathbb{C}$ we have

$$\mathcal{L}_z^\omega g = \lambda_\omega(0) h_{\theta\omega}(0) \tilde{\mathcal{L}}_z^\omega \left(\frac{g}{h_\omega(0)} \right).$$

Therefore,

$$\|\mathcal{L}_z^\omega\|_{\alpha,\xi} \leq 9 |\lambda_\omega(0)| \|h_{\theta\omega}(0)\|_{\alpha,\xi} \|(h_\omega(0))^{-1}\|_{\alpha,\xi} \|\tilde{\mathcal{L}}_z^\omega\|_{\alpha,\xi}$$

where we used that

$$\|g_1 g_2\|_{\alpha,\xi} \leq 3 \|g_1\|_{\alpha,\xi} \|g_2\|_{\alpha,\xi}$$

for any two functions. Next, for any non-vanishing $h \in \mathcal{H}_\omega^{\alpha,\xi}$ we have

$$\|h^{-1}\|_{\alpha,\xi} \leq \left(\min_x |h(x)| \right)^{-1} + \left(\min_x |h(x)| \right)^{-2} v_{\alpha,\xi}(h) \qquad (5.12.6)$$

where $h^{-1} = 1/h$. Taking $h = h_\omega(0)$, we conclude that Assumption 4.6.2 holds true with some constant $C > 0$, taking into account that the random variables Q_ω, K_ω and L_ω are bounded and that (see Section 4.4.1 and (5.9.6)),

$$\hat{h}_\omega(0) \geq L_\omega^{-1} C_0^{-1}, \quad \|l_\omega\|_{\alpha,\xi} = L_\omega$$

and

$$\|h_\omega(0)\|_{\alpha,\xi} \leq \frac{2 K_\omega}{L_\omega |\alpha_\omega(0)|} \quad \text{and} \quad \frac{1}{C_0 L_\omega} \leq |\alpha_\omega(0)| \leq v_\omega \qquad (5.12.7)$$

where we recall that $v_\omega = \frac{K_\omega M_\omega}{\|l_\omega\|_{\alpha,\xi} \|\kappa_\omega\|_{\alpha,\xi}} = \frac{K_\omega M_\omega}{L_\omega}$. $\qquad \square$

5.13   Extension to the unbounded case for minimal systems

Let $j_0, d_0 > 1$ and $\Gamma \subset \Omega$ be a measurable set such that $P(\Gamma) > 0$ and $n_\omega \leq j_0$ and $d_\omega \leq d_0$ for any $\omega \in \Gamma$. A slight modification of our arguments shows that Corollary 5.4.2 holds true when there exists a natural $J \geq 1$ so that

$$\Omega = \bigcup_{j=1}^{J} \theta^{-j}\Gamma$$

i.e., there exists a bounded measurable function $\mathcal{J} = \mathcal{J}_\Gamma : \Omega \to \mathbb{N}$ such that $\theta^{\mathcal{J}\omega}\omega \in \Gamma$ and $\theta^{j}\omega \notin \Gamma$ for any $1 \leq j < \mathcal{J}\omega$.

We will describe shortly a situation in which such a function $\mathcal{J}$ exists. Let $(\Omega, \mathcal{F}, P, \theta)$ be an irrational rotation of the unit circle where P is the Lebesgue measure and $\mathcal{F}$ is the Borel σ-algebra. Then the system (Ω, θ) is minimal and P is the unique ergodic measure. Let $\{U_i : i \geq 1\}$ be a disjoint collection of open subsets of Ω such that $P(\cup_{i=1}^{\infty} U_i) = 1$. Consider the situation when the random variables n_ω and d_ω are constant on each U_i. Then for any sufficiently large d_0 and j_0 any set Γ with the properties described above must contain an open set. For any open set $U \subset \Omega$ there exists J so that

$$\Omega = \bigcup_{j=1}^{J} \theta^{-j}U$$

and therefore there exists a function $\mathcal{J}_\Gamma$ with the desired properties. Similar examples can be given for any minimal and uniquely ergodic invertible compact system $(\Omega, \mathcal{F}, P, \theta)$ for which $P(\cup_{i=1}^{\infty} U_i) = 1$ for a collection of open sets $\{U_i : i \geq 1\}$. For instance the map $(x, y) \to (x+\alpha, y+x)$, $\alpha \notin \mathbb{Q}$ on the two dimensional torus together with the Lebesgue measure can be considered, and we refer the readers to Sections 6.5 and 6.6 in [63] for more examples.

Chapter 6

Application to random complex integral operators

6.1 Integral operators

Let $(\Omega, \mathcal{F}, P, \theta)$, $(\mathcal{X}, \rho)$, $\mathcal{E} \subset \Omega \times \mathcal{X}$ and $\mathcal{E}_\omega$ be as at the beginning of Section 5.1. For any $\omega \in \Omega$ denote by B_ω the Banach space of all bounded Borel functions $g : \mathcal{E}_\omega \to \mathbb{C}$ together with the supremum norm $\| \cdot \|_\infty$. For any $g : \mathcal{E} \to \mathbb{C}$ consider the functions $g_\omega : \mathcal{E}_\omega \to \mathbb{C}$ given by $g_\omega(x) = g(\omega, x)$. Then by Lemma 5.1.3 the norm $\omega \to \|g_\omega\|_\infty$ is a $\mathcal{F}$-measurable function of ω, for any measurable $g : \mathcal{E} \to \mathbb{C}$. Consider the σ-algebra $\mathcal{G}$ on B^* defined exactly as in Section 5.1 with the above B_ω's. Then $\mathcal{G}$-measurability of families of members of B_ω and B_ω^* have the same meaning as in Section 5.1.

Next, let $r_\omega = r_\omega(x, y) : \mathcal{E}_\omega \times \mathcal{E}_{\theta\omega} \to [0, \infty)$, $\omega \in \Omega$ be a family of integrable in y Borel measurable functions, m_ω, $\omega \in \Omega$ be a family of Borel probability measures on $\mathcal{E}_\omega$ and $u : \mathcal{E} \to \mathbb{R}$ be a measurable function so that $u_\omega \in B_\omega$, P-a.s. Consider the family of random operators R_z^ω, $z \in \mathbb{C}$ which map (bounded) Borel functions g on $\mathcal{E}_{\theta\omega}$ to Borel measurable functions on $\mathcal{E}_\omega$ by the formula

$$R_z^\omega g(x) = \int_{\mathcal{E}_{\theta\omega}} r_\omega(x, y) e^{z u_{\theta\omega}(y)} g(y) dm_{\theta\omega}(y) \qquad (6.1.1)$$

where $u_\omega(\cdot) = u(\omega, \cdot)$ for any $\omega \in \Omega$.

Assumption 6.1.1. For P-almost any $\omega \in \Omega$,

$$\|R_0^\omega \mathbf{1}\|_\infty < \infty$$

where $\mathbf{1}$ is the function which takes the constant value 1 on $\mathcal{E}_{\theta\omega}$.

Observe that

$$\|R_0^\omega \mathbf{1}\|_\infty = \sup_{g \in B_{\theta\omega} : \|g\|_\infty \leq 1} \|R_0^\omega g\|_\infty := \|R_0^\omega\|_\infty$$

and therefore under Assumption 6.1.1 for P-a.a. ω we have $\|R_z^\omega\|_\infty < \infty$ for any $z \in \mathbb{C}$, namely, R_z^ω is a continuous linear operator between the Banach spaces $B_{\theta\omega}$ and B_ω. Moreover, under this assumption R_z^ω is analytic in z when considered as a map between $B_{\theta\omega}$ to B_ω since the series $\sum_{k=0}^\infty \frac{(u_{\theta\omega})^k}{k!} z^k$ converges in $B_{\theta\omega}$ for any $z \in \mathbb{C}$.

Assumption 6.1.2. The maps $\omega \to \int_{\mathcal{E}_\omega} g_\omega(x) dm_{\theta\omega}(x)$ and $(\omega, x) \to R_0^\omega g_{\theta\omega}(x)$, $(\omega, x) \in \mathcal{E}$ are measurable for any bounded measurable function $g : \mathcal{E} \to \mathbb{C}$.

Under Assumption 6.1.1, for any $\omega \in \Omega$, $n \in \mathbb{N}$ and $z \in \mathbb{C}$ consider the n-th order iterates $R_z^{\omega,n} : B_{\theta^n\omega} \to B_\omega$ given by

$$R_z^{\omega,n} = R_z^\omega \circ R_z^{\theta\omega} \circ \cdots \circ R_z^{\theta^{n-1}\omega}. \tag{6.1.2}$$

Then we can write

$$R_0^{\omega,n} g(x) = \int_{\mathcal{E}_{\theta^n\omega}} r_\omega(n, x, y) g(y) dm_{\theta^n\omega}(y)$$

for some family $r_\omega(n, \cdot, \cdot) = r_\omega(n, x, y) : \mathcal{E}_\omega \times \mathcal{E}_{\theta^n\omega} \to [0, \infty)$ of integrable in y Borel measurable functions. We will assume that the following two sided random Doeblin condition holds true.

Assumption 6.1.3. There exist random variables $j_\omega \in \mathbb{N}$ and $\alpha_m(\omega) \geq 1$, $m \in \mathbb{N}$ such that P-a.s.,

$$\alpha_m(\omega) \leq r_\omega(m, x, y) \leq \big(\alpha_m(\omega)\big)^{-1}, \tag{6.1.3}$$

for any $m \geq j_\omega$, $x \in \mathcal{E}_\omega$ and $y \in \mathcal{E}_{\theta^m\omega}$.

Next, in the case of integral operators we say that a random triplet $(\lambda_\omega(z), h_\omega(z), \nu_\omega(z))$, $\omega \in \Omega$ consisting of a random variable $\lambda_\omega(z) \neq 0$, a bounded Borel function $h_\omega(z)$ on $\mathcal{E}_\omega$ and a complex measure $\nu_\omega(z)$ on $\mathcal{E}_\omega$ is an RPF triplet for the family R_z^ω, $\omega \in \Omega$ if for P-a.a. ω,

$$R_z^\omega h_{\theta\omega}(z) = \lambda_\omega(z) h_\omega(z), \quad (R_z^\omega)^* \nu_\omega(z) = \lambda_\omega(z) \nu_{\theta\omega}(z), \tag{6.1.4}$$

$$\nu_\omega(z)\mathbf{1} = 1 \quad \text{and} \quad \nu_\omega(z)(h_\omega(z)) = 1.$$

The main result in this chapter is the existence of measurable in ω and analytic in z RPF triplets (under certain conditions). Before formulating it we explain how the setup of Chapter 4 will be used. Consider the measure preserving system $(\Omega, \mathcal{F}, P, \vartheta)$, where $\vartheta = \theta^{-1}$. For any $z \in \mathbb{C}$ consider the family of operators A_z^ω, $\omega \in \Omega$ given by

$$A_z^\omega = R_z^{\theta^{-1}\omega} = R_z^{\vartheta\omega}. \tag{6.1.5}$$

If $(\tilde{\lambda}_\omega(z), h_\omega(z), \nu_\omega(z))$, $\omega \in \Omega$ is an RPF triplet of the family A_z^ω, $\omega \in \Omega$ with respect to the map ϑ in the sense of Section 4.1 then $(\lambda_\omega(z), h_\omega(z), \nu_\omega(z))$, $\omega \in \Omega$, where $\lambda_\omega(z) = \tilde{\lambda}_{\theta\omega}(z)$, is an RPF triplet of the family R_z^ω, $\omega \in \Omega$ with respect to the map θ (i.e. it satisfies (6.1.4)). Furthermore, observe that for any $z \in \mathbb{C}$ and $n \in \mathbb{N}$,

$$A_z^{\vartheta^{-n}\omega, n} = R_z^{\omega, n} \quad \text{and} \quad A_z^{\omega, n} = R_z^{\theta^{-n}\omega, n} \tag{6.1.6}$$

where the random iterates $A_z^{\omega, n}$ are given by (4.1.3) with ϑ in place of θ. We see then that when the results from Chapter 4 with the system $(\Omega, \mathcal{F}, P, \vartheta)$ and the families of operators A_z^ω, $\omega \in \Omega$ hold true then corresponding results for the families R_z^ω, $\omega \in \Omega$ will follow.

6.2 Real and complex cones

Denote by $B_{\omega,+}$ the set of all nonnegative members of B_ω. For any $\omega \in \Omega$ and $L > 1$ consider the real proper closed convex cone (see Appendix A),

$$\mathcal{K}_{\omega,L,\mathbb{R}} = \{g \in B_{\omega,+} : g(x) \leq Lg(y) \ \forall x, y \in \mathcal{E}_\omega\} \subset B_\omega. \tag{6.2.1}$$

Then any nonzero member of $\mathcal{K}_{\omega,L,\mathbb{R}}$ is strictly positive. Let $\mathcal{K}_{\omega,L} \subset B_\omega$ be the canonical complexification of $\mathcal{K}_{\omega,L,\mathbb{R}}$ (see Appendix A). Then the first part of Assumption 4.1.1 holds true with the family $b_\omega \equiv \mathbf{1}$, where $\mathbf{1}$ is the function which takes the constant value 1.

Theorem 6.2.1. *Let $\omega \in \Omega$ and $L > 1$.*
(i) The complex cone $\mathcal{K}_{\omega,L}$ is linearly convex,

$$\|g\|_\infty \leq 2\sqrt{2}L|\mu_\omega(g)| \quad \text{for any } g \in \mathcal{K}_{\omega,L} \tag{6.2.2}$$

and

$$\|\mu\|_\infty \leq 12(1 - L^{-1})^{-1}|\mu(\mathbf{1})| \quad \text{for any } \mu \in \mathcal{K}_{\omega,L}^* \tag{6.2.3}$$

namely the cones $\mathcal{K}_{\omega,L}$ and $\mathcal{K}_{\omega,L}^*$ satisfy Assumption 4.1.1 with $K_\omega = 2\sqrt{2}L$, $M_\omega = 12(1 - L^{-1})^{-1}$, the functional $l_\omega = \mu_\omega$ and the functional $\kappa_\omega \in (B_\omega^*)^*$ given by $\kappa_\omega\mu = \mu\mathbf{1}$, which is $\mathcal{G}$-measurable where the σ-algebra $\mathcal{G}$ was defined in Section 6.1.

(ii) For any $f \in B_\omega$ there exist $f_1, ..., f_8 \in \mathcal{K}_{\omega,L}$ so that $f = f_1 + f_2 + ... + f_8$ and

$$\|f_1\|_\infty + \|f_2\|_\infty + ... + \|f_8\|_\infty \leq r(L)\|f\|_\infty \tag{6.2.4}$$

where $r(L) = \frac{16(L+1)}{L-1}$, i.e. Assumption 4.1.3 holds true with $k_\omega = 8$ and $r_\omega = r(L)$.

6.3   The RPF theorem for integral operators

Theorem 6.3.1. *Under Assumption 6.1.1 the following holds true.*

(i) The random operators $A_z^\omega = R_z^{\theta^{-1}\omega}$ map B_ω to $B_{\vartheta\omega}$, they are continuous with respect to the supremum norms and are analytic in z. As a consequence, when also Assumption 6.1.2 holds true, the family A_z^ω, $\omega \in \Omega$ is $\mathcal{G}$-measurable in the sense of Section 4.1 where $\mathcal{G}$ is defined in Section 6.1.

(ii) Suppose, in addition, that the random variables j_ω and $\|u_\omega\|_\infty$ are bounded and let $n_0 \in \mathbb{N}$ be so that $j_\omega \leq n_0$, P-a.s. Furthermore, assume that there exists a constant $\alpha \in (0,1)$ such that $\alpha_\omega(n) \geq \alpha$, P-a.s. for any $n_0 \leq n \leq 2n_0$. Then Assumption 4.1.2 holds true with $A_z^\omega = R_z^{\theta^{-1}\omega}$, the system $(\Omega, \mathcal{F}, P, \vartheta)$, $\vartheta = \theta^{-1}$ and the cones $\mathcal{C}_\omega = \mathcal{K}_{\omega,2\alpha^{-2}}$ in some deterministic neighborhood U of 0.

Corollary 6.3.2. *Suppose that Assumptions 6.1.1 and 6.1.2 hold true, that the random variable $\|u_\omega\|_\infty$ is bounded and that there exist n_0 and α as in Theorem 6.3.1 (ii). Let U be the neighborhood of 0 from there. Then P-a.s. for any $z \in U$ there exists a unique triplet $\lambda_\omega(z)$, $h_\omega(z)$ and $\nu_\omega(z)$ consisting of a nonzero complex number $\lambda_\omega(z)$, a complex function $h_\omega(z) \in \mathcal{K}_{\omega,2\alpha^{-2}}$ and a complex continuous linear functional $\nu_\omega(z) \in \mathcal{K}_{\omega,2\alpha^{-2}}^*$ satisfying (6.1.4). For any $z \in U$ the maps $\omega \to \lambda_\omega(z)$ and $(\omega, x) \to h_\omega(z)(x)$, $(\omega, x) \in \mathcal{E}$ are measurable and the family $\nu_\omega(z)$ is measurable in ω. When $z = t \in \mathbb{R}$ then $\lambda_\omega(t) > 0$, $h_\omega(t) \in \mathcal{K}'_{\omega,2\alpha^{-2},\mathbb{R}}$ and $\nu_\omega(t)$ is a probability measure.*

Moreover, this triplet is analytic and uniformly bounded around 0. Namely, the maps

$$\lambda_\omega(\cdot) : U \to \mathbb{C}, \quad h_\omega(\cdot) : U \to B_\omega \quad and \quad \nu_\omega(\cdot) : U \to B_\omega^*$$

are analytic, where B_ω^* is the dual space of B_ω, and there exists a deterministic neighborhood $U_0 \subset U$ of 0 such that for any $k \geq 0$ there is a constant $C_k > 0$ so that

$$\max\left(\sup_{z \in U_0} |\lambda_\omega^{(k)}(z)|, \; \sup_{z \in U_0} \|h_\omega^{(k)}(z)\|_\infty, \; \sup_{z \in U_0} \|\nu_\omega^{(k)}(z)\|_\infty \right) \leq C_k, \; P\text{-}a.s.$$

$$(6.3.1)$$

where $g^{(k)}$ stands for the k-th derivative of a function on the complex plane which takes values in some Banach space and $\|\nu\|_\infty$ is the operator norm of a linear functional $\nu : B_\omega \to \mathbb{C}$.

Furthermore, there exists a constant $C > 0$ such that P-a.s. for any $n \geq n_0$, $z \in U_0$, $q \in B_{\theta^n \omega}$ and $\mu \in B_\omega^*$,

$$\left\| \frac{R_z^{\omega,n} q}{\lambda_{\omega,n}(z)} - \left(\nu_{\theta^n \omega}(z)q\right)h_\omega(z) \right\|_\infty \leq C\|q\|_\infty \cdot c^{\frac{n}{n_0}} \qquad (6.3.2)$$

and

$$\left\| \frac{(R_z^{\omega,n})^* \mu}{\lambda_{\omega,n}(z)} - \left(\mu h_\omega(z)\right)\nu_{\theta^n \omega}(z) \right\|_\infty \leq C\|\mu\|_\infty \cdot c^{\frac{n}{n_0}} \qquad (6.3.3)$$

where $\lambda_{\omega,n}(z) = \prod_{j=0}^{n-1} \lambda_{\theta^j \omega}(z)$.

6.4   Properties of cones: proof of Theorem 6.2.1

Lemma 6.4.1. *The complex cones $\mathcal{K}_{\omega,L}$, $\omega \in \Omega$ satisfy (6.2.2) and (6.2.4). In particular they are linearly convex.*

Proof. We begin with showing that the cones $\mathcal{K}_{\omega,L,\mathbb{R}}$ and $\mathcal{K}_{\omega,L}$ have bounded apertures. First, it is clear that for any $g \in \mathcal{K}_{\omega,L,\mathbb{R}}$ we have

$$\|g\|_\infty \leq L \inf g \leq L m_\omega(g) \qquad (6.4.1)$$

and therefore by Lemma 5.3 in [58],

$$\|g\|_\infty \leq 2\sqrt{2}L|m_\omega(g)| \text{ for any } g \in \mathcal{K}_{\omega,L}. \qquad (6.4.2)$$

As in Corollary 5.5.3, we deduce that the cones $\mathcal{K}_{\omega,L}$ are linearly convex.

Next, we show that the cones $\mathcal{K}_{\omega,L}$ are reproducing. Indeed, let $g \in B_{\omega,+}$. Then the function $h = g + c_L(g)$ is a member of $\mathcal{K}_{\omega,L}$, where $c_L(g) = \|g\|_\infty(L+1)/(L-1)$. Any Borel measurable bounded function $g : \mathcal{E}_\omega \to \mathbb{C}$ can be written as a linear combination of four nonnegative elements of B_ω whose supremum does not exceed $\|g\|_\infty$. As a consequence, for any $g \in B_\omega$ there exist $g_1, g_2, ..., g_8 \in \mathcal{K}_{\omega,L}$ such that $g = \sum_{i=1}^8 g_i$ and

$$\sum_{i=1}^8 \|g_i\|_\infty \leq r(L)\|g\|_\infty$$

where $r(L) = \frac{16(L+1)}{L-1}$. $\qquad \square$

Next, we will prove here that the dual cones $\mathcal{K}_{\omega,L}^* = (\mathcal{K}_{\omega,L})^*$ (see Appendix A) have bounded apertures.

Lemma 6.4.2. *For any $\omega \in \Omega$, $L > 1$ and $\mu \in \mathcal{K}_{\omega,L}^*$,*

$$\|\mu\|_\infty \leq A_L|\mu(\mathbf{1})| \qquad (6.4.3)$$

where $A_L = \frac{12}{(1-L^{-1})^2}$.

Before proving the lemma we need some notations. Write

$$\mathcal{K}_{\omega,L,\mathbb{R}} = \{g \in B_\omega : s(g) \geq 0 \ \forall s \in \mathbf{S}_\omega(L)\} \tag{6.4.4}$$

where $\mathbf{S}_\omega(L)$ is the set of all continuous linear functionals on B_ω which either have the form $s(g) = g(x)$ for some $x \in \mathcal{E}_\omega$ or the form $s(g) = g(x) - L^{-1}g(y)$ for some $x, y \in \mathcal{E}_\omega$. Then by (A.2.3),

$$\mathcal{K}_{\omega,L} = \{g \in B_\omega : \Re(\bar{s}_1(g)s_2(g)) \geq 0 \ \forall s_1, s_2 \in \mathbf{S}_\omega(L)\} \tag{6.4.5}$$

where $\bar{s}_1(g)$ stands for the complex conjugate of $s_1(g)$.

Proof of Lemma 6.4.2. By Lemma A.2.7 it is sufficient to prove that

$$B(\mathbf{1}, A_L^{-1}) := \{g \in B_\omega : \|g - \mathbf{1}\|_\infty < A_L^{-1}\} \subset \mathcal{K}_{\omega,L}.$$

Let $g \in B(\mathbf{1}, A_L^{-1})$ and write $g = \mathbf{1} + h$. By (6.4.5) it will follow that $g \in \mathcal{K}_{\omega,L}$ if we show that

$$\Re(\bar{s}_1(g)s_2(g)) = \Re(\bar{s}_1(\mathbf{1}+h)s_2(\mathbf{1}+h)) \geq 0$$

for any $s_1, s_2 \in \mathbf{S}_\omega(L)$. Consider first the case when s_i has the form $g \to g(x_i) - L^{-1}g(y_i)$ for $i = 1, 2$ and some $x_1, x_2, y_1, y_2 \in \mathcal{E}_\omega$. Since $\|h\|_\infty < 1$,

$$\Re(\bar{U}(x_1, y_1)U(x_2, y_2)) \geq (1 - L^{-1})^2 - 12\|h\|_\infty$$

where

$$U(x, y) = 1 + h(x) - L^{-1}(1 + h(y)).$$

Similarly, in the case when one of the functionals has the above form while the other is an evaluation functional e of the form $e(g) = g(x)$ we derive that

$$\Re(\bar{s}_1(g)s_2(g)) \geq 1 - L^{-1} - 6\|h\|_\infty \geq (1 - L^{-1})^2 - 12\|h\|_\infty$$

while in the case when $s_i(g) = g(x_i)$, $i = 1, 2$ we obtain

$$\Re(\bar{s}_1(g)s_2(g)) \geq 1 - 3\|h\|_\infty \geq (1 - L^{-1})^2 - 12\|h\|_\infty$$

and (6.4.3) follows. $\square$

6.5   Real cones: invariance and diameter of image estimates

We will obtain here the following

Lemma 6.5.1. *For any $\omega \in \Omega$ and $1 < L < M$ we have*

$$diam_{\mathcal{K}_{\omega,M,\mathbb{R}}}(\mathcal{K}'_{\omega,L,\mathbb{R}}) := \sup_{f,g \in \mathcal{K}'_{\omega,L,\mathbb{R}}} d_{\mathcal{K}_{\omega,M,\mathbb{R}}}(f,g) \leq 2\ln\left(\frac{ML}{M-L}\right) \quad (6.5.1)$$

where $d_{\mathcal{K}_{\omega,M,\mathbb{R}}}$ is the real Hilbert metric associated with the cone $\mathcal{K}_{\omega,M,\mathbb{R}}$ (see Appendix A) and $\mathcal{K}'_{\omega,L,\mathbb{R}} = \mathcal{K}_{\omega,L,\mathbb{R}} \setminus \{0\}$.

Proof. Let $f,g \in \mathcal{K}'_{\omega,M,\mathbb{R}}$, $x,y \in \mathcal{E}_\omega$ and set

$$Q(f,g,x,y) = \frac{Mg(x) - g(y)}{Mf(x) - f(y)}$$

assuming that the denominator does not vanish, while when it vanishes we set $Q = 0$. Set

$$Q(f,g) = \sup\left\{Q(f,g,x,y) : x,y \in \mathcal{E}_\omega\right\}.$$

We first claim that

$$d_{\mathcal{K}_{\omega,M,\mathbb{R}}}(f,g) = \ln\left(R(f,g)R(g,f)\right) \quad (6.5.2)$$

where

$$R(f,g) = \max\left(Q(f,g), \left\|\frac{g}{f}\right\|_\infty\right)$$

and we use the convention $\ln 0 := -\infty$. Indeed, let $t > 0$. Then the function $tf - g$ is a member of the cone $\mathcal{K}_{\omega,M,\mathbb{R}}$ if and only if for any $x,y \in \mathcal{E}_\omega$ we have $t \geq \frac{g(x)}{f(x)}$ and

$$t\left(Mf(y) - f(x)\right) \geq Mg(y) - g(x). \quad (6.5.3)$$

We note that when $f(x) = Mf(y)$ then for any $t > 0$ the inequality (6.5.3) holds true if and only if $g(x) = Mg(y)$, since $g \in \mathcal{K}_{\omega,M,\mathbb{R}}$. As a consequence,

$$\beta_{\mathcal{K}_{\omega,M,\mathbb{R}}}(f,g) = \ln\left(R(f,g)\right)$$

where $\beta_{\mathcal{K}_{\omega,M,\mathbb{R}}}(f,g)$ is given by (A.1.1). Replacing t by t^{-1} and f by g and using the formula (A.1.2) of the real Hilbert metric yields (6.5.2).

Next, suppose that $f,g \in \mathcal{K}'_{\omega,L,\mathbb{R}}$. In order to prove (6.5.1), it is sufficient to show that

$$R(f,g)R(g,f) \leq \frac{M^2 L^2}{(M-L)^2}.$$

Indeed, we first notice that the denominator in the definition of $Q(f, g, x, y)$ does not vanish for any x and y since $L < M$. We distinguish between four cases. Suppose first that $R(f, g) = Q(f, g)$ and $R(g, f) = Q(g, f)$. For any $x_1, y_1, x_2, y_2 \in \mathcal{E}_\omega$ set $Q = Q(f, g, x_1, y_1)Q(g, f, x_2, y_2)$. Then

$$Q = \frac{\left(M - \frac{g(y_1)}{g(x_1)}\right)\left(M - \frac{f(y_2)}{f(x_2)}\right)}{\left(M - \frac{f(y_1)}{f(x_1)}\right)\left(M - \frac{g(y_2)}{g(x_2)}\right)} \cdot \frac{g(x_1)f(x_2)}{g(x_2)f(x_1)} := I_1 \cdot I_2.$$

Clearly $I_2 \leq L^2$, since $f, g \in \mathcal{K}_{\omega, L, \mathbb{R}}$. The numerator of I_1 is nonnegative (since $M > L$) and bounded from above by M^2, while the denominator is bounded from below by $(M - L)^2$ and hence $Q = I_1 \cdot I_2 \leq M^2 L^2/(M - L)^2$. Consider now the case when $R(f, g) = \|g/f\|_\infty$ and $R(g, f) = \|f/g\|_\infty$. Then clearly,

$$R(f, g)R(g, f) \leq L^2 \leq \frac{M^2 L^2}{(M - L)^2}$$

since $I_2 \leq L^2$ for any $x_1, y_1, x_2, y_2 \in \mathcal{E}_\omega$. Suppose next that $R(f, g) = Q(f, g)$ whilst $R(g, f) = \|f/g\|_\infty$. For any $x_1, y_1, x \in \mathcal{E}_\omega$ we have

$$Q(f, g, x_1, y_1)\frac{f(x)}{g(x)} \leq Q(f, g, x_1, y_1)\frac{f(x_1)}{g(x_1)}L^2$$

$$= \frac{M - \frac{g(y_1)}{g(x_1)}}{M - \frac{f(y_1)}{f(x_1)}}L^2 \leq \frac{ML^2}{M - L} \leq \frac{M^2 L^2}{(M - L)^2}$$

and by taking supremum over all possible choices of x_1, y_1 and x we deduce that

$$R(f, g)R(g, f) \leq \frac{M^2 L^2}{(M - L)^2}.$$

Changing the roles of f and g, the same inequality holds true when $R(f, g) = \|g/f\|_\infty$ and $R(g, f) = Q(g, f)$, and the proof of the lemma is complete. $\qquad\square$

We note that the above proof implies that, in fact, $d_{\mathcal{K}_{\omega, M, \mathbb{R}}}(f, g) = \ln(Q(g, f)Q(f, g))$ for any $f, g \in \mathcal{K}'_{\omega, L, \mathbb{R}}$.

Next, we will prove the following result.

Lemma 6.5.2. *Suppose that there exist constants $n_0 \in \mathbb{N}$ and $\alpha > 0$ such that P-a.s. $j_\omega \leq n_0$ and $\alpha_n(\omega) \geq \alpha$ for any $n_0 \leq n \leq 2n_0$ where j_ω and $\alpha_n(\omega)$ appear in Assumption 6.1.3. Then, P-a.s.,*

$$A_0^{\omega, n} g \in \mathcal{K}'_{\vartheta^n \omega, \alpha^{-2}, \mathbb{R}} \tag{6.5.4}$$

for any $n_0 \leq n \leq 2n_0$ and a strictly positive bounded Borel function g : $\mathcal{E}_\omega \to \mathbb{R}$ and, in particular,

$$A_0^{\omega,n} \mathcal{K}'_{\omega,2\alpha-2,\mathbb{R}} \subset \mathcal{K}'_{\vartheta^n\omega,\alpha-2,\mathbb{R}} \subset \mathcal{K}'_{\vartheta^n\omega,2\alpha-2,\mathbb{R}}. \tag{6.5.5}$$

Moreover,

$$D_\omega(n) := \sup_{g_1,g_2 \in \mathcal{K}'_{\omega,\alpha-2,\mathbb{R}}} d_{\mathcal{K}_{\vartheta^n\omega,2\alpha-1,\mathbb{R}}}(A_0^{\omega,n} g_1, A_0^{\omega,n} g_2) \leq 4 \ln 2\alpha^{-1}. \tag{6.5.6}$$

Proof. Since $j_{\theta^{-n}\omega} \leq n_0$ and $\alpha_n(\theta^{-n}\omega) \geq \alpha$ P-a.s. for any natural $n_0 \leq n \leq 2n_0$, it is clear from (6.1.3) and the positivity of $R_0^{\theta^{-n}\omega,n}$ that

$$A_0^{\omega,n} g = R_0^{\theta^{-n}\omega,n} g \in \mathcal{K}'_{\theta^{-n}\omega,\alpha-2} = \mathcal{K}'_{\vartheta^n\omega,\alpha-2}$$

for any strictly positive bounded Borel function $g : \mathcal{E}_\omega \to \mathbb{R}$. The diameter estimate (6.5.6) is a consequence of (6.5.5) and Lemma 6.5.1. $\qquad\square$

6.6 Comparison of real and complex operators

Let $\mathbf{S}_\omega = \mathbf{S}_\omega(2\alpha^{-2})$ be the set of linear functionals $s \in B_\omega^*$ defined right after (6.4.4) with $L = L_\alpha = 2\alpha^{-2}$, where α is specified in Theorem 6.3.1. For any ω and $n \in \mathbb{N}$, let $S_n^\omega u : \mathcal{E}_{\theta\omega} \times \mathcal{E}_{\theta^2\omega} \times \cdots \times \mathcal{E}_{\theta^n\omega} \to \mathbb{C}$ be given by

$$S_n^\omega u(x_1,...,x_n) = \sum_{k=1}^n u_{\theta^k\omega}(x_k)$$

and for any $z \in \mathbb{C}$ set

$$\tilde{\varepsilon}_\omega(n,z) = |z| \|S_n^\omega u\|_\infty.$$

Proposition 6.6.1. *Let $n \in \mathbb{N}$, $z \in \mathbb{C}$ and a strictly positive function $g : \mathcal{E}_\omega \to \mathbb{R}$ satisfy $R_0^{\omega,n} g \in \mathcal{K}_{\omega,\alpha-2,\mathbb{R}}$ and $\tilde{\varepsilon}_\omega(n,z) \leq \ln 2$ for some $\omega \in \Omega$. Then for any $s \in \mathbf{S}_\omega$,*

$$|s(R_z^{\omega,n} g) - s(R_0^{\omega,n} g)| \leq (12\alpha^{-4}\tilde{\varepsilon}_\omega(n,z)) s(R_0^{\omega,n} g). \tag{6.6.1}$$

Proof. Let ω, n, z and g be as in the statement of the proposition. We first consider the case when s has the form $s(g) = g(x_0)$ for some $x_0 \in \mathcal{E}_\omega$. For any $x \in \mathcal{E}_\omega$ and $\zeta \in \mathbb{C}$ we have

$$R_\zeta^{\omega,n} g(x) = \int r_\omega(x,x_1) r_{\omega,n}(x_1,...,x_n) e^{\zeta S_n^\omega u(x_1,...,x_n)} g(x_n) \prod_{j=1}^n dm_{\theta^j\omega}(x_j) \tag{6.6.2}$$

where

$$r_{\omega,n}(x_1,...,x_n) = \prod_{j=1}^{n-1} r_{\theta^j\omega}(x_j, x_{j+1}).$$

As a consequence,

$$|s(R_z^{\omega,n}g) - s(R_0^{\omega,n}g)| = |R_z^{\omega,n}g(x_0) - R_0^{\omega,n}g(x_0)| \qquad (6.6.3)$$
$$\leq \|e^{zS_n^\omega u} - 1\|_\infty R_0^{\omega,n}g(x_0) = \|e^{zS_n^\omega u} - 1\|_\infty s(R_0^{\omega,n}g)$$

where we used that g is nonnegative. By the mean value theorem,

$$\|e^{zS_n^\omega u} - 1\| \leq 2|z|\|S_n^\omega u\|_\infty e^{|z|\|S_n^\omega u\|_\infty} \leq 4|z|\|S_n^\omega u\|_\infty = 4\tilde{\varepsilon}_\omega(n,z) \quad (6.6.4)$$

where in the second inequality we used that $\tilde{\varepsilon}_\omega(n,\omega) \leq \ln 2$, and the proof of the proposition in this case is complete.

Next, set $a = a_\alpha = \alpha^{-2}$ and consider the case when s has the form $s(g) = g(x_0) - (2a)^{-1}g(x_0')$ for some $x_0, x_0' \in \mathcal{E}_\omega$. Since $R_0^{\omega,n}g \in \mathcal{K}_{\omega,a,\mathbb{R}}$,

$$s(R_0^{\omega,n}) = R_0^{\omega,n}g(x_0) - (2a)^{-1}R_0^{\omega,n}g(x_0') \qquad (6.6.5)$$
$$\geq (a^{-1} - (2a)^{-1})R_0^{\omega,n}g(x_0') = (2a)^{-1}R_0^{\omega,n}g(x_0').$$

By (6.6.2) and (6.6.3) we have

$$|s(R_z^{\omega,n}g) - s(R_0^{\omega,n}g)|$$
$$= |R_z^{\omega,n}g(x_0) - (2a)^{-1}R_z^{\omega,n}g(x_0') - (R_0^{\omega,n}g(x_0) - (2a)^{-1}R_0^{\omega,n}g(x_0'))|$$
$$= |(R_z^{\omega,n}g(x_0) - R_0^{\omega,n}g(x_0)) - (2a)^{-1}(R_z^{\omega,n}g(x_0') - R_0^{\omega,n}g(x_0'))|$$
$$\leq \|e^{zS_n^\omega u} - 1\|_\infty (R_0^{\omega,n}(x_0) + (2a)^{-1}R_0^{\omega,n}g(x_0'))$$
$$\leq \|e^{zS_n^\omega u} - 1\|_\infty (aR_0^{\omega,n}g(x_0') + (2a)^{-1}R_0^{\omega,n}g(x_0'))$$
$$= \|e^{zS_n^\omega u} - 1\|_\infty(a + (2a)^{-1})R_0^{\omega,n}g(x_0') \leq \|e^{zS_n^\omega u} - 1\|_\infty(a + \frac{a}{2})R_0^{\omega,n}g(x_0')$$

where in the second inequality we used that $R_0^{\omega,n}g \in \mathcal{K}_{\omega,a,\mathbb{R}}$. The inequality (6.6.1) follows now from (6.6.4) and (6.6.5), and the proof of the proposition is complete. □

6.7 Complex image diameter estimates

We will prove in this section Theorem 6.3.1 and Corollary 6.3.2. Suppose that Assumption 6.1.1 holds true and that the conditions of the second statement of Theorem 6.3.1 hold true with some constants n_0 and α. Set $L = L_\alpha = 2\alpha^{-2} > 1$, $\mathcal{C}_\omega = \mathcal{K}_{\omega,\alpha^{-2}}$,

$$d_1 = 2\ln L_\alpha, \ d_0 = d_1 + 6\ln 2,$$
$$a_1 = \min\left(\frac{1}{4}(1 + \cosh(\frac{1}{2}d_0))^{-1}, \ln 2\right) \text{ and } a_0 = \frac{a_1}{3L_\alpha^2}.$$

Let $n_0 \leq n \leq 2n_0$. Then by Lemma 6.5.2, Proposition 6.6.1 and Theorem A.2.4, taking into account (6.1.6), we see that if

$$|z| \|S_n^\omega u\|_\infty = \tilde{\varepsilon}_{\theta^{-n}\omega}(n, z) < a_0$$

then

$$A_z^{\omega,n} \mathcal{C}_\omega' \subset \mathcal{C}_{\vartheta^n\omega}' \tag{6.7.1}$$

and the corresponding diameter of the image satisfies

$$\Delta_\omega(n, z) := \sup_{g_1, g_2 \in \mathcal{C}_\omega'} \delta_{\mathcal{C}_{\vartheta^n\omega}}(A_z^{\omega,n} g_1, A_z^{\omega,n} g_2) < d_0. \tag{6.7.2}$$

Next, the first statement of the theorem follows from Assumption 6.1.1 as explained after its statement. In order to prove the second assertion of Theorem 6.3.1, let $D < \infty$ be so that $\|u_\omega\|_\infty \leq D$, P-a.s. Then for P-a.a. ω and all $n \geq 1$ we have $\tilde{\varepsilon}_{\theta^{-n}\omega}(n, z) < n|z|D$. Hence (6.7.1) and (6.7.2) hold true for any $n_0 \leq n \leq 2n_0$ and z with $|z| < \frac{a_0}{2n_0 D}$ and the family A_z^ω satisfies Assumption 4.1.2 with the cones $\mathcal{C}_\omega$ and the open ball $U = B(0, \frac{a_0}{2n_0 D})$.

Now we derive Corollary 6.3.2. Applying Theorem 4.2.1 with the operators A_z^ω and the cones $\mathcal{C}_\omega = \mathcal{K}_{\omega, L_\alpha}$ we obtain that P-a.s. there exist triplets $(\tilde{\lambda}_\omega(z), h_\omega(z), \nu_\omega(z))$, $z \in U$ satisfying (4.2.1) and $\nu_\omega(z)\mathbf{1} = 1$ which is analytic in z and measurable in ω. As explained in Section 6.1, the triplet

$$(\dot{\lambda}_\omega(z), h_\omega(z), \nu_\omega(z)) = (\tilde{\lambda}_{\theta\omega}(z), h_\omega(z), \nu_\omega(z))$$

has the properties specified in the first part of Corollary 6.3.2. When $z = t \in \mathbb{R}$ taking $f_n = \mathbf{1}$ in (4.3.10) it follows that $\hat{h}_\omega(t)$ is a nonnegative function belonging to the cone $\mathcal{K}_{\omega, L_\alpha}$ and therefore by (A.2.2) it is a member of the real punctured cone $\mathcal{K}_{\omega, L_\alpha, \mathbb{R}}' = \mathcal{K}_{\omega, L_\alpha, \mathbb{R}} \setminus \{0\}$ and, in particular, it is strictly positive. Next, making the choice of $\mu_n = m_{\vartheta^n\omega}$ in (4.3.25) we derive that $\nu_\omega(t)$ is a positive measure and since $\nu_\omega(t)\mathbf{1} = 1$ it is a probability measure. Note that m_ω is indeed a member of the dual cone $\mathcal{C}_\omega^*$ by Theorem 6.2.1. Since $h_\omega(t) = \frac{\hat{h}_\omega(t)}{\nu_\omega(t)\hat{h}_\omega(t)}$ it follows that $h_\omega(t) \in \mathcal{K}_{\omega, L_\alpha, \mathbb{R}}'$. The random variable $\lambda_\omega(t)$ is strictly positive since it does not vanish and $\lambda_{\theta\omega}(t) = \int R_t^\omega \mathbf{1} d\nu_\omega(t)$.

In order to show that these triplets are uniformly bounded around 0 in the sense of the second part of Corollary 6.3.2, observe that the neighborhood V_ω defined before (4.4.4) contains a deterministic neighborhood U_0 of 0 when the random variables $|\alpha_\omega(0)|^{-1}$ and

$$v_\omega = \frac{K_\omega M_\omega}{\|m_\omega\|_\infty \|\kappa_\omega\|_\infty} = K_\omega M_\omega$$

are bounded, where $K_\omega, M_\omega, l_\omega$ and κ_ω are specified in Theorem 6.2.1 *(i)* with $L = L_\alpha = 2\alpha^{-2}$. Since M_ω and K_ω are deterministic in our situation v_ω does not depend on ω. Recall that $\alpha_\omega(0) = \nu_\omega(0)\hat{h}_\omega(0)$. The function $\hat{h}_\omega(0)$ is a member of the cone $\mathcal{K}_{\omega,2\alpha^{-2},\mathbb{R}}$ and $\int \hat{h}_\omega(0)dm_\omega = 1$. Therefore, $\alpha_\omega(0) \geq \alpha^2$ and the proof of (6.3.1) proceeds exactly as the proof of (5.4.3) in Corollary 5.4.2. Finally, applying Theorem 4.2.2, taking into account (6.3.1) and Theorem 6.2.1 we derive the exponential convergences stated in the third part of Corollary 6.3.2. □

Remark 6.7.1. In [40] and in other papers the the cones $\mathcal{K}_{\omega,L,\mathbb{R}}$ were considered as subcones of the cone $\mathcal{C}_+(\mathcal{E}_\omega)$ consisting of all nonnegative continuous functions on $\mathcal{E}_\omega$, using that their diameter with respect to the real Hilbert metric $d_{\mathcal{C}_+(\mathcal{E}_\omega)}$ does not exceed $2\ln L$ (see Lemma 3.4 in [40]). The reason we do not use these cones is that (under certain conditions) they only have bounded sectional aperture (see [58]) with respect to some L^p norm (see Theorem 7.2 and example 4.2 from [58]). Still, under certain (strong) assumptions, some of our results can be obtained when the cones only have bounded sectional aperture using similar arguments to the ones in the proof of Theorem 3.6 in [58] (in which the deterministic case is considered).

Chapter 7

Limit theorems for processes in random environment

7.1 The "conventional" case: preliminaries and main results

Let $(\Omega, \mathcal{F}, P, \theta)$ be an invertible ergodic measure preserving system. In the setup of Chapter 5, we consider the sequence of functions $\{S_n^\omega u : n \geq 1\}$ given by

$$S_n^\omega u(x) = \sum_{k=0}^{n-1} u_{\theta^k \omega}(T_\omega^n x) \tag{7.1.1}$$

as a sequence of random variables, where x is distributed according the (Gibbs) measure $\mu_\omega := \mu_\omega(0) = h_\omega(0)\nu_\omega(0)$, where $\nu_\omega(0)$ is specified in Corollary 5.4.2. For the sake of convenience, we assume that $\mathcal{L}_0^\omega 1 = 1$ which means that $\lambda_\omega(0) = 1$, $h_\omega(0) \equiv 1$ and $\mu_\omega(0) = \nu_\omega(0)$. Otherwise, we consider the normalized transfer operator $\tilde{\mathcal{L}}_0^\omega$ defined by (5.12.4), namely replace ϕ_ω with $\tilde{\phi}_\omega$ given by (5.12.3). Under the assumptions of Corollary 5.4.2 it follows from Theorem 2.3 in [38] that

$$\sigma^2 = \lim_{n \to \infty} \frac{1}{n} \text{Var}_{\mu_\omega} S_n^\omega u \tag{7.1.2}$$

exists P-a.s. and it does not depend on ω, and an appropriate coboundary characterization of positivity of σ^2 holds true. Moreover, P-a.s. the normalized sums $n^{-\frac{1}{2}}\left(S_n^\omega u - \mathbb{E}_{\mu_\omega} S_n^\omega u\right)$ converge in distribution as $n \to \infty$ towards a normal distribution with zero mean and variance σ^2.

In the setup of Chapter 6, we assume first that $R_0^\omega 1 \equiv 1$ which again means that $\lambda_\omega(0) = 1$, $h_\omega(0) \equiv 1$. Set $\mu_\omega = \nu_\omega(0)$ and consider the non-homogeneous Markov chains $\{\zeta_n^{\theta^n \omega} : n \geq 0\}$ with transition operators $R_0^{\theta^n \omega}$, $n \geq 0$ and initial distributions μ_ω. In this situation we write

$$S_n^\omega u = \sum_{j=1}^{n} u_{\theta^j \omega}(\zeta_j^{\theta^j \omega}) \tag{7.1.3}$$

and consider the sequence of random variables $\{S_n^\omega u : n \geq 1\}$. Under the assumptions of Corollary 6.3.2 it follows from Theorem 2.4 in [38] that the limit σ^2 (defined similarly to the transfer operator case) exists P-a.s., it does not depend on ω and an appropriate central limit theorem holds true.

7.1.1 *Self normalized Berry-Esseen theorem*

Assume $\sigma^2 > 0$ and for any $n \in \mathbb{N}$ set

$$\alpha_{\omega,n} = \mathrm{Var}_{\mu_\omega} S_n^\omega u$$

when this variance is strictly positive while otherwise, for the sake of convenience, we set $\alpha_{\omega,n} = n\sigma^2$. The first result here shows that there is an optimal convergence rate in the self-normalized version of the above CLT's (when $\sigma^2 > 0$).

Theorem 7.1.1. *Suppose that $\sigma^2 > 0$. Then for P-almost all ω the sequence of random variables $\hat{S}_n^\omega u = (\alpha_{\omega,n})^{-\frac{1}{2}}(S_n^\omega u - \mathbb{E}_{\mu_\omega} S_n^\omega u)$, $n \geq 1$ converges in distribution as $n \to \infty$ to the standard normal distribution $\mathcal{N}$, and there exists a random variable c_ω such that for any $n \in \mathbb{N}$,*

$$d_K(\mathcal{M}(\hat{S}_n^\omega u), \mathcal{N}) \leq c_\omega n^{-\frac{1}{2}}$$

where the law of a random variable X is denoted by $\mathcal{M}(X)$, and the Kolmogorov distance $d_K(\cdot, \cdot)$ between two laws is given by (1.1.2).

Note that in general we can obtain convergence rate of the form

$$d_K\big(\mathcal{M}(n^{-\frac{1}{2}}(S_n^\omega u - \mathbb{E}_{\mu_\omega} S_n^\omega u)), \mathcal{N}\big) \leq c_\omega n^{-\frac{1}{2}} + d_\omega |n^{-1}\alpha_{\omega,n} - \sigma^2|$$

but there is no general convergence rate in the limit $\sigma^2 = \lim n^{-1}\alpha_{\omega,n}$ since $n^{-1}\alpha_{\omega,n} - \sigma^2$ behaves as a difference between an ergodic average and its corresponding limit (see [38]).

7.1.2 *Local limit theorem*

We first consider the case when $S_n^\omega u(x)$ is given by (7.1.1) where x is distributed according to $\mu_\omega = \mu_\omega(0)$. We will rely on the following

Assumption 7.1.2.

(i) Ω is a topological space, $\mathcal{F}$ is the Borel σ-algebra and the probability measure P assigns positive measure to open sets. The map θ is continuous and there exist $\omega_0 \in \Omega$ and $m_0 \in \mathbb{N}$ such that $\theta^{m_0}\omega_0 = \omega_0$.

(ii) The maps $\omega \to \mathcal{E}_\omega$ and $\omega \to T_\omega$ are locally constant around the points $\omega = \theta^i\omega_0$, $0 \leq i < m_0$, where $\mathcal{E}_\omega$ were introduced in Chapter 5.

(iii) The maps $\omega \to u_\omega$ and $\omega \to \phi_\omega$ are continuous at the points $\omega = \theta^i \omega_0$, $0 \leq i < m_0$ when considered as maps from Ω to $\left(\mathcal{H}_\omega^{\alpha,\xi}, \|\cdot\|_{\alpha,\xi}\right)$ and the random variables H_ω and $(\gamma_\omega - 1)^{-1}$ introduced in Section 5.1 are bounded.

Note that the third condition makes sense when the second condition holds true since then $\mathcal{E}_\omega = \mathcal{E}_{\theta^i \omega_0}$ when ω lies in an appropriate neighborhood of $\theta^i \omega_0$, where $i = 0, 1, ..., m_0 - 1$. The first condition holds true, for instance, when θ is a locally distance expanding map introduced in Section 2.5.1 (denoted there by T) and P is a Gibbs measure associated with a Hölder continuous function (by Lemma 2.5.4 periodic points are dense and by Proposition 2.5.9 P assigns positive mass to open sets).

Next, under Assumption 7.1.2, we call the case a non-lattice one if the function $S_{m_0}^{\omega_0} u$ is non-arithmetic (aperiodic) with respect to the map $T_{\omega_0}^{m_0}$ in the classical sense of [28], namely if for any $t \in \mathbb{R} \setminus \{0\}$ there exist no nonzero $g \in \mathcal{H}_{\omega_0}^{\alpha,\xi}$ and $\lambda \in \mathbb{C}$, $|\lambda| = 1$ such that

$$e^{it S_{m_0}^{\omega_0} u} g = \lambda g \circ T_{\omega_0}^{m_0}, \quad \mu_{\omega_0}\text{-a.s.} \tag{7.1.4}$$

We call the case a lattice one if there exists $h > 0$ such that P-a.s. the function u_ω takes values on the lattice $h\mathbb{Z} = \{hk : k \in \mathbb{Z}\}$ and the function $S_{m_0}^{\omega_0} u$ cannot be written in the form

$$S_{m_0}^{\omega_0} u = a + \beta - \beta \circ T_{\omega_0}^{m_0} + h'\mathbf{k}; \quad \mu_{\omega_0}\text{-a.s.} \tag{7.1.5}$$

for some $h' > h$, $a \in \mathbb{R}$, $\beta : \mathcal{X} \to \mathbb{R}$ such that $e^{i\beta} \in \mathcal{H}_{\omega_0}^{\alpha,\xi}$ and an integer valued function $\mathbf{k} : \mathcal{E}_{\omega_0} \to \mathbb{Z}$.

Now we can state the local limit theorem.

Theorem 7.1.3. *Suppose that the assumptions of Corollary 5.4.2 hold true and that $\sigma^2 > 0$. Assume, in addition, that $\int u_\omega d\mu_\omega = 0$, P-a.s. Then for P-almost all ω the sequence of random variables $S_n^\omega u$, $n \geq 1$ satisfies Assumption 2.2.2. When also Assumption 7.1.2 is satisfied then Assumption 2.2.1 holds true in both the non-lattice and lattice cases. Moreover, P-a.s. for any real continuous function g on $\mathbb{R}$ with compact support,*

$$\lim_{n \to \infty} \sup_{s \in supp\, \nu_h} \left| \sigma \sqrt{2\pi n} \mathbb{E}_{\mu_\omega} g(S_n^\omega u - s) - e^{-\frac{s^2}{2n\sigma^2}} \int g d\nu_h \right| = 0 \tag{7.1.6}$$

where in the lattice case ν_h is the measure assigning mass h to each point of the lattice $h\mathbb{Z} = \{hk : k \in \mathbb{N}\}$ while in the non-lattice case we set $h = 0$ and then ν_0 is the Lebesgue measure.

Next, we consider the case when $S_n^\omega u$ is given by (7.1.3). We will rely on the following

Assumption 7.1.4. The equality $j_\omega = j_{\theta\omega} = 1$ holds true with positive probability, where j_ω is the random variable appearing in (6.1.3).

Similarly to Section 2.4 in Chapter 2, we consider the following particular lattice and non-lattice cases. For P-almost all ω, consider the set

$$B_\omega = \{h \geq 0 : u_\omega(x) - u_\omega(y) \in h\mathbb{Z} \text{ for } \mu_\omega \times \mu_\omega\text{-a.a. } (x,y)\} \qquad (7.1.7)$$

where $h\mathbb{Z} = \{hk : k \in \mathbb{Z}\}$. We call the case a lattice one if there exists $h_0 > 0$ such that P-a.s. we have $u_\omega(x) \in h_0\mathbb{Z}$ for m_ω-a.a. x and

$$h_0 = \sup\{h \geq 0 : h \in B_\omega\}.$$

We call the case non-lattice if

$$P\{\omega : B_\omega = \emptyset\} > 0.$$

Note that as in Section 2.4 there are other cases beyond what we designated as a lattice and a non-lattice case.

Theorem 7.1.5. *Suppose that the assumptions of Corollary 6.3.2 hold true and that $\sigma^2 > 0$. Assume, in addition, that $\int u_\omega d\mu_\omega = 0$, P-a.s. Then for P-almost all ω the sequence of random variables $S_n^\omega u$, $n \geq 1$ satisfies Assumption 2.2.2. When also Assumption 7.1.4 is satisfied then Assumption 2.2.1 is satisfied in both the non-lattice and lattice cases. Moreover, P-a.s. for any real continuous function g on $\mathbb{R}$ with compact support,*

$$\lim_{n\to\infty} \sup_{s\in supp\,\nu_h} \left|\sigma\sqrt{2\pi n}\mathbb{E}_{\mu_\omega} g(S_n^\omega u - s) - e^{-\frac{s^2}{2n\sigma^2}} \int g d\nu_h\right| = 0 \qquad (7.1.8)$$

where in the lattice case ν_h is the measure assigning mass h to each point of the lattice $h\mathbb{Z} = \{hk : k \in \mathbb{N}\}$ while in the non-lattice case we set $h = 0$ and then ν_0 is the Lebesgue measure.

7.2   Pressure near 0

In the deterministic case the analyticity of RPF triplets with respect to the complex parameter z yields optimal convergence rate in the CLT of order $n^{-\frac{1}{2}}$ (see [19], [28], [23], [51] and references therein) and is the key to show that Assumption 2.2.2 holds true (see [23], [28] and [51]). In this section we provide the estimates of the pressure needed in order to derive the Berry-Esseen and local limit theorems stated in Section 7.1.

We first consider the situation when $S_n^\omega u$ is given by (7.1.1) and assume that the conditions of Corollary 5.4.2 hold true. In Section 5.12 we showed that in these circumstances the assumptions of Theorem 4.6.3 hold true with $A_z^\omega = \mathcal{L}_z^\omega$ and therefore by Proposition 4.6.5 there exists a deterministic neighborhood V_2 of 0 so that the pressure functions $\bar{\Pi}_\omega$, $\omega \in \Omega$ obtained in Proposition 4.6.5 are defined on V_2. For any natural k consider the analytic function $\bar{\Pi}_{\omega,k} : V_2 \to \mathbb{C}$ given by

$$\bar{\Pi}_{\omega,k}(z) = \sum_{i=0}^{k-1} \bar{\Pi}_{\theta^j \omega}(z). \qquad (7.2.1)$$

Since we have assumed that $\lambda_\omega(0) = 1$, by Proposition 4.6.5 we have $\bar{\Pi}_{\omega,k}(0) = 0$,

$$e^{\bar{\Pi}_{\omega,k}(z)} = \lambda_{\omega,k}(z) \text{ and } |\bar{\Pi}_{\omega,k}(z)| \leq k(\ln 2 + \pi) \qquad (7.2.2)$$

for any $z \in V_2$. We begin with computing the derivative of the functions $\bar{\Pi}_{\omega,k}(z)$ at the point $z = 0$.

Lemma 7.2.1. *Suppose that the assumptions of Corollary 5.4.2 hold true. Then P-a.s. for any $k \in \mathbb{N}$,*

$$\bar{\Pi}'_{\omega,k}(0) = \int S_k^\omega u \, d\mu_\omega(0) \quad and \quad \bar{\Pi}''_{\omega,k}(0) = Var_{\mu_\omega(0)} S_k^\omega u + \mathcal{R}_{\omega,k}(0) \quad (7.2.3)$$

where with $F_\omega(0) = \nu_\omega(0)(h''_\omega(0))$,

$$\mathcal{R}_{\omega,k}(0) = F_\omega(0) - F_{\theta^k \omega}(0) + 2Cov_{\mu_\omega(0)}\big(S_k^\omega u, h'_\omega(0)\big).$$

Proof. Let $k \in \mathbb{N}$. Since $\bar{\lambda}_{\omega,k}(z) = e^{\bar{\Pi}_{\omega,k}(z)}$ (around 0) and $\bar{\lambda}_{\omega,k}(0) = 1$, P-a.s., we have

$$\bar{\Pi}'_{\omega,k}(0) = \bar{\lambda}'_{\omega,k}(0) \text{ and } \bar{\Pi}''_{\omega,k}(0) = \bar{\lambda}''_{\omega,k}(0) - (\bar{\lambda}'_{\omega,k}(0))^2. \qquad (7.2.4)$$

We first show that

$$\bar{\lambda}'_{\omega,k}(0) = \int S_k^\omega u d\mu_\omega(0) \text{ and} \qquad (7.2.5)$$

$$\bar{\lambda}''_{\omega,k}(0) = \int (S_k^\omega u)^2 d\mu_\omega(0) + \mathcal{R}_{\omega,k}(0).$$

The formulas in (7.2.3) clearly follow from (7.2.4) and (7.2.5).

In order to prove (7.2.5), we first notice that P-a.s. for any $z \in V_2$ we have

$$\lambda_{\omega,k}(z)h_{\theta^k \omega}(z) = \mathcal{L}_z^{\omega,k} h_\omega(z). \qquad (7.2.6)$$

Differentiating both sides with respect to z yields

$$\lambda'_{\omega,k}(z)h_{\theta^k\omega}(z) + \lambda_{\omega,k}(z)h'_{\theta^k\omega}(z) = \frac{d\mathcal{L}_z^{\omega,k}(h_\omega(z))}{dz} \qquad (7.2.7)$$
$$= \mathcal{L}_z^{\omega,k}\big(S_k^\omega u \cdot h_\omega(z) + h'_\omega(z)\big).$$

Similarly to (7.2.6), we have

$$\big(\mathcal{L}_z^{\omega,k}\big)^* \nu_{\theta^k\omega}(z) = \lambda_{\omega,k}(z)\nu_\omega(z), \quad P\text{-a.s.}$$

for any $z \in V_2$, and we recall that $\nu_{\theta^k\omega}(z)(h_{\theta^k(\omega)}(z)) = 1$. Taking $z = 0$, integrating the left and right hand sides of (7.2.7) with respect to $\nu_{\theta^k\omega}(0)$ and using (5.10.1) we arrive at

$$\lambda'_{\omega,k}(0) = \lambda_{\omega,k}(0)\big(\nu_\omega(0)(S_k^\omega u \cdot h_\omega(0)) + G_\omega(0) - G_{\theta^k\omega}(0)\big) \qquad (7.2.8)$$

where $G_\omega(0) = \nu_\omega(0)(h'_\omega(0))$. We claim now that $G_\omega(0) = 0$, P-a.s. Indeed, differentiating the equality $\nu_\omega(z)h_\omega(z) = 1$ with respect to z and taking into account that $h_\omega(0) \equiv 1$ we have

$$G_\omega(0) = -\nu'_\omega(0)\mathbf{1} = -\frac{d\nu_\omega(z)\mathbf{1}}{dz}\bigg|_{z=0} = -\frac{d\mathbf{1}}{dz}\bigg|_{z=0} = 0 \qquad (7.2.9)$$

where we used that $\nu_\omega(z)\mathbf{1} = 1$ for any $z \in U$.

Dividing both sides of (7.2.8) by $\lambda_{\omega,k}(0)$ and using that $G_\omega(0) = 0$ we derive that

$$\bar{\lambda}'_{\omega,k}(0) = \nu_\omega(0)(S_k^\omega u \cdot h_\omega(0)) = \mu_\omega(0)(S_k^\omega u)$$

where we used that $\nu_\omega(0) = \mu_\omega(0)$ and that $h_\omega(0) \equiv \mathbf{1}$. Next, differentiating both sides of (7.2.7) we obtain

$$\lambda''_{\omega,k}(z)h_{\theta^k\omega}(z) + 2\lambda'_{\omega,k}(z)h'_{\theta^k\omega}(z) + \lambda_{\omega,k}(z)h''_{\theta^k\omega}(z)$$
$$= \mathcal{L}_z^{\omega,k}\big((S_k^\omega u)^2 \cdot h_\omega(z) + 2S_k^\omega u \cdot h'_\omega(z) + h''_\omega(z)\big).$$

By the same reason as before, taking $z = 0$ and then integrating both sides with respect to $\nu_{\theta^k\omega}(0)$ we arrive at

$$\bar{\lambda}''_{\omega,k}(0) = \int (S_k^\omega u)^2 d\mu_\omega(0) + \mathcal{R}_{\omega,k}$$

where we used that $\nu_\omega(0) = \mu_\omega(0)$ and that $h_\omega(0) \equiv \mathbf{1}$, and the proof of the lemma is complete. $\square$

In order to develop effectively the function $\bar{\Pi}_{\omega,k}(\cdot)$ around 0 it is necessary to estimate the term $\mathcal{R}_{\omega,k}(0)$ defined in Lemma 7.2.1.

Lemma 7.2.2. *Under the assumptions of Corollary 5.4.2, there exists a constant $Q_2 > 0$ so that P-a.s. for any natural k,*

$$|\mathcal{R}_{\omega,k}(0)| \leq Q_2.$$

Proof. We first note that $F_\omega(0)$ is a real valued random variable since $\nu_\omega(0)$ is a probability measure and $h_\omega(t)$, $t \in U \cap \mathbb{R}$ are real valued functions. By Corollary 5.4.2 there exists a constant $C_2 > 0$ such that P-a.s. $\|h_\omega''(0)\|_{\alpha,\xi} \leq C_2$ and therefore

$$|F_\omega(0)| \leq \|h_\omega''(0)\|_\infty \leq C_2, \quad P\text{-a.s.} \qquad (7.2.10)$$

It remains to approximate the term $\mathrm{Cov}_{\mu_\omega(0)}(S_k^\omega u, h_\omega'(0))$. We begin with writing

$$\mathrm{Cov}_{\mu_\omega(0)}(S_k^\omega u, h_\omega'(0)) = \sum_{i=0}^{k-1} \mathrm{Cov}_{\mu_\omega(0)}(u_\omega \circ T_\omega^i, h_\omega'(0)).$$

Let $D > 0$ be such that $\|u_\omega\|_\infty \leq D$. Applying Lemma 5.10.4 *(ii)*, we obtain that for any $i = 0, 1, 2, ...$,

$$|\mathrm{Cov}_{\mu_\omega(0)}(u_\omega \circ T_\omega^i, h_\omega'(0))| \leq D\|\mathcal{L}_0^{\omega,i} h_\omega'(0) - (\mu_\omega(0) h_\omega'(0)) \mathbf{1}\|_\infty$$

where we used that $h_\omega(0) \equiv \mathbf{1}$ and $\lambda_\omega(0) = 1$, P-a.s. By (5.4.3) in Corollary 5.4.2 there exists a constant C_1 so that $\|h_\omega'(0)\|_{\alpha,\xi} \leq C_1$. Combining this with the exponential convergence obtained in Corollary 5.4.2 it follows that there exists a constant $Q_1 > 0$ so that

$$|\mathrm{Cov}_{\mu_\omega(0)}(S_k^\omega u, h_\omega'(0))| \leq Q_1.$$

We conclude that there exists a constant Q_2 such that

$$|\mathcal{R}_{\omega,k}(0)| \leq Q_2, \quad P\text{-a.s.} \qquad (7.2.11)$$

for any $k \in \mathbb{N}$. $\qquad \square$

Corollary 7.2.3. *Suppose that the assumptions of Corollary 5.4.2 hold true and that $\sigma^2 > 0$. Then there exist constants $t_0, c_0, q > 0$ such that P-a.s. for any z with $|z| \leq t_0$, $t \in [-t_0, t_0]$ and $k \geq 1$,*

$$\left|\bar{\Pi}_{\omega,k}(z) - z\bar{\Pi}_{\omega,k}'(0) - \frac{z^2}{2}\bar{\Pi}_{\omega,k}''(0)\right| \leq c_0|z|^3 k \qquad (7.2.12)$$

and when k is sufficiently large,

$$\Re(\bar{\Pi}_{\omega,k}(it)) \leq -qt^2 k. \qquad (7.2.13)$$

Proof. Let $\delta_0 > 0$ be so that the closed ball $\bar{B}(0, 2\delta_0)$ is contained in V_2. Applying Lemma 2.8.2 and taking into account the inequality in (7.2.2), we obtain that P-a.s. for any $z \in B(0, \delta_0)$ and $k \geq 1$,

$$\left|\bar{\Pi}_{\omega,k}(z) - z\bar{\Pi}_{\omega,k}'(0) - \frac{z^2}{2}\bar{\Pi}_{\omega,k}''(0)\right| \leq c_0|z|^3 k \qquad (7.2.14)$$

where $c_0 = 8(\ln(2) + \pi)\delta_0^{-3}$. By Lemmas 7.2.1 and 7.2.2,

$$\bar{\Pi}''_{\omega,k}(0) \geq \frac{1}{2}\sigma^2 k$$

for any sufficiently large k and by Lemma 7.2.1,

$$\bar{\Pi}'_{\omega,k}(0) = \int S_k^\omega u \, d\mu_\omega(0) \in \mathbb{R}.$$

Thus, there exist constants $q > 0$ and $0 < t_0 < \delta_0$ so that for any sufficiently large k,

$$\Re(\bar{\Pi}_{\omega,k}(it)) \leq -qt^2 k \qquad (7.2.15)$$

for any $t \in [-t_0, t_0]$. $\square$

Now we consider the case when $S_n^\omega u$ is given by (7.1.3). Under the assumptions of Corollary 6.3.2, taking into account that $R_0^\omega 1 = 1$ and that $\|u_\omega\|_\infty$ is bounded, it follows that Assumptions 4.6.1 and 4.6.2 hold true with the family $A_z^\omega = R_z^{\theta^{-1}\omega}$ and the system $(\Omega, \mathcal{F}, P, \vartheta)$, $\vartheta = \theta^{-1}$. Applying Proposition 4.6.5 it follows that the pressure function $\tilde{\Pi}_\omega(\cdot)$ obtained there is defined P-a.s. on some deterministic neighborhood V_2 of 0. Therefore, we can define the pressure functions $\tilde{\Pi}_{\omega,k} : V_2 \to \mathbb{C}$ by

$$\tilde{\Pi}_{\omega,k}(z) = \sum_{j=1}^n \bar{\Pi}_{\vartheta^{-j}\omega}(z)$$

which satisfy (7.2.2). The following result is derived similarly to Lemma 7.2.1.

Lemma 7.2.4. *Suppose that the assumptions of Corollary 6.3.2 hold true. Then P-a.s. for any natural k,*

$$\tilde{\Pi}'_{\omega,k}(0) = \mathbb{E} \sum_{j=1}^n u_{\theta^j \omega}(\zeta_j^{\theta^j \omega})$$

and

$$\tilde{\Pi}''_{\omega,k}(0) = Var \sum_{j=1}^k u_{\theta^j \omega}(\zeta_j^{\theta^j \omega}) + 2\mathbb{E}h'_\omega(0)(\zeta_0^\omega) \sum_{j=1}^k u_{\theta^j \omega}(\zeta_j^{\theta^j \omega})$$

$$+ F_{\theta^k \omega}(0) - F_\omega(0) - \left(\bar{\Pi}'_{\omega,k}(0)\right)^2$$

where $F_\omega(z) = \nu_\omega(z)(h''_\omega(z))$.

Remark that in the above formula we used that $\int h'_\omega(0)d\nu_\omega(0) = 0$ which follows exactly as in the transfer operator case since we assumed that $h_\omega(0) \equiv 1$. Next, the random Doeblin condition (6.1.3) together with the assumptions of Corollary 6.3.2 imply that the strong mixing condition from [38] holds true uniformly in ω, and therefore the term $|\mathbb{E}h'_\omega(0)(\zeta_0^\omega)\sum_{j=1}^k u_{\theta^j\omega}(\zeta_j^{\theta^j\omega})|$ is bounded by some constant which does not depend on k and ω. The following proposition follows similarly to Corollary 7.2.3.

Proposition 7.2.5. *Suppose that the assumptions of Corollary 6.3.2 hold true and that $\sigma^2 > 0$. Then there exist constants $t_0, c_0, q > 0$ such that P-a.s. for any z with $|z| \leq t_0$, $t \in [-t_0, t_0]$ and $k \geq 1$,*

$$|\tilde{\Pi}_{\omega,k}(z) - z\tilde{\Pi}'_{\omega,k}(0) - \frac{z^2}{2}\tilde{\Pi}''_{\omega,k}(0)| \leq c_0|z|^3 k \qquad (7.2.16)$$

and when k is sufficiently large,

$$\Re(\tilde{\Pi}_{\omega,k}(it)) \leq -qt^2 k. \qquad (7.2.17)$$

7.3 A fiberwise (self normalized) Berry-Esseen theorem: proof

In this section we will prove Theorem 7.1.1. We will only consider the case when $S_n^\omega u$ are given by (7.1.1), and the proof in the Markov chain case is analogous.

Assume without loss of generality that $\int u_\omega d\mu_\omega = 0$, P-a.s. which together with Lemma 5.10.4 *(i)* implies that $\mathbb{E}_{\mu_\omega} S_n^\omega = 0$, P-a.s. for any $n \geq 1$. Since $\lambda_\omega(0) = 1$ and $\mu_\omega = \nu_\omega(0)$ for any $n \geq 1$ we have $(\mathcal{L}_0^{\omega,n})^* \mu_{\theta^n\omega} = \mu_\omega$ and therefore P-a.s. for any $z \in \mathbb{C}$,

$$\mu_\omega(e^{zS_n^\omega u}) = \mu_{\theta^n\omega}(\mathcal{L}_0^{\omega,n}e^{zS_n^\omega u}) = \mu_{\theta^n\omega}(\mathcal{L}_z^{\omega,n}1). \qquad (7.3.1)$$

Consider the analytic function $\varphi_{\omega,n} : U \to \mathbb{C}$ given by

$$\varphi_{\omega,n}(z) = \frac{\mu_{\theta^n\omega}(\mathcal{L}_z^{\omega,n}1)}{\lambda_{\omega,n}(z)} = \int \frac{\mathcal{L}_z^{\omega,n}1(x)}{\lambda_{\omega,n}(z)}d\mu_{\theta^n\omega}(x) \qquad (7.3.2)$$

where U is the neighborhood of 0 specified in Corollary 5.4.2. Then by (7.3.1) for any $z \in U$,

$$\mu_\omega(e^{zS_n^\omega u}) = \lambda_{\omega,n}(z)\varphi_{\omega,n}(z) = e^{\bar{\Pi}_{\omega,n}(z)}\varphi_{\omega,n}(z) \qquad (7.3.3)$$

where $\bar{\Pi}_{\omega,n}(z)$ is given by (7.2.1). Next, by (7.2.5), $\lambda'_{\omega,n}(0) = \mathbb{E}_{\mu_\omega} S_n^\omega u$ and therefore by (7.3.2),

$$\varphi'_{\omega,n}(0) = 0. \qquad (7.3.4)$$

Next, let $U_0 \subset U$ be the neighborhood of 0 specified in Corollary 5.4.2. By taking a ball which is contained in U_0 we can always assume that $U_0 = B(0, r_0)$ is a ball around 0 with radius $r_0 > 0$. We claim that there exists a constant $A > 0$ such that P-a.s. for any $n \in \mathbb{N}$,

$$|\varphi_{\omega,n}(z)| \leq A \tag{7.3.5}$$

for any $z \in U_0$. By the exponential convergence statement in Corollary 5.4.2 there exist constants $A_1, k_1 > 0$ and $c \in (0,1)$ such that P-a.s. for any $z \in U_0$ and $n \geq k_1$,

$$\left\| \frac{\mathcal{L}_z^{\omega,n} 1}{\lambda_{\omega,n}(z)} - h_{\theta^n \omega}(z) \right\|_{\alpha,\xi} \leq A_1 c^{\frac{n}{j_0}} \tag{7.3.6}$$

where we replaced ω with $\theta^n \omega$. By Corollary 5.4.2, $\|h_{\theta^n \omega}(z)\|_{\alpha,\xi} \leq C_1$, P-a.s. for any $n \geq 0$ and $z \in U_0$, and (7.3.5) follows.

Next, applying Lemma 2.8.2 with $k = 1$ we deduce from (7.3.4) and (7.3.5) that there exists a constant $B > 0$ such that

$$|\varphi_{\omega,n}(z) - \varphi_{\omega,n}(0)| = |\varphi_{\omega,n}(z) - 1| \leq B|z|^2$$

for any $z \in U_1 = B(0, \frac{1}{2} r_0) = \frac{1}{2} U_0$. Moreover, by Corollary 7.2.3 there exist constants $t_0, c_0, q > 0$ such that P-a.s. for any $s \in [-t_0, t_0]$ and a sufficiently large n,

$$\left| \bar{\Pi}_{\omega,n}(is) + \frac{s^2}{2} \mathbb{E}_{\mu_\omega}(S_n^\omega u)^2 \right| \leq c_0 |s|^3 n$$

and

$$\Re(\Pi_{\omega,n}(is)) \leq -qns^2$$

where in the first inequality we also used that (see Lemma 7.2.1),

$$\bar{\Pi}'_{\omega,n}(0) = \mathbb{E}_{\mu_\omega} S_n^\omega u = 0.$$

Next, by the Esseen inequality (see [60]) for any two distribution functions $F_1 : \mathbb{R} \to [0,1]$ and $F_2 : \mathbb{R} \to [0,1]$ with characteristic functions ψ_1, ψ_2, respectively, and $T > 0$,

$$\sup_{x \in \mathbb{R}} |F_1(x) - F_2(x)| \leq \frac{2}{\pi} \int_0^T \left| \frac{\psi_1(t) - \psi_2(t)}{t} \right| dt + \frac{24}{\pi T} \sup_{x \in \mathbb{R}} |F_2'(x)| \tag{7.3.7}$$

assuming that F_2 is a function with bounded derivatives. Set $a_n = n^{-1} \alpha_{\omega,n}$ and $r_n = \sqrt{a_n}$, where $\alpha_{\omega,n}$ is defined before Theorem 7.1.1. Then a_n converges to σ^2 P-a.s. as $n \to \infty$ and r_n converges to $\sigma > 0$ P-a.s. when

$n \to \infty$. Set $t_1 = \frac{1}{2}\min(t_0, r_0)$. By (7.3.3), the mean value theorem and the above estimates, P-a.s. for any sufficiently large n and $t \in [0, t_1 r_n n^{\frac{1}{2}}]$,

$$\left|\mu_\omega\left(e^{\frac{it}{r_n\sqrt{n}}S_n^\omega u}\right) - e^{-\frac{1}{2}t^2}\right| \leq e^{\Re\bar{\Pi}_{\omega,n}\left(\frac{it}{r_n\sqrt{n}}\right)}\left|\varphi_{\omega,n}\left(\frac{it}{r_n\sqrt{n}}\right) - 1\right|$$

$$+\left|e^{\bar{\Pi}_{\omega,n}\left(\frac{it}{r_n\sqrt{n}}\right)} - e^{-\frac{1}{2}t^2}\right| \leq \left(Br_n^{-2}\right)n^{-1}e^{-q_n t^2}t^2$$

$$+\left|e^{\bar{\Pi}_{\omega,n}\left(\frac{it}{r_n\sqrt{n}}\right)} - e^{-\frac{a_n t^2}{2r_n^2}}\right| + \left|e^{-\frac{a_n t^2}{2r_n^2}} - e^{-\frac{1}{2}t^2}\right| \leq \left(Br_n^{-2}\right)n^{-1}e^{-q_n t^2}t^2$$

$$+\left(2c_0 r_n^{-3}\right)n^{-\frac{1}{2}}e^{-b_n t^2}t^3$$

where $q_n = qr_n^{-2}$, $b_n = \frac{1}{2}r_n^{-2}\min(q, a_n)$ and we used that $r_n^2 = a_n$ and that $na_n = \mathbb{E}_{\mu_\omega}(S_n^\omega u)^2$ for any sufficiently large n. Applying (7.3.7) with the distribution function of $r_n^{-1}n^{-\frac{1}{2}}S_n^\omega u = \hat{S}_n^\omega u$ and the standard normal distribution function we obtain that for any sufficiently large n,

$$d_K\left(\mathcal{M}(r_n^{-1}n^{-\frac{1}{2}}S_n^\omega u), \mathcal{N}\right)$$

$$\leq \int_0^{t_1 n^{\frac{1}{2}}r_n} \frac{\left|\mu_\omega\left(e^{\frac{it}{r_n\sqrt{n}}S_n^\omega u}\right) - e^{-\frac{1}{2}t^2}\right|}{t}dt + \left(6t_1^{-1}r_n^{-1}\right)n^{-\frac{1}{2}}$$

$$\leq \left(Br_n^{-2}\int_0^\infty te^{-q_n t^2}dt\right)n^{-1} + \left(2c_0 r_n^{-3}\int_0^\infty e^{-b_n t^2}t^2 dt\right)n^{-\frac{1}{2}} + \left(6t_1^{-1}r_n^{-1}\right)n^{-\frac{1}{2}}.$$

Theorem 7.1.1 follows now from the above estimates, taking into account that

$$\int_0^\infty \max(t, t^2)e^{-t^2}dt < \infty$$

and that r_n, q_n and b_n converge P-a.s. as $n \to \infty$ to positive limits. $\quad\square$

7.4 A fiberwise local (central) limit theorem: proof

After establishing that a non-degenerate fiberwise CLT holds true, by Theorem 2.2.3 the corresponding fiberwise LLT in both lattice and non-lattice cases holds true when Assumptions 2.2.1 and 2.2.2 are satisfied, namely when we can estimate appropriately the characteristic functions $\mathbb{E}_{\mu_\omega}e^{itS_n^\omega u}$. In this section we will show that these assumptions hold true (in the circumstances of Section 7.1.2).

7.4.1 *Characteristic functions estimates for small t's*

We prove the following result in both cases.

Lemma 7.4.1. *Suppose that the assumption of Corollary 5.4.2 hold true (or Corollary 6.3.2 in the Markov chain case) and that $\sigma^2 > 0$. Then* *for P-almost all ω the sequence of random variables $S_n^\omega u$, $n \geq 1$ satisfies Assumption 2.2.2.*

Proof. We will consider only the case of transfer operators while the case of Markov chains in random environment can be treated in a similar way. First, P-a.s. for any $n \in \mathbb{N}$ and $t \in \mathbb{R}$,

$$\mu_\omega(0)(e^{itS_n^\omega u}) = \mu_\omega(0)(\mathcal{L}_{it}^{\omega,n}\mathbf{1})$$

and therefore for any $t \in \mathbb{R}$,

$$|\mu_\omega(0)(e^{itS_n^\omega u})| \leq \|\mathcal{L}_{it}^{\omega,n}\mathbf{1}\|_\infty.$$

By (7.3.6) and the estimate $\|h_{\theta^n\omega}'(z)\| \leq C_1$ following it, there exists a constant A_2 such that

$$\|\mathcal{L}_{it}^{\omega,n}\mathbf{1}\|_\infty \leq A_2|\lambda_{\omega,n}(it)|$$

whenever $it \in U_0$. Combining this with Corollary 7.2.3 we see that there exist constants $t_1, q > 0$ such that P-a.s. for any sufficiently large n and $t \in [-t_1, t_1]$,

$$\|\mathcal{L}_{it}^{\omega,n}\mathbf{1}\|_\infty \leq A_2|\lambda_{\omega,n}(it)| = A_2 e^{\Re(\bar{\mathcal{P}}_{\omega,n}(it))} \leq A_2 e^{-qt^2 n}$$

and therefore Assumption 2.2.2 is satisfied. $\qquad\square$

7.4.2 Characteristic functions estimates for large t's: covering maps

We will prove next the following

Lemma 7.4.2. *Suppose that Assumption 7.1.2 and the assumptions of Corollary 5.4.2 hold true. Then for P-almost all ω the sequence of random variables $S_n^\omega u$, $n \geq 1$ satisfies Assumption 2.2.1 in both non-lattice and lattice cases.*

Proof. We first observe that Assumption 7.1.2 *(i)* means that Assumption 2.10.1 is satisfied. Moreover, when Assumption 7.1.2 *(iii)* is satisfied then by Lemma 5.6.1 we obtain (2.10.1), no matter which type of interval $\mathcal{I}$ is considered. This together with continuity of the maps $\omega \to u_\omega$ and $\omega \to \phi_\omega$ at the points $\theta^i\omega_0$, $i = 0, 1, ..., m_0 - 1$ guarantees that Assumption 2.10.2 holds true. Therefore, relying on Lemma 2.10.4, it is sufficient to show that Assumption 2.10.3 holds true with $\mathcal{I} = \mathbb{R}$ in the non-lattice case and with $\mathcal{I} = (-\frac{2\pi}{h}, \frac{2\pi}{h})$ in the lattice case. This is proved similarly to Section 2.9.5 in Chapter 2, namely we apply the results from [28] with the transfer operators $\mathcal{L}_{it}^{\omega_0,m_0}$, i.e. with the space $\mathcal{E}_{\omega_0}$, the map $T_{\omega_0}^{m_0}: \mathcal{E}_{\omega_0} \to \mathcal{E}_{\omega_0}$ and the function $S_{m_0}^{\omega_0}u$. $\qquad\square$

7.4.3 Characteristic functions estimates for large *t*'s: Markov chains

Lemma 7.4.3. *Suppose that Assumption 7.1.4 and the assumptions of Corollary 5.4.2 hold true. Then for P-almost all ω the sequence of random variables $S_n^\omega u$, $n \geq 1$ satisfies Assumption 2.2.1 in both non-lattice and lattice cases.*

Proof. Set $\tilde{\omega} = \theta^{j_\omega}\omega$. Then for any $y_0 \in \mathcal{X}$, $t \in \mathbb{R}$ and $g : \mathcal{X} \to \mathbb{C}$ such that $\|g\|_\infty \leq 1$ we have

$$|R_{it}^{\omega,j_\omega} \circ R_{it}^{\tilde{\omega},j_{\tilde{\omega}}} g(y_0)| \qquad (7.4.1)$$

$$= \left| \int e^{it\sum_{k=1}^{j_\omega} u_{\theta^k\omega}(y_k)} \cdot e^{it\sum_{k=1}^{j_{\tilde{\omega}}} u_{\theta^k\tilde{\omega}}(z_k)} \Big(\prod_{j=1}^{j_\omega} r_{\theta^{j-1}\omega}(y_{j-1}, y_j)\Big) r_{\tilde{\omega}}(y_{j_\omega}, z_1) \right.$$

$$\left. \times \prod_{k=1}^{j_{\tilde{\omega}}-1} r_{\theta^k\tilde{\omega}}(z_k, z_{k+1}) g(z_{j_{\tilde{\omega}}}) \prod_{j=1}^{j_\omega} dm_{\theta^j\omega}(y_{j_\omega}) \prod_{k=1}^{j_{\tilde{\omega}}} dm_{\theta^k\tilde{\omega}}(z_k))) \right|$$

$$\leq \int \prod_{k=1}^{j_{\tilde{\omega}}} dm_{\theta^k\tilde{\omega}}(z_k) \prod_{k=1}^{j_{\tilde{\omega}}-1} r_{\theta^k\tilde{\omega}}(z_k, z_{k+1})$$

$$\times \left| \int \Big(\prod_{j=0}^{j_\omega-1} r_{\theta^j\omega}(y_j, y_{j+1})\Big) r_{\tilde{\omega}}(y_{j_\omega}, z_1) e^{it\sum_{k=1}^{j_\omega} u_{\theta^k\omega}(y_k)} \prod_{j=1}^{j_\omega} dm_{\theta^j\omega}(y_j) \right|$$

$$= \int dm_{\theta\tilde{\omega}}(z_1) |R_{it}^{\omega,j_\omega} r_{\tilde{\omega}}(\cdot, z_1)(y_0)|.$$

By (2.4.10) from Section 2.4 in Chapter 2 for any z_1 we have

$$\delta_1^{-1}|R_{it}^{\omega,j_\omega} r_{\tilde{\omega}}(\cdot, z_1)(y_0)|$$

$$\leq 1 - \frac{1}{4}\delta_1^{-2} \int \int \left| e^{it\sum_{k=1}^{j_\omega} u_{\theta^k\omega}(y_k)} - e^{it\sum_{k=1}^{j_\omega} u_{\theta^k\omega}(x_k)} \right|^2$$

$$\times \prod_{j=0}^{j_\omega} r_{\theta^j\omega}(y_j, y_{j+1}) \prod_{j=0}^{j_\omega} r_{\theta^j\omega}(x_j, x_{j+1}) \prod_{j=1}^{j_\omega} dm_{\theta^j\omega}(y_j) \prod_{j=1}^{j_\omega} dm_{\theta^j\omega}(x_j)$$

where

$$\delta_1 = \delta_1(\omega, z_1, y_0) = R_0^{\omega,j_\omega} r_{\tilde{\omega}}(\cdot, z_1)(y_0) = r_\omega(j_\omega + 1, y_0, z_1) > 0$$

and we set $x_{\theta^{j_\omega+1}\omega} = y_{\theta^{j_\omega+1}\omega} = z_1$ and $x_0 = y_0$. Suppose now that the function $e^{itu_{\theta\omega}(\cdot)}$ is not constant $m_{\theta\omega}$-a.s. and that $j_\omega = j_{\theta\omega} = 1$. Then similarly to (2.4.13) in Section 2.4 (relying on (2.4.11) and (2.4.12)), there exist $\epsilon_\omega(t)$ and $\delta_\omega(t) > 0$ such that

$$\delta_1^{-1}|R_{it}^{\omega,j_\omega} r_{\tilde{\omega}}(\cdot, z_1)(y_0)| \leq 1 - \frac{(\alpha_1(\omega)\alpha_1(\theta\omega))^2 \epsilon_\omega^2(t)\delta_\omega^2(t)}{4\delta_1^2}.$$

From this we deduce similarly to Section 2.4 that

$$\|R_{it}^{\omega,2}\|_\infty < 1.$$

When this holds true for t's belonging to a compact set J then by continuity in t there exists $c_J(\omega) < 1$ such that

$$\|R_{it}^{\omega,2}\|_\infty \le c_J(\omega) < 1 \qquad (7.4.2)$$

for any $t \in J$. As in Section 2.4, the above estimate holds true for any compact subset J of $\mathbb{R} \setminus \{0\}$ in the non-lattice case and for any compact subset of $[-\frac{\pi}{h}, \frac{\pi}{h}] \setminus \{0\}$ in the lattice case. Consider the sets

$$\Gamma_\delta = \{\omega : j_\omega = j_{\theta\omega} = 1 \text{ and } c_J(\omega) < 1 - \delta\}, \ \delta > 0$$

and let $\delta > 0$ be sufficiently small so that $p_\delta = P(\Gamma_\delta) > 0$. Applying the mean ergodic theorem with the indicator function of Γ_δ we see that P-a.s. there exists an infinite sequence $l_1 < l_2 < l_3 < \dots$ of natural numbers so that for any $m \ge 1$,

$$\theta^m \omega \in \Gamma_\delta \text{ if and only if } m \in \{l_1, l_2, l_3, \dots\},$$

$\lim_{k \to \infty} \frac{l_k}{k} = \frac{1}{p_\delta}$ and

$$\lim_{n \to \infty} \frac{\max\{k : l_k \le n\}}{n} = p_\delta.$$

Now, set $m_i = l_{2i}$, $i = 1, 2, \dots$ and $k_n = \max\{k : m_k \le n\}$, $n \ge 1$. Then $k_n = \max\{k : m_k \le n\}$ grows linearly in n. Let $t \in J$ and consider the "block partition",

$$R_{it}^{\omega,n} = R_{it}^{\omega,m_1-1} \circ R_{it}^{\theta^{m_1}\omega,2} \circ R_{it}^{\theta^{m_1+2}\omega,m_2-m_1-2} \circ R_{it}^{\theta^{m_2}\omega,2} \circ \cdots$$

$$\circ R_{it}^{\theta^{m_{k_n}-1}\omega,2} \circ R_{it}^{\theta^{m_{k_n}-1+2}\omega,n-m_{k_n}-1-2}.$$

Then P-a.s. the operator norms (with respect to the supremum norm) of the blocks $R_{it}^{\theta^{m_k}\omega,2}$, $k = 1, 2, \dots$ do not exceed $1 - \delta$, while the norms of the other blocks do not exceed 1. Since k_n grows linearly in n we conclude that $\sup_{t \in J} \|R_{it}^{\omega,n}\|_\infty$ decays exponentially fast to 0 as $n \to \infty$, and the proof of the lemma is complete since we can always choose the sets J described after (7.4.2). $\qquad\square$

7.4.4 An extension

Inequality (7.4.1) suggests that a more general situation than the one of Assumption 7.1.4 can be considered. Let $\varepsilon_\omega > 0$ be a random variable and consider the sets

$$\Delta_\omega = \left\{ (y_1, ..., y_{j_\omega}) : \prod_{i=0}^{j_\omega} r_{\theta^i \omega}(y_i, y_{i+1}) \geq \varepsilon_\omega, \ \forall y_0, \ y_{j_\omega+1} \right\} \subset \mathcal{X}_{\theta\omega} \times \cdots \times \mathcal{X}_{\theta^{j_\omega}\omega}$$

and the probability measures $\mathcal{M}_\omega = \prod_{j=1}^{j_\omega} m_{\theta^j \omega}$.

Assumption 7.4.4. P-a.s. the set Δ_ω has positive $\mathcal{M}_\omega$-probability.

Under this assumption let the conditional probabilities $\bar{\mathcal{M}}_\omega$ be given by $\bar{\mathcal{M}}_\omega(\mathcal{Q}) = \mathcal{M}_\omega(\mathcal{Q}|\Delta_\omega)$. Consider lattice and non-lattice cases similar to the ones described after Assumption 7.1.4 but with the function $(x_1, ..., x_{j_\omega}) \to \sum_{k=1}^{j_\omega} u_{\theta^k \omega}(x_k)$ in place of u_ω and the measures $\bar{\mathcal{M}}_\omega \times \bar{\mathcal{M}}_\omega$ in place of $\mu_{\theta\omega} \times \mu_{\theta\omega}$. For any $a > 0$ and $j_0 \in \mathbb{N}$ set

$$\Gamma_{j_0, a} = \{ \omega \in \Omega : j_\omega, j_{\tilde{\omega}} \leq j_0, \alpha_{j_\omega}(\omega) > a \}.$$

Taking $a > 0$ and $j_0 \in \mathbb{N}$ so that $P(\Gamma_{j_0, a}) > 0$ and considering visiting sequences to the set $\Gamma_{j_0, a}$ we obtain the desired characteristic function estimates in the appropriate lattice and non-lattice cases described above essentially in the same way as in the proof of Lemma 7.4.3.

7.5 Nonconventional limit theorems for processes in random environment

7.5.1 Central limit theorem

Our setup here consists of an ergodic measure preserving system $(\Omega, \mathcal{F}, P, \theta)$, a family $\xi^\omega = \{\xi_n^\omega : n \geq 0\}$, $\omega \in \Omega$ of $\wp$-dimensional stochastic processes on probability spaces $(\mathcal{E}_\omega, \mathcal{F}^\omega, P_\omega)$ and families of sub-σ-algebras $\mathcal{F}_{k,l}^\omega$, $\omega \in \Omega$, $-\infty \leq k \leq l \leq \infty$ such that $\mathcal{F}_{k,l}^\omega \subset \mathcal{F}_{k',l'}^\omega \subset \mathcal{F}^\omega$ if $k' \leq k$ and $l' \geq l$. We define the mixing coefficients

$$\phi_n^\omega = \sup\{\phi(\mathcal{F}_{-\infty,k}^\omega, \mathcal{F}_{k+n,\infty}^\omega) : k \in \mathbb{Z}\} \tag{7.5.1}$$

using the definition (1.2.15) of $\phi(\cdot, \cdot)$.

In order to ensure some applications, in particular, to random dynamical systems we will not assume that ξ_n^ω is measurable with respect to $\mathcal{F}_{n,n}^\omega$ but instead introduce the approximation rates

$$\beta_{q,r}^\omega = \sup_{k \geq 0} \|\xi_k^\omega - \mathbb{E}_{P_\omega}[\xi_k^\omega | \mathcal{F}_{k-r,k+r}^\omega]\|_{L^q}. \tag{7.5.2}$$

We will obtain results here under two different types of assumptions concerning the distribution of the process ξ^ω.

Assumption 7.5.1. All the ξ_n^ω's have the same distribution and the distribution of $(\xi_n^\omega, \xi_m^\omega)$ depends only on ω and $n - m$, which we write for further reference by

$$\xi_n^\omega \overset{d}{\sim} \mu \quad \text{and} \quad (\xi_n^\omega, \xi_m^\omega) \overset{d}{\sim} \mu_{m-n}^\omega \qquad (7.5.3)$$

where $Y \overset{d}{\sim} \nu$ means that Y has ν for its distribution.

This assumption holds true in the situation of Section 7.1, where $\xi_n^\omega = u_{\theta^n \omega} \circ T_\omega^n$, assuming that $\mu_\omega(0)$ does not depend on ω. For instance, the case when T_ω and ϕ_ω do not depend on ω can be considered. The situation of Assumption 7.5.1 does not include the case when $\mu_\omega(0)$ depends on ω and the Markov chains in random environments considered in Section 7.1. In order to allow applications to these processes we will also obtain results under the following

Assumption 7.5.2. The distribution of ξ_n^ω depends only on $\theta^n \omega$ and the distribution of the pair $(\xi_n^\omega, \xi_m^\omega)$, $m \geq n$ depends only on $\theta^n \omega$ and $m - n$ which we write for further reference by

$$\xi_n^\omega \overset{d}{\sim} \mu^{\theta^n \omega} \quad \text{and} \quad (\xi_n^\omega, \xi_m^\omega) \overset{d}{\sim} \mu_{m-n}^{\theta^n \omega}, \ m \geq n. \qquad (7.5.4)$$

Next, let $F_\omega = F_\omega(x_1, ..., x_\ell)$, $\ell \geq 1$ be a random function on $(\mathbb{R}^\wp)^\ell$ which is measurable in $(\omega, x_1, ..., x_\ell)$. We assume that there exist non-random $K, \iota > 0$ and $\kappa \in (0, 1]$ such that P-a.s. for all $x_i, z_i \in \mathbb{R}^\wp$, $i = 1, ..., \ell$,

$$|F_\omega(x) - F_\omega(z)| \leq K[1 + \sum_{i=1}^{\ell}(|x_i|^\iota + |z_i|^\iota)] \sum_{i=1}^{\ell} |x_j - z_j|^\kappa \qquad (7.5.5)$$

and

$$|F_\omega(x)| \leq K[1 + \sum_{i=1}^{\ell} |x_i|^\iota] \qquad (7.5.6)$$

where $x = (x_1, ..., x_\ell)$ and $z = (z_1, ..., z_\ell)$. For any $N \geq 1$ set

$$S_N^\omega = \sum_{n=1}^{N} F_{\theta^n \omega}(\xi_n^\omega, \xi_{2n}^\omega, ..., \xi_{\ell n}^\omega).$$

The main goal of this section is to prove a central limit theorem (CLT) for the normalized sums

$$Z_N^\omega = N^{-\frac{1}{2}} \left(S_N^\omega - c_{N,\omega}\right)$$

for P-a.a. ω, where $c_{N,\omega}$ are appropriate centralizing constants.

For each $\theta > 0$ and ω, set

$$\gamma^{\theta}_{\theta,\omega} = \sup_{n \geq 0} \|\xi^{\omega}_n\|^{\theta}_{L^{\theta}} = \sup_{n \geq 0} \mathbb{E}_{P_{\omega}} |\xi^{\omega}_n|^{\theta}. \tag{7.5.7}$$

Our results will rely on the following assumptions.

Assumption 7.5.3. There exist $b \geq 2$, $q \geq 1$ and $m > 0$ such that

$$\frac{1}{b} > \frac{\iota}{m} + \frac{\kappa}{q}$$

and P-a.s.

$$\gamma_{m,\omega} < \infty \quad \text{and} \quad \gamma_{\iota b,\omega} < \infty.$$

We will also need

Assumption 7.5.4. There exist $d \geq 1$ and $\theta > 2$ such that P-a.s. for any $n \in \mathbb{N}$,

$$\phi^{\omega}_n + (\beta^{\omega}_{q,n})^{\kappa} \leq dn^{-\theta} \tag{7.5.8}$$

where q is specified in Assumption 7.5.3.

Theorem 7.5.5. *Suppose that Assumptions 7.5.1, 7.5.3 and 7.5.4 hold true. Then P-a.s. the limit*

$$D^2_{\omega} = \lim_{N \to \infty} \frac{1}{N} \mathbb{E}_{P_{\omega}} \left(S^{\omega}_N - \sum_{n=1}^{N} \bar{F}_{\theta^n \omega} \right)^2$$

exists, where

$$\bar{F}_{\omega} = \int F_{\omega}(x_1, x_2, ..., x_{\ell}) d\mu(x_1) d\mu(x_2) ... d\mu(x_{\ell})$$

and μ is the measure specified in Assumption 7.5.1. Moreover, for P-almost all ω the sequence of random variables

$$N^{-\frac{1}{2}} \left(S^{\omega}_N - \sum_{n=1}^{N} \bar{F}_{\theta^n \omega} \right), \ N \geq 1$$

converges in distribution as $N \to \infty$ towards a centered normal random variable with variance D^2_{ω}.

In contrast to the case when ξ^ω and F_ω do not depend on ω (considered in Chapter 1), the proof that the limit D_ω^2 exists does not rely only on approximations of expectations such as the ones obtained in Lemma 3.3.2 and Corollary 1.3.14, and we explain here how to prove its existence in our situation. After this is proved, the above theorem will follow by Stein's method. We begin with the following similar to [45] decomposition (see also Chapter 3). Let μ be the measure specified in Assumption 7.5.1 and write

$$F_{\theta^n\omega}(x_1,...x_\ell) - \bar{F}_{\theta^n\omega} = \sum_{i=1}^\ell F_{\theta^n\omega,i}(x_1,...,x_i)$$

where

$$F_{\theta^n\omega,\ell}(x_1,...,x_\ell) = F_{\theta^n\omega}(x_1,...,x_\ell) - \int F_{\theta^n\omega}(x_1,...,x_\ell)d\mu(x_\ell)$$

and for all $j = \ell - 1, \ell - 2, ..., 1$,

$$F_{\theta^n\omega,j}(\hat{x}_j) = \int F_{\theta^n\omega}(x_1,...,x_\ell)d\mu(x_\ell)d\mu(x_{\ell-1})\cdots d\mu(x_{j+1})$$

$$- \int F_{\theta^n\omega}(x_1,...,x_\ell)d\mu(x_\ell)\cdots d\mu(x_j),$$

where $\hat{x}_j = (x_1,...,x_j)$. Then relying on Assumptions 7.5.3 and 7.5.4, the arguments in [45] imply that D_ω^2 exists when the following limits

$$\lim_{N\to\infty} \frac{1}{N} \sum_{\substack{1\le n,m\le N \\ in-jm=u}} \int F_{\theta^n\omega,i}(\hat{x}_i)F_{\theta^m\omega,j}(\hat{y}_j)d\mu_{i,j,u}^\omega(\hat{x}_i,\hat{y}_j)$$

exist P-a.s. for any $1 \le i,j \le \ell$ and $u \in \mathbb{Z}$ which is divisible by $\nu_{i,j} = \gcd(i,j)$, where the measure $\mu_{i,j,u}^\omega$ is constructed as follows. Let $(s_k,t_k), k = 1,...,r$ be all the pairs (s,t) of natural numbers so that $1 \le s \le i, 1 \le t \le j$ and $it = js$ and consider the measures μ_k^ω specified in Assumption 7.5.1. Then

$$d\mu_{i,j,u}^\omega(\hat{x}_i,\hat{y}_j) \qquad (7.5.9)$$

$$= \prod_{k=1}^r d\mu_{\frac{us_k}{i}}^\omega(x_{s_k},y_{t_k}) \prod_{v\ne s_k \,\forall k} d\mu(x_v) \prod_{v'\ne t_k \,\forall k} d\mu(y_{v'})$$

where note that $\frac{us_k}{i} = s_k n - t_k m, k = 1,...,r$. Notice that the variables $|\hat{x}_i|^\iota$ and $|\hat{y}_j|^\iota$ are square integrable under the law $\mu_{i,j,u}^\omega$. Next, write $\nu := \nu_{i,j} = \gcd(i,j)$, $u = \nu u_0$, $i = i'\nu$ and $j = j'\nu$. When $u = 0$ then the equality $in = jm$ means that n has the form $n = kj'$ and then $m = ki'$.

Suppose that $u \neq 0$ and let (n_0, m_0) satisfy $i'n_0 - j'm_0 = u_0$. Such n_0 and m_0 exist since $\gcd(i', j') = 1$. Then any solution (n, m) of the equation $in - jm = u = \nu u_0$ can be written in the form $n = n_0 + \tilde{n}$ and $m = m_0 + \tilde{m}$ where $i\tilde{n} = j\tilde{n}$, namely $(\tilde{n}, \tilde{m}) = k(j', i')$ is a solution of the homogeneous equation. Therefore, we can write $\theta^n = \theta^{n_0} \circ \theta^{kj'}$ and $\theta^m = \theta^{m_0} \circ \theta^{ki'}$, and so the above limits exist if the limits

$$\lim_{N \to \infty} \frac{1}{N} \sum_{k=1}^{N} \int G_{\theta^{kj'}\omega, i}(\hat{x}_i) H_{\theta^{ki'}\omega, j}(\hat{y}_j) d\mu_{i,j,u}^{\omega}(\hat{x}_i, \hat{y}_j)$$

exist, P-a.s., where $G_{\omega, i} = F_{\theta^{n_0}\omega, i}$ and $H_{\omega, j} = F_{\theta^{m_0}\omega, j}$. Note that the random functions $G_{\omega, i}$ and $H_{\omega, j}$ also satisfy (7.5.5) and (7.5.6), perhaps with some other constants. Changing the order of summation and integration, using the latter uniform (over ω) continuity in $\hat{x}_i$ and $\hat{y}_j$ and then the dominated convergence theorem we see that such limits exist when almost sure limits of the form

$$\lim_{N \to \infty} \frac{1}{N} \sum_{k=1}^{N} g_{\theta^{j'k}\omega} h_{\theta^{i'k}\omega}$$

exist, where g and h are bounded functions. Existence of such limits follows from the double recurrence theorem in [9], and we conclude that the limit D_ω^2 exists P-a.s.

Next, suppose that Assumptions 7.5.2, 7.5.3 and 7.5.4 hold true. For any ω and n set

$$\bar{F}_{\omega, n} = \int F_{\theta^n \omega}(x_1, x_2, ..., x_\ell) d\mu^{\theta^n \omega}(x_1) d\mu^{\theta^{2n}\omega}(x_2)...d\mu^{\theta^{\ell n}\omega}(x_\ell).$$

As in Theorem 7.5.5, when the limit

$$D_\omega^2 = \lim_{N \to \infty} \frac{1}{N} \mathbb{E}\left(S_N^\omega - \sum_{n=1}^{N} \bar{F}_{\omega, n}\right)^2$$

exists P-a.s. then using Stein's method we obtain that the sequence

$$N^{-\frac{1}{2}}\left(S_N^\omega - \sum_{n=1}^{N} \bar{F}_{\omega, n}\right), \ N \geq 1$$

converges in distribution for P-a.a. ω as $N \to \infty$ towards a centered normal random variable with variance D_ω^2. In order to avoid a tedious presentation we will not formulate here a theorem stating that D_ω^2 exists under certain conditions, but instead we will explain how to prove its existence in several particular situations which will be easier to describe after the following.

Consider the functions $F_{\omega,j,n}$, $1 \leq j \leq \ell$ given by

$$F_{\omega,\ell,n}(x_1, ..., x_\ell) = F_{\theta^n \omega}(x_1, ..., x_\ell) - \int F(x_1, ..., x_\ell) d\mu^{\theta^{\ell n} \omega}(x_\ell)$$

and for $j = \ell - 1, \ell - 2, ..., 1$,

$$F_{\omega,j,n}(x_1, ..., x_j)$$

$$= \int F_{\theta^n \omega}(x_1, ..., x_\ell) d\mu^{\theta^{\ell n} \omega}(x_\ell) d\mu^{\theta^{(\ell-1)n} \omega}(x_{\ell-1}) \cdots d\mu^{\theta^{(j+1)n} \omega}(x_{j+1})$$

$$- \int F_{\theta^n \omega}(x_1, ..., x_\ell) d\mu^{\theta^{\ell n} \omega}(x_\ell) d\mu^{\theta^{(\ell-1)n} \omega}(x_{\ell-1}) \cdots d\mu^{\theta^{j n} \omega}(x_j)$$

where the measures $\mu^{\theta^m \omega}$ are specified in Assumption 7.5.2. Then we can write

$$F_{\theta^n \omega}(x_1, ..., x_\ell) - \bar{F}_{\omega,n} = \sum_{j=1}^{\ell} F_{\omega,j,n}(x_1, ..., x_j).$$

For any $1 \leq i, j \leq \ell$ consider again the pairs (n, m) satisfying $in - jm = u$ for some fixed u which is divisible by $\nu = \gcd(i, j)$. Relying on Assumptions 7.5.3 and 7.5.4 the proof that D_ω^2 exists will proceed similarly to the situation of Theorem 7.5.5 assuming that limits

$$\lim_{N \to \infty} \frac{1}{N} \sum_{\substack{1 \leq n,m \leq N \\ in - jm = u}} \int F_{\omega,i,n}(\hat{x}_i) F_{\omega,j,m}(\hat{y}_j) d\mu_{i,j,u}^{\omega,n,m}(\hat{x}_i, \hat{y}_j)$$

exist P-a.s, where the family of laws

$$\mu_{i,j,u}^{\omega,n,m} = \mu_{i,j,u}^{(\theta^n \omega, \theta^2 \omega, ..., \theta^{\ell n} \omega, \theta^m \omega, \theta^{2m} \omega, ..., \theta^{\ell m} \omega)}$$

is given by

$$d\mu_{i,j,u}^{\omega,n,m}(\hat{x}_i, \hat{y}_j)$$

$$= \prod_{k=1}^{r} d\mu_{u s_k}^{\theta^{s_k n} \omega}(x_{s_k}, y_{t_k}) \prod_{v \neq s_k \, \forall k} d\mu^{\theta^{v n} \omega}(x_v) \prod_{v' \neq t_k \, \forall k} d\mu^{\theta^{v' m} \omega}(y_{v'}),$$

the pairs (s_k, t_k), $k = 1, ..., r$ are described before (7.5.9) and $\mu_{a-b}^{\theta^b \omega}$, $a < b$ is the distribution of the pair $(\xi_b^\omega, \xi_a^\omega)$ which equals $\mathbf{i} \mu_{b-a}^{\theta^a \omega}$ where $\mathbf{i}(x, y) = (y, x)$ for any $x, y \in \mathbb{R}^\wp$. Note that the variables $|\hat{x}_i|^\iota$ and $|\hat{y}_j|^\iota$ are square integrable under the measures $\mu_{i,j,u}^{\omega,n,m}$. As in the previous case, we can parametrize the pairs (n, m) satisfying $in - jm = u$ in the form $n = n_0 + kj'$ and $m = m_0 + ki'$, and so existence of the above limits would follow from an appropriate almost sure nonconventional ergodic theorem for averages of the form

$$\lim_{N\to\infty} \frac{1}{N} \sum_{n=1}^{N} G(\theta^n \omega, \theta^{2n} \omega, ..., \theta^{\ell n} \omega)$$

where G is a bounded function. Existence of these limits in general is an open problem even when the function G has the form $G(\omega_1, ..., \omega_\ell) = g_1(\omega_1) g_2(\omega_2) \cdots g_\ell(\omega_\ell)$. Still, the above limits exist P-a.s. in the situation when θ is sufficiently well mixing and G satisfies appropriate regularity conditions (see [42] and Section 3.2). Moreover, when G is continuous and Ω is compact then it can be approximated by sums of products of the above form, and then for certain ergodic systems such almost sure limits exist (see [25], [32] and references therein), while when $\ell = 2$ then we can use again [9]. Such regularity of the resulting functions G can be obtained when F_ω, μ^ω and μ_k^ω satisfy some regularity conditions with respect to ω such as uniform in k continuity in ω.

7.5.2 *Local limit theorem*

Let $(\Omega, \mathcal{F}, P, \theta)$ be an invertible ergodic measure preserving system and let $(\mathcal{X}, \rho, T)$ be a topologically mixing one sided subshift of finite type (see Section 2.11). As in Section 2.5.1, let $f \in \mathcal{H}^{\alpha,\xi}(\mathcal{X}) = \mathcal{H}^{\alpha,\xi}$ be a real valued function and consider the transfer operator $\mathcal{L}$ given by

$$\mathcal{L}g(x) = \sum_{y \in T^{-1}\{x\}} e^{f(y)} g(y)$$

which acts on $\mathcal{H}^{\alpha,\xi}$. We assume here that $\mathcal{L}1 = 1$. Then by Theorem 2.5.6 there exists a (unique) probability measure μ on $\mathcal{X}$ so that $\mathcal{L}^*\mu = \mu$. Let $\zeta = \{\zeta_n : n \geq 0\}$ be a Markov chain on $\mathcal{X}$ with initial distribution μ and transition operator $\mathcal{L}$. Let $F : \Omega \times \mathcal{X}^\ell \to \mathbb{R}$ be a measurable function such that with some constant $C_F > 0$ for P-a.a. ω,

$$|F(\omega, x) - F(\omega, y)| \leq C_F \big(\rho_{\ell,\infty}(x, y)\big)^\alpha \qquad (7.5.10)$$

for all $x_{\bullet} = (x_1, ..., x_\ell)$ and $y = (y_1, ..., y_\ell)$ in $\mathcal{X}^\ell$ such that

$$\rho_{\ell,\infty}(x, y) := \max_{1 \leq i \leq \ell} \rho(x_i, y_i) < \xi.$$

We assume in addition that

$$\bar{F}_\omega := \int F_\omega(x_1, x_2, ..., x_\ell) d\mu(x_1) d\mu(x_2) ... d\mu(x_\ell) = 0, \quad P\text{-a.s.}$$

where $F_\omega(\cdot) = F(\omega, \cdot)$ for each ω. This is not really a restriction since we can always replace F_ω with $F_\omega - \bar{F}_\omega$. For any $N \in \mathbb{N}$ consider the random variable

$$\mathcal{S}_N^\omega = \sum_{n=1}^{N} F_{\theta^n \omega}(\zeta_n, \zeta_{2n}, ..., \zeta_{\ell n}).$$

In this section we will explain how to prove an LLT for the sequences of random variables $\mathcal{S}_N^\omega$, $N \geq 1$ for P-a.a. ω. By Theorem 2.2.3, once a (non-degenerate) CLT is established the appropriate LLT follows in both lattice and non-lattice cases when Assumptions 2.2.1 and 2.2.2 hold true, and we will focus on showing that the CLT holds true and providing conditions for the latter assumptions to be satisfied.

Let Z_0 be a random element of $\mathcal{X}$ which is distributed according to μ and set $Z_m = T^m Z_0$ for any $m \in \mathbb{N}$. Let $\mathcal{F}_{m,n}, -\infty \leq m \leq n \leq \infty$ be the family of σ-algebras generated by cylinder sets with coordinates indexed by $\max(0, m), ..., \max(0, n)$. Then, in our circumstances (see [10]) there exist constants $d > 0$ and $c \in (0, 1)$ so that for any $n, r \geq 0$,

$$\phi_n \leq dc^n \tag{7.5.11}$$

and

$$\beta_\infty(r) := \sup_{n \geq 0} \|\rho(Z_n, Z_{n,r})\|_{L^\infty} \leq dc^r \tag{7.5.12}$$

where $Z_{n,r}$ is an $\mathcal{F}_{n-r,n+r}$-measurable random element of $\mathcal{X}$ whose coordinates with indexes $n - r \leq i \leq n + r$ coincide with the coordinates of Z_n with these indexes.

Relying on (7.5.11) and (7.5.12) instead of (7.5.4) and taking into account (2.5.11), we obtain similarly to Theorem 7.5.5 that the limit

$$\mathcal{D}_\omega^2 = \lim_{N \to \infty} \frac{1}{N} \mathbb{E}(\mathcal{S}_N^\omega)^2$$

exists P-a.s. and $N^{-\frac{1}{2}} \mathcal{S}_n^\omega$ converges in distribution as $N \to \infty$ towards a centered normal random variable with variance $\mathcal{D}_\omega^2$.

Next, relying on (7.5.11) and (7.5.12) and the arguments in Sections 2.3 and 2.7, in order to show that Assumptions 2.2.1 and 2.2.2 are satisfied it is sufficient to approximate for P-a.a. ω the expectations

$$\mathbb{E}_\Theta \Big\| \prod_{n=M+1}^{N} \mathbf{L}_{it}^{(\theta^n \omega, \Theta_n)} \Big\|_\infty \tag{7.5.13}$$

where $M = M_\ell(N) = N - 2[\frac{N-N_\ell}{2}]$, $\mathbf{L}_{it}^{(\omega, \bar{x})} g = \mathcal{L}^\ell(e^{it F_\omega(\bar{x}, \cdot)} g)$ and the process $\Theta = \{\Theta_n : n \geq 0\}$ is defined in Section 2.7.1. Let $M(\Theta) =$

$(\Omega_\Theta, \mathcal{B}_\Theta, P_\Theta, \vartheta)$ be the invertible MPS corresponding to the process Θ (see Section 2.7.1) and consider the product system $\tau = \theta \times \vartheta$. Then the expression in (7.5.13) can be written in the form

$$\int \Big\| \prod_{n=M+1}^{N} \mathcal{L}_{it}^{\tau^n(\omega,\bar{\omega})} \Big\|_\infty dP_\Theta(\bar{\omega}) \qquad (7.5.14)$$

where $\mathcal{L}_z^{(\omega,\bar{\omega})} g = \mathcal{L}^\ell(g e^{z F_\omega(p_0(\bar{\omega}),\cdot)})$ for any complex z and the map $p_0 : \Omega_\Theta \to \mathcal{X}^{\ell-1}$ is defined in Section 2.7.1.

Let $\mathcal{D}_{\omega,\ell}^2$ be the limit defined similarly to $\mathcal{D}_\omega^2$ but with the function $F_{\omega,\ell}$ (defined after Theorem 7.5.5) in place of F_ω (whose existence follows similarly to $\mathcal{D}_\omega^2$'s).

Proposition 7.5.6. *Suppose that $\mathcal{D}_{\omega,\ell}^2 > 0$. Then, for P-almost all ω the sequence of random variables S_N^ω, $N \geq 1$ satisfies Assumption 2.2.2.*

We sketch here the proof of the above proposition. Observe that we can apply in these circumstances the RPF theorem with the map τ^{-1} and the random transfer operators $\mathcal{L}_z^{\tau^{-1}(\omega,\bar{\omega})}$. The main modification of the arguments in Section 2.8 is in the proof of an appropriate version of (2.8.16). In our situation we first consider the random variables $V_k : \Omega \times \Omega_\Theta \to \mathbb{R}$, $k \geq 1$ given by

$$V_k(\omega, \bar{\omega}) = \mathrm{Var}_\mu \sum_{j=0}^{k-1} F_{\tau^j(\omega,\bar{\omega}),\ell}(\zeta_{\ell j}) = \mathbb{E}_\mu \Big(\sum_{j=0}^{k-1} F_{\tau^j(\omega,\bar{\omega}),\ell}(\zeta_{\ell j}) \Big)^2$$

where

$$F_{(\omega,\bar{\omega}),\ell}(\cdot) = F_{\omega,\ell}(p_0(\bar{\omega}),\cdot).$$

Let $\zeta^{(i)} = \{\zeta_n^{(i)} : n \geq 0\}$, $i = 1,2,...,\ell$ be independent copies of ζ and for each $j \geq 0$ set

$$\tilde{\Theta}_j = \big(\zeta_j^{(1)}, \zeta_{2j}^{(2)}, ..., \zeta_{\ell j}^{(\ell)}\big).$$

Then using arguments similar to the ones proceeding Theorem 7.5.5 (or by Theorem 2.3 in [38]) the limit

$$\tilde{\mathcal{D}}_\ell^2 = \lim_{k \to \infty} \frac{1}{k} \mathbb{E}_{\tilde{\Theta}} \Big(\sum_{j=0}^{k-1} F_{\theta^j \omega,\ell}(\tilde{\Theta}_j) \Big)^2$$

exists P-a.s. and it does not depend on ω. Using the arguments in Section 4 of [30] we see that $\tilde{\mathcal{D}}_\ell^2 > 0$ if and only if $\mathcal{D}_{\omega,\ell}^2 > 0$. We conclude that

$$\lim_{k \to \infty} \int \frac{1}{k} V_k(\omega, \bar{\omega}) dP_\theta(\bar{\omega}) = \tilde{\mathcal{D}}_\ell^2 > 0.$$

Therefore, there exists a constant $b_1 > 0$ so that for a sufficiently large k,

$$P\{\omega : \int \frac{1}{k} V_k(\omega, \bar{\omega}) dP_\theta(\bar{\omega}) > b_1\} > 0.$$

Fixing a sufficiently large k and considering the jk-th hitting times, $j = 1, 2, ...$ by $\theta^n, n \geq 1$ to the above set of positive probability, we can repeat the argument in the proof of (2.8.16) using the blocks generated by these hitting times instead of the blocks generated by the time intervals $[ik + 1, (i+1)k], i = 0, 1, 2, ...$, taking into account that by the mean ergodic theorem the number of such hitting times until time n grows linearly in n.

Next, we will provide conditions which guarantee that Assumption 2.2.1 holds true.

Assumption 7.5.7. The space Ω is a topological space, $\mathcal{F}$ is the Borel σ-algebra and P assigns positive mass to open sets. The map θ is continuous and there exist $\omega_0 \in \Omega$ and $n_0 \in \mathbb{N}$ such that $\theta^{n_0} \omega_0 = \omega_0$. Moreover, the map $\omega \to F_\omega$ is continuous at the points $\omega = \theta^i \omega_0$, $0 \leq i < n_0$ when considered as a map from Ω to the space $\mathcal{H}^{\alpha,\xi}(\mathcal{X}^\ell)$.

Let $\bar{x}_0$ be a periodic point of $\hat{T} = T \times T^2 \times \cdots \times T^{\ell-1}$ and let m_0 be such that $\hat{T}^{m_0} \bar{x}_0 = \bar{x}_0$. When Assumption 7.5.7 holds true then we can always assume that $n_0 = m_0$ since otherwise we can replace both with $m_0 n_0$.

Assumption 7.5.8. The functions $F_{\theta^i \omega_0}(\cdot)$, $0 \leq i < m_0$ satisfy Assumption 2.6.1.

Consider next the function $F_{\omega_0, \bar{x}_0, m_0} : \mathcal{X} \to \mathbb{R}$ given by

$$F_{\omega_0, \bar{x}_0, m_0}(x) = \sum_{j=0}^{m_0-1} F_{\theta^j \omega_0}(\hat{T}^j \bar{x}_0, T^{\ell j} x).$$

We distinguish between lattice and non-lattice cases similarly to Section 2.6. Namely, we call the case a non-lattice one when the function $F_{\omega_0, \bar{x}_0, m_0}$ is non-arithmetic in the sense of [28], i.e. if for any $t \in \mathbb{R} \setminus \{0\}$ there exist no nonzero $g \in \mathcal{H}^{\alpha,\xi}$ and $\lambda \in \mathbb{C}$, $|\lambda| = 1$ such that

$$e^{itF_{\omega_0, \bar{x}_0, m_0}} g = \lambda g \circ T^{\ell m_0}, \quad \mu\text{-a.s.} \tag{7.5.15}$$

We call the case a lattice one when the function $F_{\omega_0, \bar{x}_0, m_0}$ cannot be written in the form

$$F_{\omega_0, \bar{x}_0, m_0} = a + \beta - \beta \circ T^{\ell m_0} + h'\mathbf{k}, \quad \mu\text{-a.s.} \tag{7.5.16}$$

for some $h' > h$, $a \in \mathbb{R}$, $\beta : \mathcal{X} \to \mathbb{R}$ such that $e^{i\beta} \in \mathcal{H}^{\alpha,\xi}$ and an integer valued function $\mathbf{k} : \mathcal{X} \to \mathbb{Z}$.

Proposition 7.5.9. *Suppose that Assumptions 7.5.7 and 7.5.8 hold true. Then in the above lattice and non-lattice cases, for P-almost all ω the sequence of random variables $\tilde{S}_N^\omega$, $N \geq 1$ satisfies Assumption 2.2.1.*

The main difference in the proof of this proposition in comparison to Section 2.9 is as follows. Similarly to Section 2.10, given a compact interval $J \subset \mathbb{R}$ not containing the origin, we consider the visiting sequence $n_1 < n_2 < n_3 < ...$ with respect to the map θ to a neighborhood of ω_0 for which relations similar to the ones in (2.10.2) hold true (in lattice and non-lattice cases). Now, instead of counting the number of disjoint blocks belonging to some Bowen ball which are contained in the word $\Theta_1\Theta_2...\Theta_n$ and estimate the probability that there is proportional to n amount of blocks (see Corollary 2.9.3 together with (2.5.11)), we count the number of blocks beginning with indexes j of the form $j = n_i$. By the mean ergodic theorem n_i grows linearly in i, and an appropriate version of Corollary 2.9.3 follows.

Next, consider sums of the form

$$W_N^\omega(x) = \sum_{n=1}^N F_{\theta^n \omega}(T^n x, T^{2n} x, ..., T^{\ell n} x)$$

where x is distributed according to μ. Then the LLT under appropriate conditions follows essentially in the same way as in Section 2.11.3. The situation of a two sided subshift of finite type can also be considered, observing that an appropriate version of Lemma 2.11.2 follows essentially in the same way.

PART 3
Appendix

Appendix A

Real and complex cones

In this appendix we describe the theory of complex projective Hilbert metrics associated with complex cones, and the contraction properties of linear maps between such cones with respect to the corresponding Hilbert metrics. We introduce the basic notations and tools which were developed first in [58] and then in [18]. Still, for readers' convenience, we first recall the definition and basic properties of real cones and the real Hilbert metric associated with it (see [7], [49], [13], [18], [19] and [58]).

A.1 Real cones and real Hilbert metrics

Let X be a real vector space. A subset $\mathcal{C}_\mathbb{R} \subset X$ is called a proper real convex cone (or, in short, a real cone) if $\mathcal{C}$ is convex, invariant under multiplication of nonnegative numbers and $\mathcal{C}_\mathbb{R} \cap -\mathcal{C}_\mathbb{R} = \{0\}$. Next, assume that X is a Banach space and let $\mathcal{C}_\mathbb{R} \subset X$ be a closed real cone. For any nonzero elements f, g of $\mathcal{C}_\mathbb{R}$ set

$$\beta_{\mathcal{C}_\mathbb{R}}(f, g) = \inf\{t > 0 : tf - g \in \mathcal{C}_\mathbb{R}\} \tag{A.1.1}$$

where we use the convention $\inf \emptyset = \infty$. We note that $\beta_{\mathcal{C}_\mathbb{R}}(f, g) > 0$ since otherwise $-g$ lays in (the closure of) $\mathcal{C}_\mathbb{R}$ which together with the inclusion $g \in \mathcal{C}_\mathbb{R}$ implies that $g = 0$. The real Hilbert (projective) metric $d_{\mathcal{C}_\mathbb{R}} : \mathcal{C}_\mathbb{R} \times \mathcal{C}_\mathbb{R} \to [0, \infty]$ associated with the cone is given by

$$d_{\mathcal{C}_\mathbb{R}}(f, g) = \ln\left(\beta_{\mathcal{C}_\mathbb{R}}(f, g)\beta_{\mathcal{C}_\mathbb{R}}(g, f)\right) \tag{A.1.2}$$

where we use the convention $\ln \infty = \infty$.

Next, let X_1 and X_2 be two real Banach spaces and let $\mathcal{C}_i \subset X_i, i = 1, 2$ be two closed real cones. Let $A : X_1 \to X_2$ be a continuous linear transformation such that $A\mathcal{C}_1 \setminus \{0\} \subset \mathcal{C}_2 \setminus \{0\}$ and set

$$D = \sup_{x_1, x_2 \in \mathcal{C}_1 \setminus \{0\}} d_{\mathcal{C}_2}(Ax_1, Ax_2).$$

271

The following theorem is a particular case of Theorem 1.1 in [49] (see [7] for the case when $C_1 = C_2$).

Theorem A.1.1. *For any nonzero $x, x' \in C_1$ we have*

$$d_{C_2}(Ax, Ax') \leq \tanh\left(\frac{1}{4}D\right) d_{C_1}(x, x')$$

where $\tanh \infty := 1$

This lemma means that any linear map between two (punctured) real closed cones weakly contracts the corresponding Hilbert metrics, and this contraction is strong if the (Hilbert) diameter of the image is finite.

A.2 Complex cones and complex Hilbert metrics

A.2.1 *Basic notions*

Let Y be a complex Banach space. We say that $C \subset Y$ is a *complex cone* if $\mathbb{C}'C \subset C$, where $\mathbb{C}' = \mathbb{C} \setminus \{0\}$. The cone C is said to be *proper* if its closure $\bar{C}$ does not contain any two dimensional complex subspaces. The *dual cone* $C^* \subset Y^*$ is the set given by

$$C^* = \{\mu \in Y^* : \mu(c) \neq 0 \ \forall c \in C'\}$$

where $C' = C \setminus \{0\}$, and, as usual, Y^* is the space of all continuous linear functionals $\mu : Y \to \mathbb{C}$ equipped with the operator norm. It is clear that C^* is a complex cone. We will say that C is *linearly convex* if for any $x \notin C$ there exists $\mu \in C^*$ such that $\mu(x) = 0$, i.e. the complement of C is the union of the kernels $Ker(\mu), \mu \in C^*$. Then the complex cone C^* is linearly convex, since for any $c \in C$ the corresponding evaluation map $\nu \to \nu(c)$ is a member of the dual cone $(C^*)^*$ of C^*.

We introduce now the notation of the complex Hilbert projective metric δ_C of a proper complex cone defined in [18]. Let $x, y \in C'$ and consider the set $E_C(x, y)$ given by

$$E_C(x, y) = \{z \in \mathbb{C} : zx - y \notin C\}.$$

Since C is proper (and $\mathbb{C}'$ invariant) the set $E_C(x, y)$ is nonempty. When x and y are collinear set $\delta_C(x, y) = 0$ and otherwise set

$$\delta_C(x, y) = \ln\left(\frac{b}{a}\right) \in [0, \infty]$$

where

$$a = \inf|E_{\mathbb{C}}(x, y)| \in [0, \infty] \quad \text{and} \quad b = \sup|E_{\mathbb{C}}(x, y)| \in [0, \infty]$$

are the "largest" and "smallest" modulus of the set $E_{\mathcal{C}}(x, y)$, respectively. Observe that $\delta_{\mathcal{C}}(x, y) = \delta_{\mathcal{C}}(c_1 x, c_2 y)$ for any $c_1, c_2 \in \mathbb{C}' = \mathbb{C} \setminus \{0\}$, i.e. $\delta_{\mathcal{C}}$ is projective. When $\mathcal{C}$ is linearly convex then $\delta_{\mathcal{C}}$ satisfies the triangle inequality and so it is a projective metric (see [18] and [19]). We remark that a different notion of a complex Hilbert metric $d_{\mathcal{C}}$ was defined in [58], which was prior to the definition of $\delta_{\mathcal{C}}$. We refer the readers to Section 5 in [18] for relations between $d_{\mathcal{C}}$ and $\delta_{\mathcal{C}}$, among them a certain equivalence between them for a wide class of complex cones, including canonical complexifications of real cones introduced in the next section. This means that for such cones it makes no difference whether we use $\delta_{\mathcal{C}}$ or $d_{\mathcal{C}}$, and in this section we will only present results concerning $\delta_{\mathcal{C}}$.

A.2.2 The canonical complexification of a real cone

Next, let X be a real Banach and let $\mathcal{C}_{\mathbb{R}} \subset X$ be a real cone. Let $Y = X_{\mathbb{C}} = X + iX$ be its complexification (see Section 5 from [58]). Following [58], we define the canonical complexification of $\mathcal{C}_{\mathbb{R}}$ by

$$\mathcal{C}_{\mathbb{C}} = \{x \in X_{\mathbb{C}} : \Re(\bar{\mu}(x)\nu(x)) \geq 0 \ \ \forall \mu, \nu \in \mathcal{C}_{\mathbb{R}}^*\} \qquad (\text{A.2.1})$$

where

$$\mathcal{C}_{\mathbb{R}}^* = \{\mu \in X^* : \mu(c) \geq 0 \ \ \forall c \in \mathcal{C}_{\mathbb{R}}\}$$

and X^* is the space of all continuous linear functions $\mu : X \to \mathbb{R}$ equipped with the operator norm. Then $\mathcal{C}_{\mathbb{C}}$ is a proper complex cone (see Theorem 5.5 of [58]) and by [58] and [18] we have the following polarization identities

$$\mathcal{C}_{\mathbb{C}} = \mathbb{C}'(\mathcal{C}_{\mathbb{R}} + i\mathcal{C}_{\mathbb{R}}) = \mathbb{C}'\{x + iy : x \pm y \in \mathcal{C}_{\mathbb{R}}\} \qquad (\text{A.2.2})$$

where we recall that $\mathbb{C}' = \mathbb{C} \setminus \{0\}$. Moreover, when

$$\mathcal{C}_{\mathbb{R}} = \{x \in X : \mu(x) \geq 0 \ \ \forall \mu \in \mathcal{S}\}$$

for some $\mathcal{S} \subset X^*$, then

$$\mathcal{C}_{\mathbb{C}} = \{x \in X_{\mathbb{C}} : \Re(\bar{\mu}(x)\nu(x)) \geq 0 \ \ \forall \mu, \nu \in \mathcal{S}\} \qquad (\text{A.2.3})$$

since $\mathcal{S}$ generates the dual cone $\mathcal{C}_{\mathbb{R}}^*$. We conclude this section with the following result which appears as Lemma 4.1 in [19].

Lemma A.2.1. *A canonical complexification $\mathcal{C}_{\mathbb{C}}$ of a real cone $\mathcal{C}_{\mathbb{R}}$ is linearly convex if there exists a continuous linear functional which is strictly positive on $\mathcal{C}_{\mathbb{R}}' = \mathcal{C}_{\mathbb{R}} \setminus \{0\}$.*

A.2.3   *Apertures and contraction properties*

The *aperture* of a real cone $C_\mathbb{R}$ in some real Banach space is defined (see [58]) by

$$K(C_\mathbb{R}) = \inf_{\mu \in (C_\mathbb{R}^*)'} \sup_{x \in C_\mathbb{R}'} \frac{\|\mu\|\|x\|}{\mu(x)}$$

namely it is the infimum of K-values for which there exists a continuous linear functional $\mu \in C_\mathbb{R}^*$ such that

$$\|x\|\|\mu\| \leq K\mu(x) \quad \text{for any} \ \ x \in C_\mathbb{R}$$

which (see again [58]) is also the infimum of K-values for which there exists a continuous nonzero functional $\mu \in C_\mathbb{R}^*$ such that

$$\|x\| \leq \mu(x) \leq K\|x\| \quad \text{for any} \ x \in C_\mathbb{R}. \tag{A.2.4}$$

Now, following [58], the aperture of a complex cone $\mathcal{K}_\mathbb{C}$ in some complex Banach space is defined similarly by

$$K(\mathcal{K}_\mathbb{C}) = \inf_{\mu \in \mathcal{K}_\mathbb{C}^*} \sup_{x \in \mathcal{K}_\mathbb{C}'} \frac{\|\mu\|\|x\|}{|\mu(x)|}.$$

Then for any $K > 0$ we have $K(\mathcal{K}_\mathbb{C}) \leq K$ if there exists a continuous linear functional $\mu \in \mathcal{K}_\mathbb{C}^*$ such that for any $x \in \mathcal{K}_\mathbb{C}$,

$$\|x\|\|\mu\| \leq K|\mu(x)| \tag{A.2.5}$$

and we note that $K(\mathcal{K}_\mathbb{C}) < K$ implies that (A.2.5) holds true for some $\mu \in \mathcal{K}_\mathbb{C}^*$. The following result appears in [58] as Lemma 5.3.

Lemma A.2.2. *Let X be a real Banach space and let Y be its complexification. Let $C_\mathbb{R} \subset X$ be a real cone, and assume that (A.2.4) holds true with some μ and K. Then the complexification $C_\mathbb{C}$ of $C_\mathbb{R}$ satisfies (A.2.5) with $\mu_\mathbb{C}$ and $2\sqrt{2}K$, where $\mu_\mathbb{C}$ is the unique extension of μ to the complexified space Y.*

Next, the following assertion is formulated as Theorem 3.1 in [19] and it summarizes some of the main results from [18].

Theorem A.2.3. *Let $(X, \|\cdot\|)$ be a complex Banach spaces and $C \subset X$ be a complex cone.*

(i) Suppose that the cone C is linearly convex and of bounded sectional aperture. Then $(C'/ \sim, \delta_C)$ is a complete metric space, where $x \sim y$ if and only if $\mathbb{C}'x = \mathbb{C}'y$.

(ii) Let $K > 0$ and $\mu \in \mathcal{C}^*$ be such that (A.2.5) holds true for any $x \in \mathcal{C}$. Then for any $x, y \in \mathcal{C}'$,

$$\left\| \frac{x}{\mu(x)} - \frac{y}{\mu(y)} \right\| \le \frac{K}{2\|\mu\|} \delta_\mathcal{C}(x, y).$$

(iii) Let $\mathcal{C}_1$ be a complex cone in some complex Banach space X_1, and $A : X \to X_1$ be a complex linear map such that $A\mathcal{C}' \subset \mathcal{C}_1'$. Set

$$\Delta = \sup_{u, v \in \mathcal{C}'} \delta_{\mathcal{C}_1}(Au, Av)$$

and assume that $\Delta < \infty$. Then for any $x, y \in \mathcal{C}'$,

$$\delta_{\mathcal{C}_1}(Ax, Ay) \le \tanh\left(\frac{\Delta}{4}\right) \delta_\mathcal{C}(x, y).$$

Theorem A.2.3 *(ii)* means that any linear map between two (punctured) cones whose image has finite $\delta_{\mathcal{C}_1}$ (Hilbert) diameter is a weak contraction with respect to the appropriate projective metrics and that this contraction is strong.

A.2.4 *Comparison of real and complex operators*

Let X and Y be real Banach spaces and denote their complexifications by $X_\mathbb{C}$ and $Y_\mathbb{C}$. Let $\mathcal{S} \subset Y^*$, consider the real cone

$$\mathcal{C}_\mathbb{R} = \{y \in Y : s(y) \ge 0 \ \forall s \in \mathcal{S}\}$$

and denote its canonical complexification by $\mathcal{C}_\mathbb{C}$.

Theorem A.2.4. *Let $\mathcal{K} \subset X$ be a real closed cone and denote its canonical complexification by $\mathcal{K}_\mathbb{C}$. Let $P : X \to Y$ be a real linear transformation such that $P(\mathcal{K}') \subset \mathcal{C}_\mathbb{R}'$. Suppose that the real Hilbert diameter satisfies*

$$D = \sup_{u, v \in \mathcal{K}'} d_{\mathcal{C}_\mathbb{R}}(Pu, Pv) < \infty.$$

Let $A : X_\mathbb{C} \to Y_\mathbb{C}$ be a complex linear transformation and assume that there exists $\varepsilon > 0$ such that

$$\delta = \delta(\varepsilon, D) := 2\varepsilon\left(1 + \cosh\left(\frac{D}{2}\right)\right) < 1$$

and that for any $s \in \mathcal{S}$ and $x \in \mathcal{K}$,

$$|s(Ax) - s(Px)| \le \varepsilon s(Px).$$

Then $A(\mathcal{K}_\mathbb{C}') \subset \mathcal{C}_\mathbb{C}'$ and for any $y \in \mathcal{K}_\mathbb{C}'$,

$$\delta_{\mathcal{C}_\mathbb{C}}(Ay, Py) \le 3\ln \frac{1}{1 - \delta}$$

and in particular

$$\sup_{x, y \in \mathcal{K}_\mathbb{C}'} \delta_{\mathcal{C}_\mathbb{C}}(Ax, Ay) \le D + 6\ln \frac{1}{1 - \delta}.$$

In the case when $\mathcal{K} = \mathcal{C}$ this is Theorem 4.5 from [19] and the proof of Theorem A.2.4 goes exactly in the same way. Remark that a close result was first obtained in Proposition 6.3 from [58] with the aforementioned metric $d_{\mathcal{C}_{\mathbb{C}}}$.

A.2.5 Further properties of complex dual cones

We begin with a result which is proved in Lemma 2.4 from [18].

Lemma A.2.5. *Let $\mathcal{C}$ be a complex linearly convex proper cone in a complex Banach space. Then for any $x, y \in \mathcal{C}'$,*

$$\delta_{\mathcal{C}}(x, y) = \sup_{f, g \in \mathcal{C}^*} \ln \left| \frac{f(y) g(x)}{f(x) g(y)} \right|.$$

Next, let $\mathcal{C}_1$ be a complex cone in some complex Banach space. A direct calculation shows that for $f, g \in \mathcal{C}_1^*$ we have

$$E_{\mathcal{C}_1^*}(f, g) = \left\{ \frac{g(x)}{f(x)} : x \in \mathcal{C}_1' \right\}.$$

This together with the definition of $\delta_{\mathcal{C}_1^*}$ yields that

$$\delta_{\mathcal{C}_1^*}(f, g) = \ln \sup_{x, y \in \mathcal{C}_1'} \left| \frac{f(x) g(y)}{g(x) f(y)} \right| = \sup_{x, y \in \mathcal{C}_1'} \ln \left| \frac{f(x) g(y)}{g(x) f(y)} \right|.$$

Combining this with Lemma A.2.5 we derive the following.

Lemma A.2.6. *Let X_1 and X be complex Banach spaces and $\mathcal{C}_1 \subset X_1$ and $\mathcal{C} \subset X$ be complex cones such that $\mathcal{C}$ is proper and linearly convex. Let $A : X_1 \to X$ be a complex linear map so that $A\mathcal{C}_1' \subset \mathcal{C}'$. Then the dual operator $A^* : X^* \to X_1^*$ satisfies $A^* \mathcal{C}^* \subset \mathcal{C}_1^*$ (this is true in general, of course) and*

$$\sup_{x, y \in \mathcal{C}_1'} \delta_{\mathcal{C}}(Ax, Ay) = \sup_{\mu, \nu \in \mathcal{C}^*} \delta_{\mathcal{C}_1^*}(A^* \mu, A^* \nu) \qquad (A.2.6)$$

namely, both images have the same diameter with respect to the appropriate complex Hilbert metrics.

We conclude this appendix with the following simple lemma.

Lemma A.2.7. *Let $(X, \| \cdot \|)$ be a complex Banach space and let $\mathcal{C} \subset X$ be a complex cone. Let $x_0 \in X$ and $M \in (0, \infty)$ be such that $B(x_0, \frac{1}{M}) = \{x \in X : \|x - x_0\| < \frac{1}{M}\} \subset \mathcal{C}$. Then for any $\mu \in \mathcal{C}^*$,*

$$\|\mu\| \leq M |\mu(x_0)| \qquad (A.2.7)$$

namely, the aperture of the dual cone does not exceed M as can be seen by considering the evaluation functional $\mu \to \mu(x_0)$.

Proof. Let $\mu \in \mathcal{C}^*$, $\lambda \in \mathbb{C}$ and $h \in X$ be such that $|\lambda| < 1$ and $\|h\| \leq \frac{1}{M}$. Then $|\mu(x_0) + \lambda\mu(h)| > 0$, since $x_0 + \lambda h \in \mathcal{C}$. Suppose that $|\mu(h)| > |\mu(x_0)|$ and let $\lambda = -\frac{\mu(x_0)}{\mu(h)}$. Then $|\lambda| < 1$ and so we obtain that

$$|\mu(x_0) - \mu(x_0)| = |\mu(x_0) + \lambda\mu(h)| > 0$$

which is a contradiction. Thus, $|\mu(y)| \leq M|\mu(x_0)|$ for any $y \in X$ such that $\|y\| \leq 1$ and (A.2.7) follows. $\qquad\square$

Bibliography

[1] A.D. Barbour, *Stein's Method for Diffusion Approximations*, Probab. Th. Rel. Fields 84 (1990), 297-322.

[2] A.D. Barbour and S. Janson, *A functional combinatorial central limit theorem*, Electron. J. Probab. 14 (2009), 2352-2370.

[3] A.D. Barbour and L.H.Y. Chen, *An Introduction to Stein's Method*, Singapore University Press, Singapore, 2005.

[4] J.R. Blum, D.L. Hanson and L.H. Koopmans, *On the strong law of large numbers for a class of stochastic processes*, Z. Wahrsch. verw. Geb 2 (1963), 1-11.

[5] P. Billingsley, *Convergence of Probability Measures*, 2nd ed. Wiley, New York, 1999.

[6] P. Billingsley, *Convergence of Probability Measures*, Wiley, New York, 1968.

[7] G. Birkhoff, *Extension of Jentzsch's theorem*, Trans. A.M.S. 85 (1957), 219-227.

[8] E. Bolthausen, *Markov process large deviations in τ-topology*, Stoch. Proc. Appl. (1987), 95-108.

[9] J. Bourgain, *Double recurrence and almost sure convergence*, J. Reine Angew. Math. 404 (1990), 140-161.

[10] R. Bowen, *Equilibrium states and the ergodic theory of Anosov diffeomorphisms*, Lecture Notes in Mathematics, volume 470, Springer Verlag, 1975.

[11] R.C. Bradley, *Introduction to Strong Mixing Conditions*, Volume 1, Kendrick Press, Heber City, 2007.

[12] L. Breiman, *Probability*, SIAM, Philadelphia (1992).

[13] P.J. Bushell, *The Cayley-Hilbert metric and positive operators*, Linear Alg. and Appl. 84 (1986), 271-281.

[14] I.P. Cornfeld, S.V. Fomin, and Ya.G. Sinai, *Ergodic Theory*, Springer-Verlag, Berlin, 1982.

[15] L.H.Y Chen and Q.M. Shao, *Normal approximation under local dependence*, Ann. Probab. 32 (2004), 1985-2028.

[16] C. Castaing and M.Valadier, *Convex analysis and measurable multifunctions*, Lecture Notes Math., vol. 580, Springer, New York, 1977.

[17] M.J. Kasprzak, A.B. Duncan and S.J. Vollmer, *Note on A. Barbours paper*

on Steins method for diffusion approximations, Electron. Commun. Probab. 22 (2017), 1-8.

[18] L. Dubois, *Projective metrics and contraction principles for complex cones,* J. London Math. Soc. 79 (2009), 719-737.

[19] L. Dubois, *An explicit Berry-Esseen bound for uniformly expanding maps on the interval,* Israel J. Math. 186 (2011), 221-250.

[20] M.D. Donsker and S.R.S. Varadhan, *Asymptotic evaluation of certain Markov process expectations for large time,* I. Commun. Pure. Appl. Math. (1987), 1-47.

[21] H. Furstenberg, *Recurrence in Ergodic Theory and Combinatorial Number Theory,* Princeton Univ. Press, Princeton, NJ 1981.

[22] H. Furstenberg, *Nonconventional ergodic averages,* Proc. Symp. Pure Math. 50 (1990), 43-56.

[23] Y. Guivarćh and J. Hardy, *Théorèmes limites pour une classe de chaînes de Markov et applications aux difféomorphismes d'Anosov,* Ann. Inst. H. Poincaré Probab. Statist. 24 (1988), no. 1, 73-98.

[24] A.L. Gibbs and F.E. Su, *On choosing and bounding probability metrics,* Int. Statist. Rev. 70 (2002), 419-435.

[25] Y. Gutman, W. Huang, S. Shao and X. Ye, *Almost sure convergence of the multiple ergodic average for certain weakly mixing systems,* arXiv:1612.02873v2.

[26] Y. Hafouta, *Stein's method for nonconventional sums,* arXiv:1704.01094.

[27] P.G. Hall and C.C. Hyde, *Martingale central limit theory and its application,* Academic Press, New York, 1980.

[28] H. Hennion and L. Hervé, *Limit Theorems for Markov Chains and Stochastic Properties of Dynamical Systems by Quasi-Compactness,* Lecture Notes in Mathematics vol. 1766, Springer, Berlin, 2001.

[29] Y. Hafouta and Yu. Kifer, *A nonconventional local limit theorem,* J. Theoret. Probab. 29 (2016), 1524-1553.

[30] Y. Hafouta and Yu. Kifer, *Berry-Esseen type inequalities for nonconventional sums,* Stoch. Proc. Appl. 126 (2016), 2430-2464.

[31] Y. Hafouta and Yu. Kifer, *Nonconventional polynomial CLT,* Stochastics, 89 (2017), 550-591.

[32] W. Huang, S. Shao and X. Ye, *Pointwise convergence of multiple ergodic averages and strictly ergodic models,* arXiv:1406.5930v2.

[33] I.A. Ibragimov and Yu.V. Linnik, *Independent and Stationary Sequences of Random Variables,* Wolters-Noordhoff, Groningen, 1971.

[34] R.M. de Jong, *A strong law of large numbers for triangular mixingale arrays,* Stat. & Probab. Lett. 27 (1996), 1-9.

[35] J. Jacod and A.N. Shiryaev, *Limit Theorems for Stochastic Processes,* 2nd. ed., Springer-Verlag, Berlin, 2003.

[36] T. Kato, *Perturbation theory for linear operators. Classics in Mathematics.* Springer-Verlag, Berlin, 1995.

[37] Yu. Kifer, *Perron-Frobenius theorem, large deviations, and random perturbations in random environments,* Math. Z. 222(4) (1996), 677-698.

[38] Yu. Kifer, *Limit theorems for random transformations and processes in*

random environments, Trans. Amer. Math. Soc. 350 (1998), 1481-1518.

[39] Yu. Kifer, *Optimal stopping and strong approximation theorems*, Stochastics, 79 (2007), 253-273.

[40] Yu. Kifer, *Thermodynamic formalism for random transformations revisited*, Stoch. Dyn. 8 (2008), 77-102.

[41] Yu. Kifer, *Nonconventional limit theorems*, Probab. Th. Rel. Fields, 148 (2010), 71-106.

[42] Yu. Kifer, *A nonconventional strong law of large numbers and fractal dimensions of some multiple recurrence sets*, Stoch. Dynam. 12 (2012), 1150023.

[43] Yu. Kifer, *Nonconventional Poisson limit theorems*, Israel J. Math. 195 (2013), 373–392.

[44] Yu. Kifer, *Strong approximations for nonconventional sums and almost sure limit theorems*, Stochastic Process. Appl. 123 (2013), 2286-2302.

[45] Yu. Kifer and S.R.S. Varadhan. *Nonconventional limit theorems in discrete and continuous time via martingales*, Ann. Probab. 42 (2014), 649-688.

[46] Yu. Kifer and S.R.S. Varadhan, *Nonconventional large deviations theorems*, Probab. Th. Rel. Fields 158 (2014), 197-224.

[47] Yu. Kifer and A. Rapaport, *Poisson and compound Poisson approximation in conventional and nonconventional setups*, Probab. Th. Rel. Fields 160 (2014), 797-831.

[48] Yu. Kifer, *Ergodic theorems for nonconventional arrays and an extension of the Szemerédi theorem*, Discrete Cont. Dynam. Sys. 38 (2018).

[49] C. Liverani, *Decay of correlations*, Ann. Math. 142 (1995), 239-301.

[50] S.V. Nagaev, *Some limit theorems for stationary Markov chains*, Theory Probab. Appl. 2 (1957), 378-406.

[51] S.V. Nagaev, *More exact statements of limit theorems for homogeneous Markov chains*, Theory Probab. Appl. 6 (1961), 62-81.

[52] L. Kuipers and H. Niederreiter, *Uniform Distribution of Sequences*, Wiley, New York, 1974.

[53] V. Mayer, B. Skorulski and M. Urbański, *Distance expanding random mappings, thermodynamical formalism, Gibbs measures and fractal geometry*, Lecture Notes in Mathematics, vol. 2036 (2011), Springer.

[54] M. Pollicott, *A complex Ruelle-Perron-Frobenius theorem and two counterexamples*, Ergodic Theory Dynam. Systems, 4(1) (1984), 135-146.

[55] D. Pollard, *Convergence of stochastic processes*, Springer, New York, 1984.

[56] Y. Rinott, *On normal approximation rates for certain sums of dependent random variables*, J. Comput. Appl. Math. 55 (1994), 135-143.

[57] W. Rudin, *Real and Complex Analysis*, McGraw-Hill, New York, 1987.

[58] H.H. Rugh, *Cones and gauges in complex spaces: Spectral gaps and complex Perron-Frobenius theory*, Ann. Math. 171 (2010), 1707-1752.

[59] B.A. Sevast'yanov, *Poisson limit law for a scheme of sums of dependent random variables*, Theory of Probab. Appl. 17 (1972), 695-699.

[60] A.N. Shiryaev, *Probability*, Springer-Verlag, Berlin, 1995.

[61] C. Stein, *A bound for the error in the normal approximation to the distribution of a sum of dependent random variables*, Proc. Sixth Berkeley Symp. Math. Statist. Probab. 2 (1972), 583-602. Univ. California Press, Berkeley.

[62] C. Stein, *Approximation Computation of Expectations*, IMS, Hayward, CA (1986).

[63] P. Walters, *An Introduction to Ergodic Theory*, Springer, New York, 1975.

Index

Printed in the United States
By Bookmasters